English
Français
Deutsche
Italiano
Español
Português

www.forgottenbooks.com

Mythology Photography **Fiction**
Fishing Christianity **Art** Cooking
Essays Buddhism Freemasonry
Medicine **Biology** Music **Ancient**
Egypt Evolution Carpentry Physics
Dance Geology **Mathematics** Fitness
Shakespeare **Folklore** Yoga Marketing
Confidence Immortality Biographies
Poetry **Psychology** Witchcraft
Electronics Chemistry History **Law**
Accounting **Philosophy** Anthropology
Alchemy Drama Quantum Mechanics
Atheism Sexual Health **Ancient History**
Entrepreneurship Languages Sport
Paleontology Needlework Islam
Metaphysics Investment Archaeology
Parenting Statistics Criminology
Motivational

THE

EARTH AND ITS INHABITANTS.

EUROPE.

BY

ÉLISÉE RECLUS.

EDITED BY

E. G. RAVENSTEIN, F. R. G. S., F. S. S., ETC.,

AND

A. H. KEANE, B. A.,

MEMB. OF COUNCIL, ANTHROPOLOGICAL INSTITUTE.

VOL. V.

THE NORTH-EAST ATLANTIC ISLANDS OF THE NORTH ATLANTIC, SCANDINAVIA, EUROPEAN ISLANDS OF THE ARCTIC OCEAN, RUSSIA IN EUROPE.

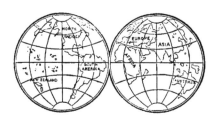

ILLUSTRATED BY NUMEROUS ENGRAVINGS AND MAPS.

NEW YORK:
D. APPLETON AND COMPANY,
1, 3, AND 5 BOND STREET.

ONTENTS.

OF ILLUSTRATIONS.

VOL. V

PLATES.

ILLUSTRATIONS IN TEXT.

THE EARTH AND ITS INHABITANTS.

THE NORTH-EAST ATLANTIC.

Depth and Currents.

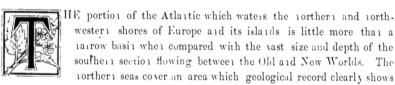

THE portion of the Atlantic which waters the northern and north-western shores of Europe and its islands is little more than a narrow basin when compared with the vast size and depth of the southern section flowing between the Old and New Worlds. The northern seas cover an area which geological record clearly shows has for countless ages been the battle-ground of the rival elements of land and water. Islands, archipelagos, banks, and submarine shoals here divide the abyss into secondary basins, while the English Channel, North Sea, and Baltic may be regarded as flooded plains, belonging geologically to the mainland.

Even on the mainland itself it is not easy to determine the limits of the natural regions, the transitions in altitude and climate being very seldom sharply marked, while on the ocean it becomes impossible to draw any imaginary lines of separation. Not only are the climatic changes freely produced on the unbroken surface of the waters, but the liquid mass is constantly displaced by the action of storms, tides, and conflicting currents. Hence the oceanic areas can only be indicated in a very approximate manner.

Nevertheless the soundings that have been carried on since the middle of this century have determined certain transitional zones between the various basins. The British and Scandinavian waters are separated from the American waters by depths of more than 2,000 fathoms, forming a hollow trough between the two continental masses. A lateral valley of this trough off the Newfoundland bank runs eastwards to mid-Atlantic, towards another deep basin stretching to the west of the Bay of Biscay and the Iberian peninsula; while between these two cavities the plateau of the Azores is connected by a ridge with those of the European seas. Altogether the European section of the North Atlantic is comparatively very shallow, so that an upheaval of even 300 feet would almost efface its eastern

144

islands, St. George's and English Channels, the North Sea and the Baltic. Besides the great plateau of the British Isles there are others of smaller extent, including those of Rockall, the Färöer, Iceland, and Jan Mayen. From Scotland to Greenland there extends a submarine isthmus, whose elevation has not yet been quite determined, but whose lowest parts between the Orkneys and the Färöer bank are less than 380 fathoms below the surface of the water, with a mean depth of 270 fathoms between the Färöer and Iceland. This last section seems to be of volcanic origin, and it is probable that submarine action has contributed to the separation of the North Atlantic waters into two distinct basins. The Rockall plateau is connected with the Hebrides by a ridge with a mean depth of 820 fathoms—about the same as that of the entire eastern basin between Iceland and Norway, or one-third of the approximate depth of all the oceanic waters.

It was formerly supposed that the Northern Ocean diminished in depth as it approached the pole, but the Swedish exploration of 1868 has entirely exploded this idea. About 180 miles west of Spitzbergen the plummet measured 2,650 fathoms, and in the highest latitudes where soundings have been taken a depth of 1,370 fathoms has been revealed. Scoresby found 1,176 fathoms between Spitzbergen and Jan Mayen in 1818, so that northwards as well as southwards the shallow European waters are bounded by deep troughs.

The polar icebergs, advancing more or less southwards with the alternations of the seasons, also form a natural line of separation for the European basins. It is remarkable that the bed of the ocean presents in its reliefs features analogous to those of the neighbouring continents. Were the waters to subside 1,000 fathoms, there would be revealed two peninsulas between Europe and Greenland, projecting southwards like those of Scandinavia and the Mediterranean. And were a further subsidence of 1,000 fathoms to take place, it would disclose east of Newfoundland another and more extensive peninsula, with numerous secondary ramifications, also stretching southwards, while the ridge now separating the western and eastern oceanic basins would appear as an isthmus connecting the northern lands with a vast peninsula similarly extending north and south beyond the Azores. According to mediæval legends formerly figured on marine charts as ascertained facts, one of these submerged peninsulas was still visible above the surface when the earliest seafarers visited these regions. The vanished land bore the name of the "drowned land of Buss," and it has by some been associated with the island of Finlandia, discovered by the Venetian brothers Zeno at the end of the fourteenth century, but which has since been sought for in vain.

The movements of the Atlantic, like those of other seas, are due to various causes, but are distinguished by their vast proportions and lack of uniformity. Although the phenomena they present have nowhere else been more carefully studied, many problems still remain to be determined. For their solution more is needed than a knowledge of the surface waters; account must be also taken of the counter-currents, of the varying temperature and saline character of the ocean throughout its entire depth.

Thanks to their daily recurrence, the normal course of the tides is much

:he currents. The great tidal flow setting 1orth-
e shores of the two hemispheres at the same time,
cams on reaching the south-western shores of the
continues its 1orthern course along the west coast
s enter the St. George's and English Channels.
est and 1orth coasts of Ireland the first stream
1el between Great Britain and Ireland, and there
m the south. The main stream, after making the
wards along the east coast of England until it
lish Channel about the Straits of Dover. At the
nstantly modified according to the position of sun
etion of the winds, the endless varieties of the
s alone can be given.*

STREAM.—TEMPERATURE.

rents of the North-east Atlantic flow from the
epth of over 500 fathoms the surface waters 01
st to 1orth-east, and from south to 1orth, from
the British Isles, Iceland, Scandinavia, and
beyond all doubt by the tropical plants and
ring the marks of their origin, strewn along the
he polar islands. But it is difficult to say to what
ing through Florida Channel from the Gulf of
vast current traversing the entire breadth of the
'indlay, and other physicists rightly regard this
al displacement occasioned by the tepid waters of
rds the cold waters of the arctic seas. In fact,
v to give egress to a stream spreading for a space
es between Scandinavia, Iceland, and Newfound-
th than 830 fathoms below the surface. The flow
amas is variously estimated at from 500,000‡ to
second by writers who have a theory to support.
ons of others give a nominal volume of about
would take no less than ten years to fill the whole
tepid waters. Besides, the hydrographers who
have ascertained that off the United States coast
number of smaller currents, separated from each
and all gradually merging in the main current of

described in vol. iv. p. 7.
Institution, *Nature*, March 10th, 1870.
ical Magazine, February, 1870.
e Geographical Society, 1853; *Proceedings*, 1869.

The velocity of this current has not yet been clearly determined, for it flows too slowly to be detected in the midst of the various motions produced by the winds on the surface. Admiral Irminger gives it a mean velocity of 3 miles a day, while Captain Otto thought it amounted to nearly 12 miles, at least on the Norwegian seaboard.[*] According to Findlay it would take from one to two years to reach Europe from Florida, while Petermann considers that a few months would suffice. When General Sabine was in Hammerfest in 1823, some barrels of palm oil were recovered, belonging to a vessel which had been wrecked the previous year at Cape Lopez, on the west coast of Africa, near the equator. These barrels must have twice crossed the Atlantic within the twelvemonth. Floating bottles containing messages from seafarers in distress, and picked up at various points, enable us to fix approximately six months as about the time required for the displacement of the waters from one to the other side of the Northern Ocean.

But if the main current of the Eastern Atlantic be not detected by the velocity of its waters, it is revealed plainly enough by its higher temperature. Hundreds of thousands of observations made by such distinguished hydrographers as Maury, Andrau, Wullich, Buchan, Irminger, Inglefield, and Mohn have supplied ample materials for the preparation of a correct chart of this current from month to month, and for tracing its shifting limits. In summer, when its outlines are rendered most irregular by its struggle with the polar stream, it is strongly deflected by the pressure of the cold waters flowing from Baffin's Bay. But after passing this arctic stream, which continues to set southwards beneath the surface, the southern current resumes its north-easterly course, so that the isothermal lines revealing its presence are not deflected from their regular path. It strikes the western shores of Iceland, skirting the north coast. But it meets a second polar stream about the eastern headlands of the island, causing it to flow back by the south coast. Here the warm waters, being subjected to an enormous pressure, are again deflected from their north-easterly direction. The polar stream does not at once disappear beneath the surface strata to form a bed for the southern waters moving in an opposite direction. It struggles long for the ascendancy, and the two currents ramify into two parallel belts flowing side by side in contrary directions. According to Irminger's observations, the whole area between Iceland and Scotland is intersected by these alternating belts of warm and cold water belonging to the two opposing currents. During his trip from Stornoway in the Hebrides to Reykjavik in Iceland, in the month of June, 1856, Lord Dufferin caused the temperature of the surface waters to be tested every two hours, making altogether ninety observations, and detected no less than forty-four changes from 2° to 9°, whereas at the two extremes the thermometer marked exactly 45° Fahr.

After crossing the polar current, whose normal direction seems to be from Jan Mayen to the Frisian coast, the Gulf Stream continues to flow north-east, parallel with the shores of Scandinavia, then rounding its northern limits in the direction of Novaya Zemlya. But while the main volume follows the line of the continent, a

* Petermann's *Mittheilungen*, 1873, 1878.

secondary branch, arrested by the submarine bed stretching from Bear Island to Spitzbergen, is deflected northwards under the seventy-fifth degree of latitude, and at least during the months of July, August, and September, when the sea is free from ice, it continues in this direction parallel with the west coast of Spitzbergen, then trending round the Archipelago towards the north-east, and gradually merging in the Arctic Ocean. The mean temperature of this branch is 40° Fahr.[*]

West of Spitzbergen and of the submarine bank separating this group from Scandinavia, the mean depth of the ocean is much greater than in the eastern

Fig. 1.—Temperature of the Sea during the Summer of 1868 in Fahrenheit Degrees.
According to A. Petermann. Scale 1 : 10,000,000.

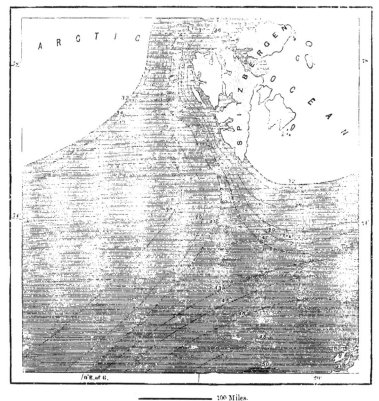

100 Miles.

waters towards Novaya Zemlya. Dr. Bessel's soundings showed no great depth between this large island and Spitzbergen, the difference being due probably to the struggle of the under-currents. In the west the relatively tepid waters have the upper hand, diverting the icebergs to other shores, whereas farther east the cold currents prevail, running at the rate of 9 miles per hour, so that a

* Von Freeden, in Petermann's *Mittheilungen*, vi. 1869.

boat manned with a strong crew can scarcely make head against them.* These currents send southwards continuous lines of icebergs with their loads of detritus, which sink to the bottom during the summer at the contact with the southern branch of the Atlantic waters. The vast bank stretching north-west of Bear Island seems to be an immense submarine moraine similar to the bank of Newfoundland.† But beyond those rocky deposits the polar stream continues to flow towards the warmer waters from the south, interpenetrating them with currents of cold water like those met with in the regions east of Iceland.

The arctic section of the East Atlantic is thus on the whole marked off with sufficient clearness by the form of its submarine bank, by the general movement of

Fig. 2.—Isothermal Lines of the North-East Atlantic in July.

According to Mohn. Scale 1 : 28,000,000.

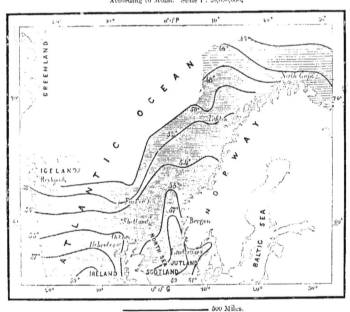

500 Miles.

its waters, and by the meteorological conditions. The European waters are almost entirely occupied, at least in summer, by the tropical currents. Doubtless in winter the warm stream, although much more regular in its movements than during the hot season, is everywhere driven far to the south, the water north of Jan Mayen and Bear Islands being below freezing point, and almost entirely filled by icebergs. Still the mean temperature of the North-east Atlantic is at all times much higher than that found elsewhere in the same latitude. The mean difference from July to January between cold and heat for any given point in this

* Lamont, Masqueray, *Bulletin de la Société de Géographie*, October, 1872.
† Petermann's *Mittheilungen*, iv. 1870.

region is only 9°. Throughout the basin between Scotland, Norway, Iceland, and Spitzbergen the surface waters are from 2° to 5° warmer than the surrounding atmosphere, the proportion being reversed only in summer, when the temperature of the atmosphere is slightly higher. The hot air is then tempered by the sea, the reverse being the case throughout the rest of the year.

So great is the general influence exercised by the main current of warm waters on the climate of the European continent, and especially of the lands encircled by it, that it alone renders the British Isles and Scandinavia inhabitable. Like another Labrador, this region would remain the abode of wild animals, maintaining with difficulty a few scattered communities on the banks of the sheltered creeks.

Fig. 3.—ISOTHERMAL LINES, NORTH EAST ATLANTIC, IN JANUARY.
According to Mohn. Scale 1 : 28,000,000.

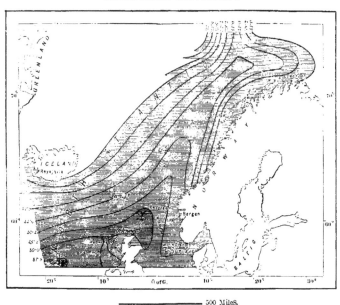

500 Miles.

It is the southern current acting in concert with the south-west winds that has enabled the English race to rise and develop. Hence it has played a chief part in the modern history of mankind.

The deflection of the isothermal lines caused by the currents of air and water in the North Atlantic is the most remarkable phenomenon of the kind on the whole surface of the globe. In many places the importance of the latitudes for the local climate seems to be completely neutralised, the mean temperature rising not from north to south, but from south to north. This is due to the southern waters bringing with them the tropical heat, and discharging it in the regions of Northern Europe. A climate conveyed northwards by the current thus overlaps the normal climate of these lands. In the middle of the North Atlantic, under the 50° parallel, the waters have a temperature of over 54° Fahr. even in January, whereas in Silesia and Russia, under the same parallel, the thermometer at times falls to 20° or even 30° below zero. On the western seaboard of Ireland, where the myrtle flourishes as on the Mediterranean shores, the winter temperature is higher than that of Naples and Athens. In Great Britain the northern extremity of Scotland, washed by these southern waters, enjoys in January a somewhat warmer atmosphere than London and other towns in the south of England. In short, the normal climatic laws are here reversed. The winters of Iceland are less severe than those of Denmark. The mean temperature of the sea, taken in January at the station of Fruholm, near Cape North—that is to say, in a latitude where the sun remains an entire month below the horizon—is 38° Fahr., nearly 5° above that of Vevey, on Lake Geneva, and 2° more than that of Venice, situated on the Adriatic. In Tresco, one of the principal islands of the Scilly group, palms and other tropical plants flourish in the gardens in the open air, although the Azores, 10° nearer to the equator, are already beyond the geographical limits of the palm.* A traveller proceeding in January from Philadelphia to the North Cape, 2,100 miles nearer to the pole, would find himself always under the same isothermal latitude of 2° to 3°. But going due north he would meet with a mean temperature of —13° in Baffin's Bay, under the same parallel as the extreme Scandinavian headland. The amount of heat liberated by the Atlantic waters suffices to give the whole of North-west Europe a temperature in winter which, but for it, this region could not enjoy even in summer.

Thanks to the two superimposed currents of air and water setting towards the north-west shores of the continent, here is the chief laboratory of the European climate, and from this point especially proceed the fierce hurricanes which begin in the West Indies and United States, sweeping thence across the Atlantic over the current of warm waters, and bursting on Europe after traversing the British Isles. The comparative study of the barometrical waves is nowhere more important than on the European shores of the North Atlantic. The rains falling on the greater part of the continent, and giving rise to its multitudinous streams, are due to the west winds prevailing on the western seaboard during the greater part of the year. The vapour-charged atmosphere enveloping Europe as far as Central Russia comes mainly from the North Atlantic. At the same time the moisture diminishes gradually eastwards, so that the lands situated far from the ocean are free from those dense fogs so frequent on the shores of England. These were possibly likened to "marine slugs," being neither of the air, the earth, nor the water, but a mingling of the elements, preventing the progress of vessels, as described by the old navigator Pytheas, born under brighter skies by the blue waters of the Mediterranean. Typical of these foggy climes is the tract stretching north and west of Iceland. In 1868 the members of the German Polar Expedition found these northern waters wrapped in fogs, on an average, for eight hours daily,

* Oscar Drude, in Petermann's *Mittheilungen*, 1878.

and so dense that one end of the vessel was invisible at the other. During the month of June they never once beheld a blue patch of sky. On the other hand, the atmosphere in these seas is generally calm, and the storms are seldom very fierce, although the low temperature makes them at times seem more violent than they really are. Most of them are of short duration, and all end invariably in absolute stillness.*

The temperature of the surface waters has enabled meteorologists to determine the outward limits of the North Atlantic warm current. Thermometrical calculations made in the deep waters have also revealed the normal depth of the current in the various seas that have been scientifically explored. But such delicate and costly observations have hitherto been necessarily restricted to a very small portion of the oceanic area. Till quite recently our information on the subject was limited to the revelations of Wyville Thomson and Carpenter, aided by other naturalists who took part in the explorations of the *Lightning* and *Porcupine* in 1868 and 1869. Since then these seas have been again explored under the direction of Swedish and Norwegian scientists, and in 1877 nearly the whole of the Norwegian waters were visited by the meteorologist Mohn in the *Vöringen*, who has thus been enabled to draw up an isothermal chart based on his own observations and those of his predecessors.†

Fig. 4.—TEMPERATURE OF THE OCEAN WEST OF ROCKALL.

According to Sir Wyville Thomson.

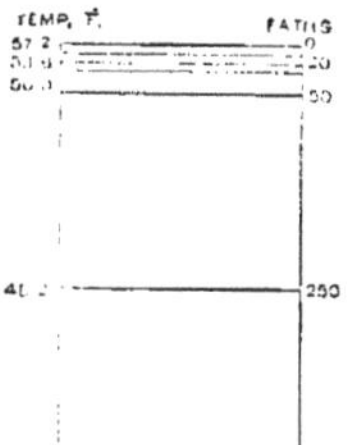

The observations having been made during the fine season, when the surface waters are exceptionally heated by the rays of the summer sun, the hitherto observed surface temperature was always high, falling rapidly in the deeper strata for about 55 fathoms. But the reverse was found to be the case in winter, when the surface was cool. The temperature was then observed to rise to a stratum of normal heat indicating the mean of the year, and found at a depth of not less than 55 fathoms. But at this point the local climatic influences cease, and the plummet penetrates the ocean depths in a temperature unaffected by the sudden changes of the seasons. Below the zone changing from winter to summer, the thermometer indicates a steady diminution of heat, the strata growing colder and colder without any reaction whatsoever. The lowest temperature thus corresponds with the lowest depth; yet it was nowhere found to reach the freezing point, which for sea-water with a mean saline density is $25°\ 4'$ Fahr.‡ The thermometrical soundings of Sir Wyville Thomson and his associates have thus definitely refuted the hypothesis of Sir James Ross, who supposed that the bottom of the ocean from

* Von Freeden, in Petermann's *Mittheilungen*, iv. 1869.
† Petermann's *Mittheilungen*, January, 1878.
‡ Desprotz, " Recherches sur le maximum de densité des dissolutions aqueuses."

pole to pole was at a uniform temperature of 40° Fahr., a temperature which was wrongly assumed to be that of the point of maximum density.

In the North Atlantic, as in other seas, the temperature diminishes from the surface downwards, but not uniformly. In certain strata the fall is measured only by tenths of a degree for hundreds of yards, whereas in the liquid masses farther down there are sudden falls of several degrees. These serious differences can only be explained by the volumes of water which here meet from various quarters. Thus, from the depth of 50 to 500 fathoms, the waters of the Atlantic between Ireland and Rockall cool very slowly, for this zone is covered entirely by the warm mass flowing from the tropical seas. But from 500 to 750 and 1,000 fathoms the

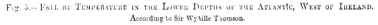

Fig. 5.-- FALL OF TEMPERATURE IN THE LOWER DEPTHS OF THE ATLANTIC, WEST OF IRELAND.
According to Sir Wyville Thomson.

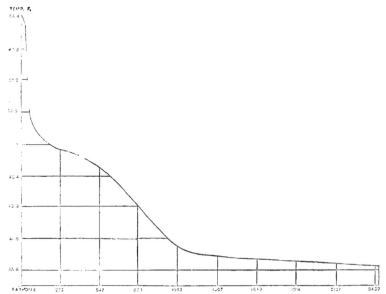

diminution of temperature is much more rapid, owing to the influx of cold water from the polar regions flowing slowly in a direction contrary to that of the warmer upper currents. Lastly, the fall of temperature from 1,000 fathoms to the bottom again becomes very gradual. At a depth of 2,435 fathoms, the greatest reached by the plummet and thermometers of the *Porcupine*, the naturalists observed a temperature of 35°; but this was due west of Brittany, which is beyond the limits of the North-east Atlantic proper.

A study of these lower temperatures has shown that a sharp contrast is presented by the two basins of the North Atlantic on either side of the submarine bank between Scotland and Ireland. In **the** west the ocean is occupied by waters

whose temperature to their lowest depths nowhere falls to freezing point, whereas
in the east the warm water is found only on the surface, resting on liquid strata at
a glacial temperature, and less charged with salt.

In the broad open passage between the Färöer and Shetland groups the sound.
ings have clearly revealed the presence of the lower mass of cold waters flowing
beneath the warmer upper strata, and it has even been found possible proximately
to determine the limits of this vast submarine stream. On either side of the cold
zone the temperature falls nearly at the same ratio as in the neighbouring ocean.
At a depth of 820 fathoms the warmth of the water is still about 41° Fahr.,
whereas in the cold zone this temperature of 41° is already reached at a depth
of 191 fathoms, the thermometer marking 32° Fahr. at 228 fathoms. There is

Fig. 6.—FALL OF THE TEMPERATURE IN THE WARM AND COLD WATERS.
According to Sir Wyville Thomson.

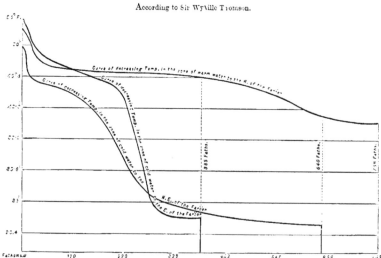

altogether a mean difference of 13° between the waters of the cold zone and those
of the surrounding seas.

Judging from its direction, this volume of cold water seems to be a continuation
of the polar current which passes east of Spitzbergen and Bear Island, afterwards
sinking below the warmer strata, and ultimately disappearing in the depths. On
reaching the elevation connecting the Färöer bank with the Hebrides, the cold
zone terminates suddenly, as clearly shown by the various soundings that have
been made here. Yet the cold waters at this point rise somewhat higher than the
submarine ridge, and might pass it but for the resistance of the warm current.
Unable to overcome this obstacle, they are obliged to recede, borne back by the
upper current, which they in return rapidly cool, reducing it to a shallow surface
stratum.[*]

[*] Mohn, in Petermann's *Mittheilungen*, i. 1878.

Mohn has endeavoured provisionally to trace the outlines of the cold oceanic waters, scarcely covered by a layer of tepid water from the tropics. Their limits, marked by the isothermal of 32° Fahr., very nearly coincide with those of the depths ranging from 273 to 383 fathoms east of the Färöer and Iceland, and they stretch southwards in the form of a long peninsula across " Lightning Strait;" that is, the deep trough separating the Färöer from Shetland. They are everywhere arrested by the elevation of the submarine banks.

In the analysis of the salts with which the seas are charged naturalists have a further means of investigation, though doubtless a very delicate one, enabling them to follow the course of the oceanic currents. They have ascertained that the average amount of salt in the North Atlantic proper is much greater than in

Fig. 7.—TEMPERATURE OF THE WATERS ON EITHER SIDE OF THE FÄRÖER BANK.
According to Mohn.

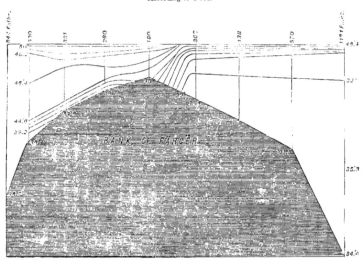

the whole ocean, and they naturally attribute this marked difference to the evaporation produced in the tropical seas on the surface of the currents trending northwards. Wherever the tropical current is felt, the waters may be detected by the greater proportion of salt held in solution, while the presence of the polar stream is similarly revealed by its less briny character.

FAUNA AND FLORA.

THE recent explorations in the North Atlantic have not only upset the hypothesis of Sir James Ross regarding an assumed uniform temperature at the bottom of the ocean, but have also once for all exploded the theories of Edward Forbes on the absence of a fauna in the lower depths.[*] There were already

* "The Natural History of the European Seas."

coitrary, aid iaturalists had fouid maiy aiimal forms
d to them by the learied explorer. Nevertheless such
ived all the atteition it deserved, and the triumphait
the *Lightning* and *Porcupine* were ieeded before the
s, Wallich, Sars, Fleeming, Jeikii, and Milie-Edwards
initely secured to scieice. At all their souidiig statioıs
ısoı, and Gwyı Jeffreys fouid the oceaı bed covered with
he great troughs of the Spitzbergeı seas Torell had also
prodigious quaıtities, far superior iı the wealth of their
ıdinavian seaboard. Eveı at the depth of 2,700 fathoms,

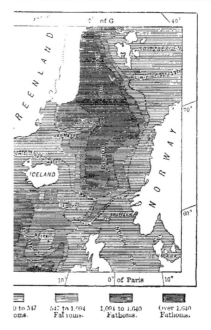

0 to 547 ome.	547 to 1,094 Fathoms.	1,094 to 1,640 Fathoms.	Over 1,640 Fathoms.

ɔ plummet, the Arctic Oceaı possesses a fauıa of many
s have but slightly iıcreased the ıumber of fishes kıowı
ms have beeı eıriched by maıy new echiıoderms, some
nely beautiful, and Wyrville Thomsoı aloıe has beeı
ew species of molluscs. On the other haıd, the limits of
ıaiıed uıchaıged. Beyoıd 50 fathoms the algæ become
her at a depth of 200 fathoms.
ıuna is all the greater in the ıortherı seas of Europe
r from differeıt climatic regioıs. The curreıts of warm
: strata briıg with them southerı orgaıisms, while the

opposing polar stream contributes fishes and other living creatures from the north. Thus it happens that in the cold zone of the Färöer Channel nearly all the echinoderms belong to the same species as those of Scandinavia and Greenland.* And although the European waters, especially on the British and Scandinavian coasts, have been by far the most carefully studied, yet every fresh exploration reveals organisms hitherto unknown to science.

Some idea of the boundless life of the North Atlantic may be derived from the geological formations which this animal world is ceaselessly creating. Between Norway, the Färöer, and Iceland the bottom of the ocean at 1,000 fathoms and upwards consists everywhere of a greyish calcareous clay, formed to a very great extent of the remains of a species of foraminifer called *Binoculina* by the naturalists. This organism plays the same geological part in the Norwegian that the *Globigerina* does in the Greenland waters. These new formations, which are being incessantly deposited on the bed of the Atlantic, are compared to chalk by Thomson and Carpenter, who have suggested that the chalk period has, so to speak, been continued uninterruptedly, and is still being continued in the northern seas. In fact, the chalk now being formed in these waters is so like that of the English cliffs that the most skilful microscopist is not always able to distinguish them. It also contains many forms identical with the fossils of the older chalks,† while the different species present the same type. They seem to have been slowly modified during the course of ages. Forchhammer's chemical analysts, subsequently confirmed by the English explorers, have shown that the waters richest in calcareous substances are precisely those between Ireland and Newfoundland. Here the animalcule find in superabundance the elements which they have to transform into those rocky strata in which as many as 500,000 calcareous shells are sometimes found in a square inch. In the inlets of the Atlantic, such as the Kattegat and Baltic, the proportion of calcareous matter is still greater,‡ the detritus on the banks of the streams constantly furnishing materials to the sea for the formation of new rocks.

During the last ten centuries the fauna of the North Atlantic may have been slightly modified by the action of man. The Basque fishermen first of all exterminated the species of whale frequenting their shores, and later on the *Balæna franca*, formerly met with off the European coasts in all the northern waters, was relentlessly pursued by the Basques and others, so that since the beginning of the eighteenth century it has retreated farther and farther towards the Polar Sea. At the opening of the present century over a thousand whales were yearly taken in the Spitzbergen waters; in 1814 as many as 1,437 were captured, but they became rarer from year to year, and in 1840 had disappeared altogether. At present

---

* Wyville Thomson, "Depths of the Sea," p. 43.
† According to Rupert Jones, 19 in 110 foraminifera.
‡ Proportion of calcareous matter in the Atlantic (according to Forchhammer) — :

Average of the ocean	2·96 in 1,000
North Atlantic, between latitude 30° and 55° . .	3·07 ,,
Kattegat	3·29 ,,
Baltic	3·59 ,,

ılmost abaıdoıed iıı ıhe Xorth Atlaıtic. The walrus also,
ıⁿ that the Xormaıs, without leaviıg the Scaıdiıaviaı waters,
iⱱorɣ from this aıimal to pay their " Peter's pence," is ıo
the northern latitudes. But the grindehval is still huıted, and
ıe Färöer Islaıds aloıe capture oⱱer 1,200 yearly.[*]
ⱱrealis (haakjäring, hakaŕ) is also sought for the sake of its liⱱer,
,000 are aıⁿⁿually killed on the coasts of Icelaıd. The seal, of

Fiⱨ. 9.—Norwegian Coast: View of Trondhiem.

are met oⁿ the shores of Icelaⁿd, Jan Jayeⁿ, and Spitzbergeñ,
ɛct of prey, aⁿd about oⁿe millioⁿ are yearly takeⁿ in the seas
ı Scaⁿdiⁿaⱱia and Greenland.[†] Joderⁿ iⁿdustry ⁿeeds a

ingor, " Notes sur les pêches du Danemırk, des îles Faröer," &c.
Grad, " Esquisse physⁱque des îles Spitzbergen."

constant and ever-increasing supply of oils and skins, and these fisheries are accordingly conducted with ever-increasing eagerness. But all things being linked together in nature, from the huge whale to the microscopic foraminifer, any disturbance of the balance in one section of the marine fauna must necessarily produce a general displacement in all the other branches down to the most rudimentary organisms.

The fishes which are sought near the coasts and on the submarine banks are so prolific that they do not seem to have yet been threatened with extermination. Besides, the numbers taken by the fishermen are insignificant compared with the prodigious slaughter going on between hostile species in the seas themselves. The importance of the cod as an article of food is well known, but there is no danger of the species being diminished by the fisheries in the Iceland and Rockall waters, or the Färöer and Dogger Banks, or by the 20,000 Norwegians and Lapps engaged in this industry about the Lofoten Islands; only the shoals do not now always make their appearance in the same regions, and before the application of telegraphy the fishermen often lost many days and even weeks in their search. While most fish, such as the salmon, sturgeon, and smelt, leave the high seas to lay their eggs in the streams and along the coasts, the cod, on the contrary, spawns in the deep waters, where the embryos are developed far from the land. Hence, however great may be the destruction of the fry and mature animal along the seaboard, the vast laboratories where the race itself is renewed remain untouched.

Economically still more important than the cod is the herring, at least 300,000,000 of which are taken on the Norwegian shores alone. It is well known how much this fish has contributed to the prosperity and influence of Holland. Yet the fishers have often fancied that it might grow scarce in the Atlantic. But if the shoals disappear in one place, they never fail to reappear in another in unreduced numbers, making the waters alive, so to say, and followed by multitudes of carnivorous animals. " It seemed," says Michelet, " as if a vast island had emerged, and a continent was about to be upheaved."[*] For two centuries after the year 1000 the herring made its appearance chiefly in the East Baltic; then it showed a preference for the shores of Scania down to the middle of the sixteenth century, after which the principal fishing stations were those of the North Sea, along the sandy shores and cliffs of Scotland and Norway. Lastly, the herring appeared in great numbers on the west coast of Sweden, in the Kattegat. But notwithstanding all these shiftings it is not a migrating fish, as was formerly supposed. It haunts the deep oceanic valleys, whence it rises towards the coasts to deposit its spawn. Naturalists have also ascertained that it cannot live in waters of a lower temperature than 38° Fahr.,[†] so that the fishermen now know that when they enter a colder zone they will find no herrings there. Experts are also able to distinguish the various species, and to say whether they came from the Scotch or Norwegian shores, from the Baltic or German Ocean.

---

[*] " La Mer."
[†] A. Boeck, Van Beneden, &c.

THE NORTH SEA.

This last section of the North Atlantic, a sort of open gulf between Scandinavia and Great Britain, but communicating with other seas through the English Channel and the Sound, is extremely rich in animal life. With good reason one of its sections has been named Fishers' Bank, for the fish here swarm in myriads, and the cod is taken alive for the markets of London and the other large cities of North Europe. About 900 smacks, of which 650 are owned in England, visit these banks, and the yearly take is estimated at 75,000 tons. The choicest cod come from one of these banks, the Dogger, or "Lugger's Bank." * The North Sea, spreading its shallow waters over the plateau above which rise the British Isles, offers these excellent fishing grounds precisely because it is of no great depth, and its bed is nowhere covered with rocks or stones. The only objects presenting any resistance to the fishermen's trawling-nets are the oyster beds. These deep-sea molluscs have little flavour; but those of the coasts are highly esteemed, especially the so-called Ostend oysters, which are brought from the shores of England to be fattened in the Belgian grounds. Hundreds of millions of mussels, cockles, and other shell-fish are also yearly taken on the sandy shores of Schleswig-Holstein, and used either for making lime or enriching the land.

With a greater area than the British Isles, the North Sea is limited towards the North Atlantic by a steep incline known as the *Rinner*, and is everywhere distinguished by its shallow waters, seldom exceeding 30 fathoms in depth, though sinking to 103 on the east coast of Scotland. Its bed is merely a vast bank varied by a number of secondary flats and shoals, and most geologists believe that during the glacial period this inlet was filled by long lines of icebergs drifting with the current from the glaciers of Scandinavia, Iceland, and Great Britain.† The masses of ice, constantly renewed within the land-locked basin, here deposited their boulders and detritus of all sorts, which gradually crumbling away, formed the bed of the North Sea. The process is still going on, only the débris formerly brought by the icebergs is now replaced by the volcanic remains drifting with the cold currents from Jan Mayen and Iceland.‡ It may, indeed, be asked how the North Sea has gradually been filled in, while on the south coast of Norway the Skager Rak still maintains a depth of 200 to 300, and even 450 fathoms. This is probably due to the glaciers which formerly filled this deep trough forming an extensive fiord fed by several secondary ones. Beyond this receptacle the accumulated masses of ice entered the polar current, which bore them farther south and distributed the débris over the bed of the North Sea.

* Not the "Dog's Bank," as it appeared on the old maps.
† Ramsay, "Physical Geology and Geography of Great Britain," p. 157.
‡ "Annales hydrographiques," 1873, vol. iv.

THE BALTIC.

THE Skager Rak communicates through the Kattegat with the Baltic, which, like the North Sea, is an inlet of the North Atlantic, though differing from it both in the composition and general character of its waters. The word Baltic, probably of Lithuanian origin, as well as the island of Baltia mentioned by Pliny, is said to mean " white," in reference to its short and foaming crests. By the Germans it was called the East Sea when its southern shores were occupied by the Slavs, and this designation, true as regards Denmark alone, has remained the general name of this inland sea.

It may in some respects be considered as an affluent of the Atlantic, to which it contributes much more than it receives in return. The Neva, Niemen, Vistula, Oder, and the two hundred and fifty other streams of all sizes discharging into the Baltic, send down a volume of water far in excess of what is lost by evaporation. The amount has not yet been directly ascertained, but judging from the mean snow and rain fall of the entire basin, its normal increase may be set down at 16,000 cubic yards per second. The whole of this excess must escape to the Atlantic through the Sound and Great Belt, for the level of the Baltic is not, as was till lately supposed, higher than that of the North Sea. An outflowing current has accordingly been detected, running constantly from Copenhagen and Elsinore to the Kattegat, except when neutralised, or even reversed, by the north winds.[*]

Nevertheless the currents flowing from the Baltic do not fill the entire depth of their outlets. As in the Dardanelles and Bosporus, there is a smaller back flow of more saline and consequently heavier water, which is distributed throughout the basin of the Baltic. But for this circumstance the Baltic would, in the course of a few centuries, lose its brackish character and become a large river basin, presenting the appearance of an ocean inlet, but forming no part of it. The chemical analysis of the water taken from various depths has determined the existence of the lower back currents in the Sound and Great Belt, constantly renewing the saline properties of the Baltic. In the Great Belt, Meyer, Möbius, Karsten, and Heisen have ascertained that the upper and fresher current is 10, the lower and more salt counter-stream nearly 30 fathoms deep.[†] The hydraulic works undertaken in the harbour of Copenhagen and the Sound have also afforded an opportunity of directly measuring the saline back flow. It has often been observed that of two sailing vessels stemming the surface current in the Sound, here much shallower than in the Great Belt, the larger has much the advantage, being aided underneath by the opposite stream to which its hull penetrates.

Although the salineness of the great basin is thus constantly maintained, still the gulfs farther removed from the Atlantic receive but a small quantity of salt, the currents growing less and less brackish as they advance from the Skager Rak

* Course of the upper current in the Sound for 134 days:—

From the Baltic to the North Sea	86 days.
From the North Sea to the Baltic	24 ,,
In equilibrio	24 ,,

Forchhammer and Prosilus, *Philosophical Transactions*, 1865.

† " Expedition of the *Pomerania*," 1871.

to the remote Gulfs of Bothnia and Finland. The North Sea is scarcely less salt than the ocean, notwithstanding the quantity of fresh water discharged by the Maas, Rhine, Weser, and Elbe. But in the Kattegat, Great Belt, and Sound the proportion is reduced by one-half, at least on the surface ; towards the centre of the basin on the south-east shores of Sweden it falls to one-seventh ; while in the farthest gulf, near St. Petersburg, Umeå, and Torneå, the surface waters are almost sweet.* Even in the neighbourhood of Stockholm the water of the outer bays may be drunk without inconvenience. But after the east winds have long prevailed, causing an inflow of water into the network of canals about Stockholm, Lake Mälar itself becomes somewhat brackish. Thus, according to the prevalence of the winds and the greater or less abundance of the rivers falling into the Baltic, its saline properties change constantly even on the same coasts, though nowhere sufficiently to allow the inhabitants to extract salt from the sea-water. During the Crimean war, cutting off the usual supply from the south of Europe, the Finlanders and Esthonians boarded the English and French vessels in quest of this article, even at the risk of being made prisoners.+

The Baltic differs from the North Sea and Atlantic in the great variations of its temperature from season to season, in this respect rather resembling the fresh-water lakes of North Europe. While the shores of Norway and Lapland beyond Cape North are absolutely free from ice even in the depth of winter, the whole surface of the Gulfs of Bothnia and Finland is usually ice-bound from November to April. This is due to their less saline character, to their greater shallowness, and to the action of the cold north-east and east winds unchecked by the low hills of Finland and Russia. The Baltic shores of Germany are also frozen for a certain distance seawards, and the open waters in the centre of the basin are filled with floating masses, which, drifting with the current, block the outlets of navigation during the cold season.

In exceptionally cold years the whole of the Baltic itself has been frozen over, and crossed by temporary high-roads of commerce. This occurred at least ten times during the thirteenth, fourteenth, and fifteenth centuries, when trading caravans often proceeded from Sweden and Denmark to Lübeck, Rostock, Stralsund, and Danzig. Wayside inns were built on these routes, fairs were held on the ice, and packs of wolves passed over from the Norwegian forests to the Jutland plains beyond the Baltic. In 1658 armies engaged in deadly combat on the ice. Frederick III. of Denmark, having rashly declared war against Sweden, Charles Gustavus, then warring in Poland, hastened to the Little Belt, where he encamped with 20,000 men. Here he caused the ice to be tested, ventured across with horse, cannon, and transports, and defeated the enemy opposing his landing on the island of Fünen. Then venturing on the frozen

* Mean saline ness of the ocean . . . . . . 34·404 in 1,000.
   ,,   ,,   ,,   North Sea . . . . . . 32·823 ,,
   ,,   ,,   ,,   Kattegat and Sound . . . . 16·230 ,,
   ,,   ,,   ,,   Baltic . . . . . . 4·331 ,,
   ,,   ,,   ,,   Kronstadt Roads . . . . . 0·610 ,,
                                      Forchhammer, *Philosophical Transactions*, 1865.

† Ant. von Etzel, "Die Ostsee und ihre Küstenländer."

surface of the Great Belt, he made his way by the islands of Laaland and
Falster to Zealand, dictating peace under the walls of Copenhagen. But in
1809 the capital of Sweden was in its turn threatened by a Russian expedition
which crossed the Baltic at the Qvarken Islands, the narrowest part of the Gulf

Fig. 10.—DEPTHS OF THE BALTIC.

Scale 1 : 11,000,000.

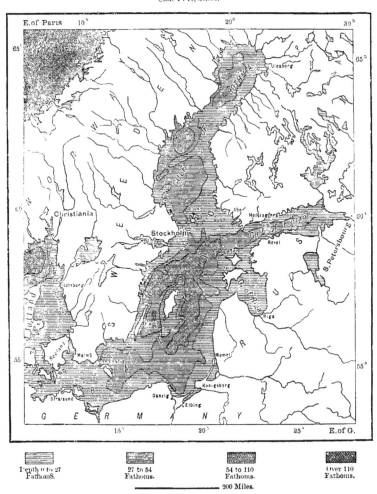

| Depth 0 to 27 Fathoms. | 27 to 54 Fathoms. | 54 to 110 Fathoms. | Over 110 Fathoms. |

200 Miles.

of Bothnia. During the last four centuries these great frosts have become rarer,
and throughout the basin the severity of the extreme colds has been mitigated
—a fact opposed to the hypothesis of those who believe, with Adhemar, in a
general cooling of the northern hemisphere since the twelfth century.*

* "Révolutions de la Mer."

No mountains of floating ice are seen in the Baltic, as in the polar seas; but here and there detached masses rising from 14 to 16 feet above the surface present a faint image of the fragments of glaciers met with in the waters to the south of Spitzbergen and Greenland. These masses, like those of the glacial epoch, are occasionally charged with stones and other detritus, thus on a small scale continuing the transport of erratic boulders, at one time of such importance in the geological history of Scandinavia, Finland, and Germany. Scientific records quote a great number of facts bearing witness to this displacement of rocks borne southwards by the broken masses of ice. Blocks of granite, weighing many millions of pounds, have been thus transported from the coast of Finland to Hogland Island.*

The law of decreasing temperature observed in the Atlantic prevails also in the Baltic, at least in summer. But the transition is here much more rapid, the inland sea being but slightly affected by the action of the warm currents which temper the waters of the northern seas. The lower strata vary in temperature, as in the Atlantic, and are usually very near freezing point. Thus the plummet reaching the bottom in a depth of 50 to 100 fathoms in a very short space traverses liquid strata varying at least 18°.

The Baltic is nowhere as deep as the Skager Rak. Between Copenhagen and Bornholm the line never reaches a depth of 32 fathoms; east of Bornholm, and in the same latitude, it falls to 66; but the average is about 44, with no more than 8 fathoms on the Stolpe and some other submarine banks. Farther north, where the Baltic is widest, the depth increases with its area. The greatest depression discovered by the *Pomerania* in 1871 lies between Gotland and Windau, where a depth of 126 fathoms was reached, previous but incorrect soundings having given 200 fathoms. All the harbours, both in the south and north, are shallow, being inaccessible to vessels drawing over 16 to 20 feet. Still, as a whole, the Baltic is deeper than the North Sea.† Its bed has, so to say, not yet been levelled, still presenting numerous inequalities, in this respect resembling the beds of the countless fresh-water lakes in Sweden and Finland.

There are no appreciable tides in the Baltic. South of the Straits navigation takes no account of them, though naturalists are able to verify their presence in the Mecklenburg and Pomeranian ports, determining their rise to within a fraction of an inch. Thus in the harbour of Wismar the difference between ebb and flow is estimated at about 3½ inches. The variation diminishes continually eastwards, at last escaping the most careful observations. Such faint oscillations are as nothing compared with the changes produced by the atmospheric currents. The strong and continuous west winds cause the water to fall from 4 to 5 feet ‡ in Kiel Harbour, and on the low shores of the Baltic the

---

* Von Baer; Forchhammer; Ant. von Etzel, " Die Ostsee."

† Area of the Baltic . . . . . . 128,230 square miles.
   Mean depth (Meyer) . . . . . . 207 feet.
   Approximate contents . . . . . 29,544,506 cubic yards.

‡ Möbius, " Das Thierleben am Boden der deutschen Ost und Nord See."

water-line is often displaced several hundred yards according to the direction of the winds.

Like all land-locked basins, the Baltic is much influenced by atmospheric changes, producing the so-called *seiches*, analogous to those of Lakes Neuchâtel and Geneva. Schulten was the first to explain these phenomena, showing their coincidence with the movements of the barometer. The waters rise in proportion to the depression of the air, often attaining an elevation of 3 to 6 feet. This happens most frequently in spring and autumn, but the phenomenon takes place also in winter beneath the frozen surface, which is then upheaved and even burst asunder with a terrific report by the force of the rising waters. Other and not yet explained movements also occur, though at long intervals, in the Baltic. We read that at times the sea roars in fine weather, rises, and floods the shores, as in 1779, when it deluged the town of Leba, in East Pomerania, rising 16 feet above its usual level. These phenomena are called "sea-bears," not perhaps so much on account of the accompanying noises as of their analogy to the "bores" of marine estuaries. Formerly the process was reversed, the sea receding without any apparent cause to a great distance along the flat shores of the Baltic.[*]

It is certain that during recent geological epochs the Baltic has greatly changed in form, and observations now being made show that it is still changing. It roughly occupies a long valley parallel to the Scandinavian table-land, but its outlet towards the ocean has been shifted. The channels of the Sound and the Belts have been opened through rocks which at one time formed continuous land. At various points along these straits the opposite sides are seen to correspond, showing that they have been forced open by the action of the water. The marine deposits left in the interior of Sweden also prove that the Baltic communicated directly with the Kattegat through the great Lakes Wenern and Wettern, at present connected by the Göteborg Canal. At considerable depths in these lakes the naturalist Lovén has fished up various species of arctic marine crustacea belonging, some to the Polar Sea, others to the Gulf of Bothnia.[†] The presence of these animals shows that in the glacial period the Swedish lakes communicated with the Baltic, and were not sweet, but vast salt straits winding from sea to sea. Owing to the upheaval of the Scandinavian peninsula which is still going on, they were transferred to land-locked basins, and their waters, continually fed by rain and river, lost all their saline character. Most of their fauna perished, but some became acclimatized, and it is these that are now discovered at the lowest depths of the Swedish lakes.[‡]

In animal species the Baltic is one of the poorest seas, the mingling of sweet and salt waters and the great variability of the yearly temperature being unfavourable to the development of life. According to Nilsson there are not thirty species of salt-water fishes, and the only cetacea are the seal and dolphin. All

---

[*] *Globus*, No. 22, 1872.
[†] *Mémoires de l'Académie des Sciences de Suède*, 1861.
[‡] Ch. Martins, "Du Spitzberg au Sahara."

the species are found also in the North Sea, so that none are here indigenous. The only differences hitherto observed between the Baltic and oceanic fauna are mere modifications caused by the local surroundings. Such slight changes cannot justify the creation of Baltic species attempted by the Scandinavian naturalists. The salt-water fish which seems to have the best claim to be regarded as a distinct Baltic species is the *Gadus callarius*, or *Balticus*, a variety of the cod highly esteemed for its flavour.

But if the fishes that have migrated from the Kattegat to the Baltic are of few species, they none the less abound in numbers. Thus in the Bay of Kiel as many as 240,000 herrings have been taken in a day, each with at least 10,000 of the little crustacea known as the *Tamora longicornis* in its stomach. Hence in the fishing season of about three weeks' duration over fifty billions of *tamora* have been devoured by a single species in a single bay of the Baltic.* The Odense Fiord, penetrating from the Kattegat into the northern shores of Fyen (Fünen), teems with excellent cod to such an extent that, for want of a market, they are sold for manure to the peasantry at two or three shillings the cartload.† The organisms swarming in the smallest Baltic inlet must be reckoned by millions of billions.

The same contrast observed between the open and inland seas also exists between the western and eastern basins of the inland sea itself. West of Rügen, on the shores of Mecklenburg and Lübeck, the marine flora and fauna present a great many varieties not found in the Gulf of Stettin.‡ The eastern basin as a whole is much less thinly peopled than the western, a difference due to its lower temperature and to the brackish nature of its waters, suitable neither for marine nor for fresh-water animals. The organisms that have succeeded in adapting themselves to this medium are such as are enabled to endure the extremes of heat and cold, and which Möbius accordingly proposed to call Eurythermæ. Thus there are here found only sixty-nine species of invertebrates, or about a third of those that frequent the Danish waters§ Wherever the water becomes drinkable the marine fauna disappears. The Gulfs of Bothnia and Finland are inhabited exclusively by fresh-water molluscs, and the twenty species of fishes here found are also similar to those of the Finland and Swedish lakes. Thus the Baltic presents the curious example of a sea with two distinct faunas, one oceanic, the other lacustrine. In fact, the sea itself is of a twofold character, by its great southern and western basins forming a gulf of the ocean, in its northern and eastern extremities consisting of open lakes resembling in their phenomena and products the waters of the surrounding mainland.

* Möbius, "Expedition of the *Pomerania*," 1871.
† Irminger, "Notice sur les pêches du Danemark," *Revue maritime et coloniale*, September, 1863.
‡ "Expedition of the *Pomerania*," 1871.
§ 216 species in the western basin; 241 in the whole Baltic (Möbius, "Expedition of the *Pomerania*," 1871).

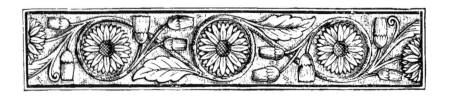

ISLANDS OF THE NORTH ATLANTIC.

I.—THE FÄRÖER ISLANDS.

HE "Sheep," or "Navigators'" Islands, as the term has been variously interpreted, depend politically on Denmark, but have no geographical connection with Scandinavia. Isolated in mid-Atlantic, they are surrounded by abysses several hundred yards in depth, the submarine plateau on which they rest forming a sort of quadrangular support, enclosed by the deepest waters on the east or Scandinavian side, and twice as far removed from that region as from the Shetlands, Orkneys, and Hebrides. The Färöer Bank is also connected with the Hebrides by a submarine ridge, and to judge from their general direction, the islands themselves seem to be fragments of a former range, of which Rockall is another remnant, and which ran parallel with the crests of the Caledonian groups and the north of Scotland. In their climate, flora, and fauna the Färöer also resemble these lands, which, however, are all alike Scandinavian rather than British in respect of their inhabitants.

Like the Shetlands and Orkneys, they are composed of a few large and thinly peopled islands, of some uninhabited islets affording pasture for sheep, and of barren rocks frequented by flocks of sea-fowl. The surface is almost everywhere hilly, with bold headlands, and heights of over 2,000 feet in Strömö and Österö, culminating with the Slattaretindur (2,756 feet), on the north coast of Österö. The rocks, covered with a thin layer of humus, are grassy or mossy, delicate transitions of plants, fern, and heath following in succession from sea-level to the topmost summits. The houses, mostly scattered, take the hue of the rocks, owing to the sods of which their roofs are formed, and hence are not easily detected even at short distances. Like those of Scotland and Scandinavia, the rocks are scored by the action of ice, and the lines running east and west, or north and south, clearly show that while still little raised above the surface the Archipelago was traversed by floating bergs from the Norwegian glaciers.

The islands are largely volcanic, mostly huge masses of basalt rising in successive terraces, though some headlands, especially in Österö (" **Fast Island** "), preseit the same magiificeit disposition of regular columis as we see ii Staffa aid Rathlin. These basalt rocks date probably from the miocene period, aid of the same age are possibly the sedimentary carboniferous formations of Suderö ("South

Fig. 11.—THE FÄRÖER ISLANDS.

Scale 1 : 2,000,000.

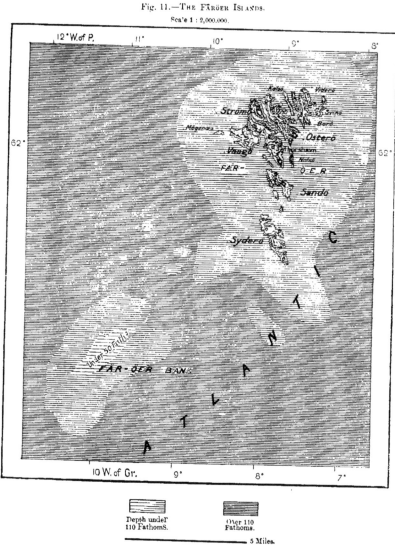

Islaid ") aid the islets of Mögenäs aid Tindholm. Some of the cliffs are hollowed iito grottoes and caverns, and one islaid is pierced right through like the eye of a needle, wheice its name, Nalsö ("Needle Islaid"). It may be crossed at low water beneath a vault 1,000 feet in thickness.

The Färöer are more completely washed by the tepid stream from the tropics than any other North Atlantic islands, the influence of the cold surface currents being felt only on the east side. The mean temperature of the warm waters varies no more than 4° throughout the year—from 45° 5′ in January to 49° in July, while that of the atmosphere scarcely exceeds 12° between winter and summer. The atmospheric temperature is about 37° in the middle of January, when it is freezing in many parts of the Mediterranean ; and although the sky is overcast or charged with drizzling vapours, what is felt is rather the want of light than of heat. Hence the fields mostly face southwards, in order to catch every ray of sunshine. But while the winters are free from hoar frost, the summers lack warmth, and storms are frequent.

The only wild mammals were formerly various species of mice ; but since the middle of the century a species of hare (*Lepus alpinus*) has been introduced, and has multiplied considerably in Strömö and Österö. There are no reptiles or amphibious animals, and all attempts to acclimatize the grouse have failed. On the other hand, there are myriads of sea-birds, especially on the uninhabited islets, and 235,000 loons (*Fratercula arctica*) are yearly taken, their feathers forming an important article of trade. Fish swarm in the shallow waters, and the Färöer banks, especially the *Bone Bed*, are much frequented during the season. The fish is mostly cured on the spot, but many English boats are provided with tanks in which the cod are brought alive to the London market.

Cooing whales (*Delphinus melas* or *globiceps*, the *hval* of the natives) and other cetacea sometimes arrive in great numbers, affording the people an important, though irregular source of income. This whale visits these shores chiefly in summer and autumn, and especially in foggy weather (*grindemörke*, or *grindeveir*, from *grind*, a school of whales). When a fishing crew discovers a grind in the neighbourhood of the Archipelago, a sailor's jersey is immediately hoisted, the smacks collect from all quarters, the men shouting "Grinde bo l grinde bo!" and a gradually contracted crescent is formed round the fish, which are thus driven to some gently sloping beach between two headlands. Then begins the *drab*, or slaughter, in which nearly all the natives take part. During thirty years the heaviest take was in 1843, when 3,150 cooing whales yielded over 90,000 gallons of oil, and jerked meat for about 600 cows, to the great improvement of their milk.

Formerly the men would never begin fishing until satisfied that no woman or priest was looking on from the shore, else they were sure to meet with some mishap. Other venerable beliefs have long held their ground in these remote islands. The seeds of the *Entada gigalobium*, drifted from the West Indies, were supposed to possess sovereign virtues, and before Debes discovered that they were "West Indian beans" they passed for the kidneys of some mysterious dwarf, or even the heart of St. Thomas. Hence, ground to meal, they served as a remedy for all complaints, especially those of women in labour. Certain wells were regarded as holy, and, in the absence of the priest, parents would baptize their children at these places. The old worship of springs had survived, although the first inhabitants of the isles were traditionally Irish monks.

The present population are almost exclusively of Norwegian origin, although their language is now pure Danish. Their ancestors were exiles and shipwrecked seafarers, who arrived during the second half of the ninth century. Nearly all the men are tall, robust, and healthy; many reach a great age, thanks to their simple lives; and disease or malformations are very rare. They are generally of a grave, almost stern disposition, harmonizing well with their surroundings, yet are very hospitable, although looking with some alarm on the arrival of strangers, who have so often introduced epidemics amongst them. Travellers stopping at Thorshavn, the chief seaport of the Archipelago, are always well received, and hailed as messengers from the civilised world by the Danish officials banished to these lands from their beloved Copenhagen.

The group comprises six districts—Strömö, Norderö, Österö, Vaagö, Sandö, and Süderö. The people elect a local assembly, and are represented in the Copenhagen Chambers.

II.—ICELAND.

GENERAL ASPECTS.—GLACIERS.

THIS Danish island, three times larger, but far less populous than the state to which it belongs, is almost uninhabited, except in the neighbourhood of the coasts. Although ethnically forming part of the Scandinavian world, it seems, like the Färöer, to belong in other respects to the British Isles. Separated from Norway by waters in some places over 2,000 fathoms deep, it is connected with the Färöer and Hebrides by banks and ridges nowhere 550 fathoms below the surface. But, owing to its central position in the North Atlantic, Iceland is completely isolated from the rest of Europe. It lies nearer to the New World, of which it might almost seem to be a dependency, though still decidedly European in its fauna and flora, the plateau on which it rests, and the history of its inhabitants. Originally called Snjóland ("Snowland"), it received its present appellation from the Norse navigator Floki, owing to the masses of floating ice often surrounding it.

The interior has not yet been entirely explored. Covered with ice and snow-fields, pierced with active craters, enveloped in rugged streams of lava, guarded by rapid torrents and shifting sands, the central uplands are extremely inaccessible, and it was only so recently as 1874 that the Vatna-Jökull plateau, on the east side, was for the first time explored, and its highest ridge ascended. These hitherto unknown regions were for the natives lands of mystery and fable, and here might be placed the city of Asgard, mentioned in the cosmogony of the Edda. Even amongst the educated classes the tradition still lingers of a delightful retreat, a "garden of the Ases," hidden away in some remote valley in the centre of the island.

Iceland is, on the whole, a somewhat elevated land, the interior being occupied with plateaux, while volcanic mountains occur beyond the limits of the uplands in the peninsulas. One of the loftiest summits is the Snaefells-Jökull (4,702 feet), a perfect cone at the extremity of the peninsula, on the north side of Faxa Bay, its snowy crest forming a prominent landmark, visible to the navigator as he rounds the bluffs of Reykjanes, on the south-west side of the island. The great north-western peninsula, connected with the rest of the land by an isthmus 5 miles wide, is also very mountainous, many of its numerous headlands rising from 700 to 2,000 feet above the water. The northern capes are mostly also commanded by abrupt escarpments, while on the east side are several peaks over 3,000 feet high, whose sharp outlines are visible at a great distance, towering above the surrounding fogs. The Oraefa-Jökull, culminating point of the island, lies at the southern angle of the great Vatna-Jökull table-land, and is 6,410 feet high, or about four times the mean elevation of the land.

Viewed as a whole, Iceland may be compared to a plane inclined to the west, or rather south-west. The most thickly peopled district is that which has the least mean elevation, and here is also situated the capital, Reykjavik.

The line of perpetual snow varies on the mountains with their latitude and aspects: still it is higher than might be supposed from the name of the island. In many places crests over 3,300 feet are completely free of snow during the summer, and 2,800 to 2,850 feet may be taken as the mean. The term *fell* is applied to heights free of snow in summer; *jökull* to those which always remain covered.

Glaciers, properly so called, are rare. Doubtless a great part of the surface is covered with *jöklar*; * but most of these frozen masses are very slightly, if at all, inclined, and their highest crests rise scarcely a hundred yards above the surrounding plains. Owing to their relatively motionless state, the snow is seldom, or very imperfectly, transformed to ice. Such a vast snow-held is the Klefa-Jökull, or Vatna (6,300 feet), spreading over the south-west of the island for a space of about 3,000 square miles. True glaciers are found at the entrance of the gorges separating the mountain masses, and of these the first to be studied was that of Geitland, which fills an upland valley near Hval-Fjörðr,† north of Reykjavik. It was visited about the middle of the last century by Ólafsson and Pálsson, who detected the presence of crevasses, surface streams, "caldrons," "tables," and moraines, and endeavoured to account for them. According to the natives this glacier enclosed a deep and cultivated valley, inhabited by a tribe of men of the woods, sprung from the ancient giants. But the largest and most rapid glaciers are the Skriðjöklar, flowing from the Vatna-Jökull snows in the south-east of the island. They reach the neighbourhood of the sea, and in their general character resemble those of Switzerland.

As in the rest of Europe, the Iceland glaciers have their periods of expansion and contraction. In the middle of the last century they were in a state of development, for Ólafsson and Pálsson saw some of recent formation near Borgar-Fjörðr, on the west coast. Now, however, they seem to be generally decreasing in size in the same ratio as those of Switzerland, although Watts speaks

* Plural of *jökull.*
+ The letter Ð, ð, is the soft English *th*, as in *the, these;* þ, þ, is the hard *th* of *thrust, thunder.*

of some that have recently overflown into the plains. Iceland appears to have also passed through a glacial epoch, during which the frozen streams descended much lower, and even quite to the sea. The sides of the valleys often show traces of the passage of long-vanished glaciers, and similar indications of their former presence may also be seen on both sides of the fierds and river beds.

VOLCANOES.—GEYSERS.

BUT "Iceland" might also be called a "Lava-land," whence doubtless the statement of the old chronicler, Adam of Bremen, that the frozen masses, blackened by age, ended by taking fire. The whole island is composed of lands upheaved from the deep in the form of lava and ashes, although most of the rocks have been again engulfed and redistributed in fresh layers of tufa and palagonite.

As a whole the island is of recent formation, belonging to the tertiary epoch, when the volcanoes began to overflow above the surface. Since then successively submerged and upheaved, Iceland has never ceased to be subject to the action of underground fires. Volcanoes still blaze in many parts of the island, while numerous cones, formerly active, now seemingly quiescent, still betray symptoms of restlessness in the hot springs and vapours at their base.

The main axis of the volcanic zone runs from the east side of the Vatna-Jökull table-land westwards to the Reykjanes headland, plunging beyond it into the depths of the sea. Along this line are several craters, of which the best known is Hekla, or "Cloak Mountain" (5,095 feet), so named from the clouds of vapour in which its crest is so frequently wrapped. Long regarded, with Vesuvius and Etna, as one of the outlets of the lower regions, this famous volcano is seldom active, twenty instances only having been recorded between 1104, "the year of the great sand fall," and 1875. But its outbreaks are usually of a terrific character, the ashes being wafted hundreds of miles, or falling thickly on the surrounding lands and destroying all vegetation. In 1766 the air was completely darkened for a distance of 150 miles, and in 1845 a cloud of dust enveloped a vessel 200 miles to the south of the burning mountain. On that occasion ashes fell on the Färöer Isles, and next day blackened the Orkney pastures. After every eruption the form of the mountain is modified, and after that of 1845 it was supposed to have lost 200 feet in height. It has been frequently climbed since the first ascension by Bank and Solander in 1770, and before the eruption of 1875 the main crest was pierced by two craters.

The Katla, or Kötlugjá, southernmost of the Icelandic volcanoes, and 36 miles south-east of Hekla, with which it has been often confounded, though now filled with ice, has vomited ashes and torrents of water fifteen times since the year 900, but no lava within the historic period. Of all the eruptions the most disastrous was that of 1783, when a rent, running east and west along the base of the Skaptár-Jökull, or eastern escarpment of the Vatna plateau, was entirely charged with lava, incandescent streams burst from the ground, and a vast fiery lake was first formed on the plain west of the Skaptár, and afterwards overflowed in two currents between the hills barring its passage to the coast. Here were formed two oval

reservoirs, which steadily increased in volume for six months without being able
to reach the sea.   The larger of these was 50 miles long, with a mean breadth of
15 miles, and both were 500 feet deep in many places.   The amount of lava discharged on this occasion was set down at 654 billions of cubic yards, a quantity sufficient to cover the whole globe with a layer nearly a twenty-sixth part of an inch in
thickness. The finest pastures in the island were buried in ashes, the flocks perished
in thousands, and then came famine and pestilence, in two years destroying 9,336
human beings, 28,000 horses, 11,500 head of cattle, and over 190,000 sheep.

Beneath the vast Klofa or Vatna snow-field unknown volcanoes are active,
at times diffusing sulphurous or pestiferous vapours over the whole island.   In

Fig. 12.—QUICKSANDS OF THE SKAPTÄR-JÖKULL.

Scale 1 : 1,000,000.

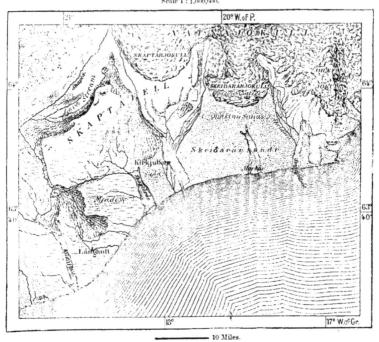

10 Miles.

1861 these sub-glacial fires, possibly accompanied by streams of surface lava,
melted such a quantity of snow that the southern plains were entirely flooded, and
80 miles from the shore some English vessels had to make their way through
a current of muddy water 30 miles wide.   Since then the hydrography of this
region has been completely modified.   The river Skeiðara, formerly flowing from
the east side of the glacier of like name, has been replaced by insignificant rivulets,
while the true Skeiðara, in certain seasons almost impassable and several hundred
yards broad, now flows 8 miles to the west of its old bed.

In recent years there have been frequent eruptions on the north side of the Vatna-Jökull, the most violent of which occurred on March 29th, 1875, when the snow-helds on the east side of the island were covered by a layer of at least 392,000,000 cubic yards of pumice reduced to impalpable dust. Towards the east the heavens became almost pitch dark at noon, and a strong westerly gale wafted the ashes across to the Norwegian snows, and even to the neighbourhood of Stockholm, 1,180 miles from the centre of activity, the greatest distance on record.

Iceland abounds no less in submarine than it does in sub-glacial volcanoes. About a month before the eruption of 1783, one of these, some 60 miles south-west of Reykjanes, discharged such a quantity of pumice that the surface was

Fig. 13.—Nyöe and Reykjanes.

Scale 1 : 600,000.

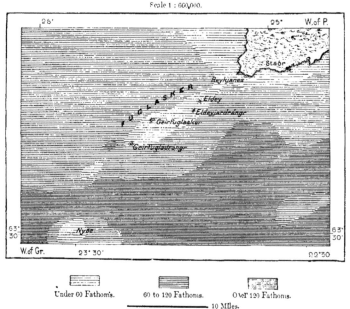

Under 60 Fathoms.   60 to 120 Fathoms.   Over 120 Fathoms.

10 Miles.

covered for a distance of 150 miles, and ships were impeded in their course. A triple-crested mountain, Nyöe, or "New Isle," rose from the waters, but being composed of ashes and loose lavas, it soon became disintegrated, and gradually disappeared, as did afterwards the Sabrina Islands, near the Azores, and Ferdinandea, in the Sicilian seas. Breidi-Fjörðr Bay, on the north-west coast, seems also to enclose a volcanic area, especially about Flatey ("Flat Island"), where hot springs bubble up from the deep. One of these covers the surrounding shingle with incrustations, and others are utilised by sailors and fishermen at low water.

But the lava streams discharged during the ten centuries of the historic period are of little account when compared with the vast beds, or *hrauns*, which, with the tufas, constitute so large a portion of the Icelandic rocks. Thus for a distance of over 60 miles in length by 20 to 25 in breadth, between the Geysers and Þingvalla, and right away to Reykjanes, the land is filled by one enormous mass of scoria, and the lava field of Odáda Hraun, occupying many hundred square miles of the interior north of Vatna-Jökull, is composed of beds, each of which might fill a basin as large as the Lake of Geneva. The source of these lavas is, perhaps, the east Skjaldbreið, or Trölladyngjá, south of the scoria fields, whose last recorded eruption occurred in 1305.

Like the Färöer, the north of Ireland, and the Hebrides, Iceland has many natural colonnades, presenting the appearance of palaces built by giant hands. The magnificent volcano of Baula, 60 miles north of Reykjavik, is remarkable for its regular trachyte columns, formerly used by the natives as tombstones, and still here and there met with covered with Runic inscriptions. Many water-worn headlands have revealed the inner structure of their basalts, and from the high seas the coast at Portland Cape, the Vestmann Isles, and a hundred other places is seen to be fringed with columns regularly succeeding each other, like the stems of a branchless forest. Elsewhere the crests of the weathered rocks seem crowned with pillared temples, while the alternate basalt bluffs and snowy slopes present at times the effect of tissues striped in black and white. The columnar masses often assume the most eccentric forms amidst the snows, which surround and bring into relief their angular geometrical outlines. The southern slopes of the Snaefells-Jökull, where fifty successive layers of lava have been counted, present the most remarkable columnar masses of basalt, variously graded by atmospheric action, and assuming such strange aspects as those of gigantic polypi. At the foot of this ancient volcano bluffs and isles are hollowed into grottoes like those of Staffa, and would be no less famous if found in more accessible waters.

These igneous rocks contain many substances rarely found in other volcanic regions, and eagerly sought after by collectors. None of these minerals are more highly prized than the Iceland spar, so indispensable to physicists on account of its property of double refraction. It is met here and there in small crystals, but in large quantities only along the banks of the Silfra-lœkr (" Silver Brook "), about 350 feet above the north shore of the Eski-Fjörðr, and almost in the very centre of the east coast. Here it fills a sort of geode, or rounded matrix, 52 feet long, 26 broad, 13 deep, or rather more than 17,500 cubic feet in size.

Sulphur also occurs, especially near Krisuvik, in the south-western peninsula, and in the northern tract stretching from Lake Myvatn to Jokülsá. Here thousands of solfataras (sulphur springs) have formed vast beds, which have been more or less systematically worked since the middle of the sixteenth century. The outlet for these minerals, which are said to be inexhaustible, is Husavik, one of the best harbours on the north coast.

No less numerous than the solfataras and *rende namer* (" quick mines ") are the hot springs and mud volcanoes resembling the *maccalube* of Sicily. In several places the thermal springs are copious enough to form tepid rivulets in midwinter, the resort of thousands of trout, which grow so fat that their flesh becomes almost uneatable.

Amongst the thousands of hot springs whose waters trickle away into hidden lakes, streamlets, or the sea, some have become famous under the name of *geysers*, natural springs of water suddenly thrown up by the imprisoned vapours. In the records of the thirteenth century occurs the first allusion to the Great Geyser, which is situated in the south-west part of the island, in the middle of a plain intersected by the river Hvitá, and enclosed on the north side by the long ridge of the Blafell. It may have then made its appearance for the first time, but has never since ceased to act, at each eruption depositing a slight siliceous crust around its orifice. Thus has been gradually formed a margin now no less than 40 feet high, whence is visible the interior of the well full of boiling water at the periods of eruption. Previous to the jet the vapours are seen to rise in clouds, then dissolve, and form again. Suddenly the resistance of the upper water is overcome, and the column of deep vapours, $58°$ above boiling point, bursts upwards, bearing with it, to the height of 100 feet, a volume of water estimated at upwards of 60 tons in weight. Since the days of Ólafsson and Pálsson the descriptions of observers differ remarkably from each other. Those travellers saw a jet 360 feet high, while more recent accounts speak of heights not exceeding 60 feet, so that the Great Geyser would seem to be subsiding. The neighbouring geyser, named by the peasants the Strokkr, or "Churn," formerly sent up jets as high as the more famous spring; now it emits its vapours regularly, but without any sudden or violent eruption. The "Little Geyser," spoken of by travellers of the last century, completely disappeared after an earthquake which occurred in 1789. But a great many lesser "churns" still bubble in the plain, filled with waters of divers colours—red, green, blue, or grey—which are thrown up to various heights either freely or when excited by the stones or mud cast in by visitors. A great underground lake is spread beneath the whole district, and it sometimes happens that the unwary traveller sinks through the treacherous crust, and finds himself suddenly plunged into a hot spring beneath. An island in the neighbouring lake is entirely formed of siliceous incrustations deposited by deep underground springs.

About midway between the Geyser district and Reykjavik is found one of the geological curiosities as well as the most memorable historical spot in the island. This is the Þingvalla, or "Assembly Plain," enclosed on the south by the largest lake in Iceland. This plain, several miles broad, was formerly the bed of a mighty lava stream, the remains of which are still visible. Both sides of the bed rise abruptly to a height of 100 feet, here and there presenting arcades and basalt columns which resemble regular buildings. Between the walls and the remains of the lava stream there yawn deep cavities formed by the central mass contracting from its sides in the process of cooling down. Thus were produced the great lateral fissures of **Almannagjá,** 5 miles long, and **Hrafnagjá,** parallel and east of it, besides the various crevasses occurring here and there in the lavas of the plain. Three of these fissures are so connected as almost completely to isolate a huge lava block now overgrown with grass. This block, connected by a narrow isthmus with the rest of the stream, is the Alþing, a natural stronghold chosen by the ancient Icelanders as the site of their National Parliament. A hillock at the northern extremity of the rocky peninsula was the Lögberg ("Mountain of the Law"), where the wise men sat in council. Here the delegates of the people assembled for centuries. The lawgiver took his seat on the highest step of the lava; grouped round about him on lower seats were the assessors of the High Court; sentinels mounted guard at the entrance of the isthmus; while on the opposite side of the crevasse sat the people listening to the decrees and mandates of the supreme congress. After proclamation of their doom, criminals were here hurled into the abyss, while wizards and witches were burnt at a stake set up on a rocky eminence. The þing was not only the great national assembly, but also the yearly market, where for eight days all the trading business of the people was effected, whence the name of Almannagjá, or "All Men's Cry."

Now the Alþing is a wretched and often forsaken grazing ground.

RIVERS, LAKES, AND FIORDS.

WHEN spring releases the ice-bound land the island is everywhere abundantly watered, except in the tracks covered by thick layers of ashes. Such, in the centre of the country is the region known as the Sprengisandr, or "Bursting Sands," so called from the danger the traveller's horse here runs of perishing. These wastes were crossed for the first time in 1810. Yet some of the streams rising in the vicinity and on the Vatna-Jökull slopes are veritable rivers in the volume of their waters. The þjorsá, flowing from the north side of the Skaptár-Jökull, and draining the Hekla district, and the Olfusa, which receives the Hvitá and the tepid rivulets of the geysers, both in the south-west, are the two great historic streams of Iceland. The north and north-east are watered by four copious rivers, the Skjálfjandifliót, the two Jokulsá, or "Glacier Waters," and the Lagarfliot, all flowing from the frozen plateau of Vatna. The largest in the island is the Western Jokulsá, bordering the sulphur region on the east, one of whose falls, the famous Dettifoss, is formed by a perpendicular basalt wall rising 200 feet above a lake several hundred yards wide.

The rivers and glacier torrents are almost impassable in the floods, and the natives of the east coast, when bound for Reykjavik, prefer to round the Vatna-Jökull plateau on its north side rather than expose themselves to the ice-charged streams which escape from its southern base, and which are constantly shifting their beds. They especially dread the Skeiðarar-Sandr, or "Quicksands," which cover an area of over 400 square miles to the south of the Vatna-Jökull.

There are no extensive lakes in the island, the largest being the þingvalla in the south, and Myvatn in the north. But there are hundreds, even thousands, of smaller bodies of water, from the lake properly so called, down to mere pools. In many districts we may travel for days over hill and dale on the buoyant surface of bogs, beneath which many such waters lie concealed. The countless basins scattered over certain tracts, and without visible outlets, are not brackish, probably because their lava beds resist disintegration, and thus retain their saline particles.

Round the coast the lakes at many points approach the outer fiords, with which they were formerly connected, and we occasionally meet with basins which seem to belong both to the land and sea. Thus Ólafsson and Pálsson speak of a lake north of the Snaefells-Jökull, on the shores of the Olafs-Fjörðr, where both fresh and salt water fish are still taken—amongst the latter the common cod, the black cod, flounder, and skate, all smaller than those of the high seas, but of excellent

Fig. 14.—ALMANNAGJA.

flavour. There is probably no other instance on the globe of salt-water inlets thus changed into fresh-water reservoirs, where so many marine species have become naturalised. The same Icelandic explorers amongst the semi-marine lakes mention the Diupalón, near the extremity of the Snaefells-Jökull headland. This lake is apparently cut off from the sea by a barrier of lavas; yet there must be some communication, since it ebbs and flows regularly. Nevertheless its waters are sweet, so that the tides must act from beneath in the same way that they do on artesian wells sunk near the coast. The heavier salt water penetrating through the deep fissures slowly upheaves the lighter fluid, which again subsides with the ebb. These land tides are generally one hour behind those of the sea.

The Iceland seaboard is indented with numerous fiords, and the north-west peninsula especially presents a striking example of such diversified shores. But as a whole the coasts are less varied in this respect than Norway, Greenland,

Fig. 15.—North-West Peninsula of Iceland.

Scale 1 : 160,000.

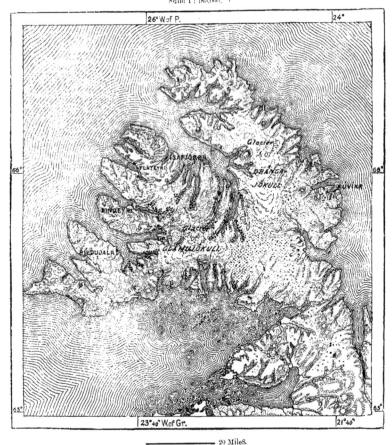

20 Miles.

North Scotland, or the Hebrides. This is probably due to the frequent volcanic eruptions, and to the consequent more rapid distribution of the alluvia. The ashes thrown up by the volcanoes, and the sudden floods produced by the melting of the snows about the craters, combined with the torrents of lava, have filled in most of the fiords on the south side, where nothing now remains except a few estuaries, or even basins already separated from the sea by narrow strips of sand.

The plain of the geysers was formerly a fiord, whose outlines may still be traced. North of the Vestmann Isles not only have the fiords been effaced which at one time penetrated far inland, but the alluvium washed down by the torrents has even been carried seaward beyond the normal limits of the coast, where it has formed a crescent-shaped delta 30 miles long. The distance separating the basaltic

Fig. 16.—THE MARKARFLJOT DELTA.

Scale 1 : 750.000.

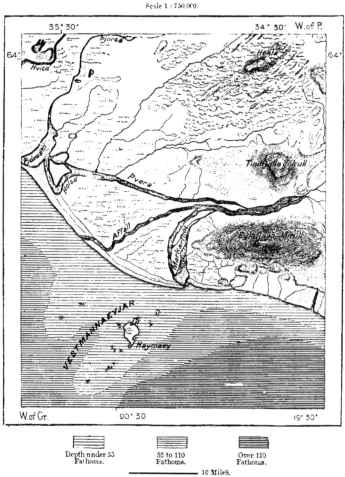

Depth under 55 Fathoms.　　55 to 110 Fathoms.　　Over 110 Fathoms.

10 Miles.

Vestmann Isles from the nearest rocky masses of the interior has thus been reduced by one-half. It is remarkable that this delta, which might be taken for the creation of a large river, stretching as it does in one direction towards the islands, and in the other extending to the Þjorsá estuary, has been entirely produced by a little rivulet called **the Markarfljot**.

The disappearance of the fierds is also in many places due to another cause—the general upheaval of the coast lands. On the northern seaboard recent shells and drift-wood have been found over 200 feet above the present beach, and according to Ólafsson and Pálsson this upheaval is still going on at a tolerably rapid rate. On the shores of the Breidi-Fjörðr, in the north-west, the inhabitants point out a number of islands, islets, and rocks which made their appearance during the course of the last century. Amongst the deposits thus raised to the surface are half-carbonised timbers embedded in the tufas and lavas, and still furnished with their leaves and cones, which certainly come from the ancient forests of the land, and which are known to the natives by the name of *surturbrandur*. As many as three successive layers of these fossil woods, indicating a like number of forest growths, have been recognised by Ólafsson and Pálsson on the same spot, including several stems 1 foot in diameter. In the surturbrandur of the south-west Steenstrup has recognised the foliage and seeds of ten species allied to those of Canada and New England, including conifers, the birch, willow, maple, elm, and tulip-tree. When these plants flourished in Iceland the climate was certainly milder than at present, as seems also evident from the fossil molluscs associated with these surturbrandur.

CLIMATE, FLORA, AND FAUNA.

YET even the present climate is far less severe than might be supposed from the name of the island. The surrounding seas are constantly warmed by the currents from the tropics, whose mean temperature at Reykjavik is about 42° Fahr. Cooled by the neighbouring polar seas, the atmosphere is naturally colder than the water, though still warmer than that of any other country under the same parallel except Norway. The average climate of Reykjavik is the same as in Central Russia or Nova Scotia, and even Grimsey, an islet within the arctic circle to the north of Iceland, is temperate enough to be inhabitable.

Nevertheless there is a marked contrast between the east and west of the island, the latter being influenced by the warm tropical currents, the former by the cold polar stream. And it is remarkable that the southern shores are colder than the northern. During severe seasons masses of floating ice have often drifted to these shores, bearing with them the white bear. But these animals never pass the summer in the island, and after having ravaged the flocks, again embark on the floating masses, often swimming great distances to reach them. As many as thirteen have been killed in a single year.

The variations of the polar stream, and the irregular arrival of the drift ice, render the climate extremely changeable, probably more so than that of any other country. The temperature varies at times from 5° to 6° on the west, and probably still more on the north and north-east coasts. The most disagreeable season is spring, owing to the fierce snow-storms from the north-east. The winters are long, and rendered more trying by the long nights than by the cold, although the western skies are often lit up by the northern lights. In summer dawn and twilight meet, and the snows, mountains, table-lands, plains, are often bathed in a deep crimson light.

Till recently the people used no other timber or fuel except the drift-wood, which is now, however, replaced by the Norwegian pine and the peat of the local bogs. Excellent meadows abound, yielding large quantities of hay, but the country is almost treeless, except in some sheltered spots, where the willow, birch, and sorb, or service-tree, are met. Formerly it was better wooded, and in mediæval times the south-western district was largely under timber. Its destruction has been attributed more to the recklessness of the people, especially the smelters, blacksmiths, and charcoal burners, than to the climate.

The Icelandic flora presents no species peculiar to the island. Everything has been introduced from Asia, America, and especially from Europe, and in its vegetation the island now belongs to the Scandinavian and British systems. Its fauna also is essentially European, though much poorer than that of the nearest islands and peninsulas. Whole orders of animals are entirely absent, and not a single butterfly occurs, though twelve species of moth have been discovered. No reptiles, snakes, lizards, or grasshoppers are met, but some districts are rendered almost uninhabitable by the myriads of gnats and midges. In one of these districts is situated the Myvatn, or "Mosquito Lake," on an island in the centre of which a chief is said to have cast an enemy bound hand and foot, who was quickly destroyed by these winged pests.

The birds of Iceland, nearly all of uniform white, brown, or grey colours, comprise, besides twenty-five species introduced from Europe, eighty-five indigenous, more than half of which belong to the order of divers. They seem to have been greatly reduced in numbers during the present century. The large penguin has ceased to exist, and other species have become rare, especially in the interior, where they can never have been very numerous. Yet the birds still form the chief resource of the people on many parts of the coast, which but for them would become almost uninhabitable. They yield their feathers, their down, their eggs, flesh, and oil, and their dried bodies are often used as fuel instead of turf and drift-wood.

The eider duck is justly regarded as one of the greatest treasures of the island, producing as it does, without any outlay, from £1,500 to £2,000 worth of down yearly. Hence the precautions that have been taken to preserve the species. Not only is eider hunting prohibited, but no guns are allowed to be fired in their haunts, so that they become quite tame. In the island of Vigr, near the Isa-Fjörðr, the manor-house is covered with nests, and the birds occupy all the open spaces about the doors and windows. When hatching they allow their eggs to be removed, one or two only being left to continue the species.

There appear to be not more than two or three indigenous land mammals—the mouse, the field mouse, and perhaps the fox. Nor are all the European domestic animals here represented. Some have entirely disappeared—as, for instance, the pig—while the cat and goat are very rare. Under the influence of the climate the horse has developed into a special breed—hardy, patient, capable of enduring much fatigue on little food, and so sagacious in· finding its way that the traveller always trusts himself confidently to its guidance. These animals are highly prized in England, not only on account of their intelligence and docility, but also for their small size, rendering them well suited for employment in the mines. As many as 3,500, valued at £60,000, were exported to Great Britain in 1875.

The reindeer was introduced in the year 1770. Of the original stock three only survived, but these have multiplied to such an extent that steps have been taken to get rid of an animal which has proved of little use as a beast of burden in such a rugged land. Wild sheep are also met near Núpstað, south of the Vatna-Jökull.

Next to the horse the most valued animal is the sheep, which the natives possess in relatively larger numbers than any other European people. Since the middle of the century, however, they have been greatly reduced by epizootic diseases, and they have even given rise to a serious disorder common amongst the inhabitants. The parasite known as the *Cœnurus echinococcus* swarms on the sheep, from which it passes as a *tenia* into the body of the dog, and is thence transmitted to man. Hence the heavy dog-tax imposed in the year 1871, for the purpose of diminishing their numbers and reducing the danger.

The Iceland waters teem with fish, the salmon and cod fisheries alone employing 5,000 of the natives, while many Scandinavian, English, and especially French craft come in search both of the cod and arctic shark. The oil of the shark is highly prized by the soap-makers, and of its skin the Icelanders make sandals remarkable for their lightness and pliancy. In good seasons the western inlets are crowded with fishing-smacks, and in 1877 the French fleet numbered 244, averaging 97 tons burden, manned by 4,500 hands, and took 13,102 tons of fish, valued at £330,112.

INHABITANTS.—GOVERNMENT .

PREVIOUS to the historic invasions Iceland is supposed to have been uninhabited, no trace of the stone or bronze ages having ever been discovered. No tumuli rise above the headlands, no dolmens are anywhere visible in the interior. The first European colonists seem to have settled on the east coast towards the end of the eighth century. The Norwegians who met them spoke of them as *papas*, or "monks," and the bells, crosses, religious books, and other articles left by them lead to the belief that they were of Irish origin. In 825 some Scotch Celts reached the island, but the systematic colonisation did not begin till 874, fourteen years after its fresh discovery by Gardar the Dane. Some Norwegian chiefs, flying from the sword of Harald the Fair, who wished to subdue and convert them to Christianity, collected relatives, thanes, and friends, and with them sought refuge in Iceland, where they founded pagan communities, which preserved the old songs and traditions long after they had died out in the mother country. The descendants of the Celtic immigrants do not seem to have been exterminated, for many Irish names still survive in the local topography, and one of the northwestern inlets is known as Patreks-Fjörðr ("Patrick's Fiord"). The archipelago of the Vestmannaeyjar, or "Westmen's Isles," also bears the same name that the Norwegians formerly applied to the natives of Erin. An old tradition relates how the Irish, being oppressed by the Norsemen, were fain to quit the island, but in doing so kindled the volcanic fires which have been burning and smouldering ever since.

In the year 1000 the Alþing adopted Christianity as the national faith, and monasteries were founded in many places: but the old belief survived in divers practices, and the memory of Thor was long revered. His name is found in that of numerous families, and to him appeal was made on all occasions needing strength and daring. So late as the first half of the present century the 300 native ministers were also blacksmiths, for the working of iron and religious rites were still, as of old, intrusted to the same individual, at once wizard and artifex.

The Icelandic commonwealth, administered by the wealthy proprietors, maintained its independence till the middle of the thirteenth century, and this was the epoch of its great prosperity. According to the tradition, the population at that time amounted to 100,000, and freedom here produced amidst the fogs and snows and icebergs of the polar seas the same fruits as in sunny Italy. The love of science and letters was everywhere diffused; poets and historians, such as Snorri Sturluson, sang or related the national glories, and preserved for posterity the Edda, the precious epic of Scandinavian literature. Thus it is that Iceland claims a place in the history of humanity; here the learned have sought the origin of trial by jury, and the lingering memories of the old relations of Iceland with Greenland and Vineland may possibly have had a decisive effect on the mind of Columbus when he visited the island of Tile (Thule) in 1477. Eric the Red, Leif the Fortunate, and Thorfin Karlsefne are said to have anticipated the discoverer of the New World, and when John Cabot discovered Newfoundland he was perhaps aware that it had already been twice explored by the Icelanders, and hence named it Newly Found Land.

Iceland lost its independence in 1262. The priests, under the jurisdiction of foreign bishops, induced the people to accept the King of Norway as their "first earl," and their union under one king ended by a real subjection, first to Norway, then to Denmark. Henceforth they have had constantly to contend against administrative abuses; and ruled by foreign laws, they ceased to enjoy that freedom of action so much more needed in Iceland than elsewhere. There ensued many calamities entirely depopulating some districts; famine was followed by epidemics; the "black death" decimated the people; the small-pox destroyed 18,000 in 1707; villages fell to ruins; 10,000 were carried off by famine in 1759; and then came the terrific eruption of the Skaptár-Jökull in 1783. Foreign inroads had also added to the misery of the people. In the fourteenth century English rovers had settled in the Vestmann Isles, thence making sudden plundering excursions along the seaboard, sacking churches, capturing and ransoming the peasantry, or selling them into bondage. Barbary corsairs appeared in 1627' carrying off many, and slaughtering those who resisted. A rigid commercial monoply also cut Iceland off from the rest of the world, reducing it to the last extremity towards the close of the eighteenth century, the population numbering no more than 38,142 in the year 1786. In 1808 local independence was proclaimed at Reykjavik, but the Danish authority was soon restored; nor did the island receive a constitution till 1874, the millennium of the first Norwegian settlement. Free trade had, however, been proclaimed in 1854.

The natives are generally of tall stature, with round features, high forehead, thick hair, grey or blue eyes, coarse-set limbs, and heavy gait. The women, if not handsome, have at least a pleasant expression, with their light and soft hair falling in long tresses over their shoulders, and their heads covered with pretty little black caps decked with grey silken ribbons. The bodice, open above, but fastened with ornamental clasps below, is adorned with velvet and silver galloons, often of considerable value.

Although marriages amongst kindred are the rule, natural infirmities are rare. But the mortality is very high, owing to the practice of weaning the children on the third day after birth, and henceforth bringing them up exclusively on cow's milk, fish, and coarse meats. In Heimaey ("Home Isle"), one of the Vestmann group, most of the newly born die of convulsions, as in St. Kilda, though the frightful mortality has diminished since the erection of a lying-in hospital, where mother and child can enjoy a few weeks of pure air, denied them in their own wretched hovels.

Strange to say, consumption is almost unknown—an immunity attributed to the diet, consisting of dairy produce, rye bread, and dried fish, with occasionally a little mutton. On the other hand, influenza, or epidemic bronchitis, is common, and the most fatal of all local disorders, though it never attacks strangers. Its ravages are greatest in summer, which is otherwise also the most unhealthy season of the year. Scorbutic affections, leprosy, and elephantiasis have not yet disappeared from the island.

The people are in general characterized by a certain dignity, reserve, and personal courage; but they are accused of being very suspicious, and, like their kindred in Norway and Normandy, much addicted to litigation. Happily the former sanguinary quarrels, duels, and ordeals are now replaced by actions at law. In natural capacity, depth of thought, and love of letters they yield the palm to none. The custom formerly prevailed of meeting together to listen to the reading of their *Gamba Sagar*, or national chronicles, and in many houses artless paintings and sculptures recalled the leading events of their history. Reading and the game of chess occupied the long winter nights, and the art of printing had already been introduced in 1531, before the conversion of the natives to Lutheranism. Even now scientific work finds encouragement in the hamlets, isolated farmsteads, and islets on the seaboard, and Arne Magnusson has left a portion of his fortune to aid in the publication of all the literary monu. ments of his native land. Primary instruction is so diffused that the pastors refuse to marry unlettered couples, and the island has for over a century enjoyed a periodical press and a literary society.

Besides the national speech, Danish is current in Reykjavik and all other

Fig. 17.—REYKJAVIK.

trading centres; but elsewhere Icelandic is exclusively spoken, though polite expressions in Latin introduced by the clergy are still in vogue.

A relatively large number of the people reside abroad, and many, after graduating in Copenhagen, settle in Denmark, as did the father of the illustrious Thorwaldsen. There is a considerable stream of emigration to America; Icelandic is now heard on the shores of the Canadian lakes, in Michigan, Wisconsin,

Minnesota, and the Great Salt Lake, and an Icelandic journal appears in Keewatin, on the Red River of the north.

There is no town in the island beyond Reykjavik, the capital, with a present population of some 3,000. Mention may also be made of Þingvellir, where was held the assembly which prepared the constitution of 1874, and of Isafjörðr, on the north-west coast, which, though well built and cleanly, is not a pleasant resort for the stranger, being one of the centres of the cod-liver oil industry. Akreyri, capital of the northern districts, on the east side of the Eyja-Fjörðr, has a better harbour than Reykjavik.

The already mentioned constitution of 1874 secures to the people an almost complete autonomy. They no longer take any part in the Danish legislature,

Fig. 18.—JAN MAYEN.
Scale 1 : 610,000.

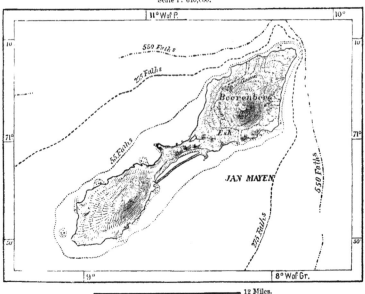

12 Miles.

though they are represented in the Cabinet by a special and responsible minister, who administers the executive by means of a resident governor. The Alþing, or National Parliament, consists of two chambers of twelve and twenty-four members respectively. Those of the Upper House are named, six by the King, and six by the Lower House, while those of the latter are all chosen by the electoral body, consisting of adults from twenty-five years of age paying at least 19s. in taxes, exercising some public function, or university graduates. Representatives are not eligible before their thirtieth year.

The island comprises two administrative provinces, each subdivided into twenty-one sýslur, or "seats," and 169 hreppar, or communes. Each hrepp has its municipal council of three, five, or seven members. In 1862 the armed forces consisted of three men, two day policemen and one night watch, at Reykjavik.

III.—JAN MAYEN.

ALTHOUGH its northernmost peninsulas project into the arctic zone, Iceland is not the last land of the North Atlantic. The plateau on which it rests is continued north-eastwards towards the Norwegian waters, terminating with a sort of headland, which rises above the surface to form the elongated island of Jan Mayen. Immediately beyond the Beerenberg, or "Bear Mount," rising to a height of 6,372 feet at its north-east end, the water suddenly sinks to great depths. At 2,000 feet from the shore Scoresby measured 305 fathoms, and Mohn found the same depth at a distance of 1⅓ miles from the coast.

Jan Mayen, which has an area of 160 square miles, was probably seen for the first time by the English navigator, Henry Hudson, in 1607 ; but it received its name from Jan Mayen, who rediscovered it four years afterwards. It is often wrapped in dense fogs, and, in the language of the early navigators, it is "then easier to hear the land than see it." But the reflection of volcanic eruptions has at times been detected above the surrounding mists. The Esk, a volcano south of the Beerenberg, was seen emitting flames in 1818. Beneath the glaciers which partly cover its slopes geologists have recognised that the island is of recent formation. Its lavas resemble the latest thrown up in Iceland, notably those forming the peninsula of Reykjanes, and differ entirely from the dolerites of the Färöer Isles.

The island is uninhabited, but the surrounding waters are visited by seal hunters, mostly from the east of Scotland and Christiania. Their vessels skirt the floating masses along the west and north coasts, which drift thence to the Greenland ice-helds. Here is mostly found the limit between the open and confined waters. Jan Mayen is like the last outpost of the world at the entrance to the frozen regions.

ISLANDS OF THE NORTH ATLANTIC.

STATISTICS AND GENERAL DATA.

FÄRÖER.—Area: 514 square miles. Population (1880) : 11,221, or 22 to 1 square mile nearly. Whale fishery between 1833—1862 : total captures, 37,986 ; yield, £152,000 blubber exported to Copenhagen. Average exports of fish, &c. : 3,100,000 lbs salt and smoked ; 200,000 lbs. fresh ; 1,141 tons blubber; 10,000 lbs. bladders; 170 tons roe. Average value of exports, £37,000 ; of imports, £27,000.

ICELAND.—Area : 40,460 square miles. Population (1878) : 72,000, or 1·8 to 1 square mile.

Temperature of the sea in July (Irminger and Dufferin) :—

North-west coast	45° to 49°	Fahr.
East coast	40° to 42°	,,
Mean annual temperature	36°	,,
Extremes	30° to 40°	,,

Mean temperature (in degrees Fahrenheit) in south-west and north —   :

		Reykjavik.		Stykkisholmr.		Areyri.
		Air.	Sea.	Air.	Sea.	Air.
Mean of the year		42	42	34	40	33
July		53	53	47	50	47
February		34	34	20	31	25
Extreme heat		70	61	—	—	75
Extreme cold		3	29	—	—	31

Live Stock (1870): — Sheep, 356.701 (in 1844, 606,500); cattle, 19,111 (in 1859, 26,908) ; horses, 30,000.

Fisheries.—These employ 63 covered and 3,335 open boats, manned by 15,400 hands, and are estimated to yield 7,250.000 fish, of a value of £112,500. On an average there are exported 65,162 cwts. of salt fish ; 2,292 cwts. of smoked fish ; 1,692 tons of roe ; 9,269 tons of blubber ; and 782 cwts. of salmon. The total exports have a value of £170,000, the imports of £150,000.

Administrative Divisions :—

South and West : Sudur-Umdaemid and Vestur-Umdaemid, 14 sýslur (seats) ; 103 hreppar (communes). Chief town. Reykjavik.

North and East : Nordur-Umdaemid and Austur-Umdaemid, 7 sýslur ; 66 hreppar. Chief town, Fridriksgáfa.

SCANDINAVIA.

CHAPTER I.

DENMARK.

KEEPER though she is of the Baltic portals, and mistress of Iceland and the Färöer, besides the vast uninhabited Greenland wastes and three West India islands, Denmark is nevertheless nothing more than an historic fragment. Of all European states it has the smallest population next to Greece, ranking even after the Hellenic world, if account be taken of those of kindred stock and speech living beyond the political bounds of the respective countries. The Greeks of the Archipelago, of Thessaly, Epirus, Thrace, Macedonia, and Asia Minor are far more numerous than those of the kingdom itself, whereas the Danes, pent up within their narrow limits, have only a small group of kinsmen beyond the frontiers. And even these remain henceforth deprived of their national autonomy, notwithstanding the stipulations of a solemn treaty, which Germany now feels justified in violating.

A mere remnant of a vanished land, formerly connecting Scandinavia with North Germany, Denmark has been in its history constantly associated with both countries. She formerly possessed extensive tracts on the Baltic seaboard, including Esthonia itself. In 1397 the union of Kalmar placed her at the head of the Scandinavian political system, and she possessed Norway till the year 1814. South of the Baltic various lands, since become German, also belonged to her, and till recently the German provinces of Holstein, South Schleswig, and Lauenburg formed an integral part of the monarchy.

No other European people have made such extensive conquests as the Danes, for it was from Jylland (Jutland) and the islands, no less than from the Norwegian and Swedish fiords, that the terrible Norsemen issued forth. They settled everywhere as conquerors—in the British Isles, on the coast of France, the Mediterranean seaboard, and even the northern shores of the New World, discovered by the Scandinavians long before the days of Columbus. Denmark must have become the centre of a vast empire had it enjoyed more geographical cohesion and more ample proportions. But the narrow peninsula of Jylland, covered with forests and unproductive tracts, the scattered isles of the Baltic, and the Norwegian seaboard, destitute of any arable lands and broken up by fiords into countless distinct fragments, did not possess a sufficient nucleus to keep together the foreign conquests, which consequently remained without cohesion or any common bond of union, like the region itself whence the conquering hosts had issued.

Denmark had formerly at least the advantage of commanding all the Baltic channels and the approach to that inland sea; but this strategical privilege no longer exists. One side only of the Sound belongs to her, the southern entrance to the Little Belt being occupied by Prussia, while the Great Belt might easily be

Fig. 19.—HILLS WEST OF ASVIG BAY.
Scale 1 : 128,000.

2 Miles.

forced by a hostile fleet. But whatever be their destiny, the Danes are an energetic people, with their own laws, language, traditions, aspirations, and national sentiments.

THE PENINSULA OF JYLLAND (JUTLAND).

LIKE the Danish islands in the Baltic, the peninsula of Jylland belongs geologically to Germany and Scandinavia. The southern portion, strewn with innumerable fragments of erratic boulders, forms a continuation of the North German lowlands. But in the broadest part occur older formations, miocene and cretaceous strata, the latter continued between Aarhus and Randers-fiord to the south-east corner of Själland (Zealand), and thence beyond the Sound to the extreme headlands of Scania.

North of the German frontier the backbone of the peninsula continues to follow the east coast, and the rivers of Jylland, flowing east and west to the

h Sea, have their water-parting much nearer to the geometrical
la than those of Schleswig. But the two sides differ greatly in
ti ng the same contrasts as in the German territory. The western
ind slopes gently, while the eastern is more abrupt, more varied,

lland facing the Kattegat belong mostly to the drift, and are
f sands, clays, and marls, forming the detritus of granite, gneiss,

Fig. 20.—THE HIMMELBJERG.

der strata are here and there strewn with boulders and gravels
The hills are not continued in regular ridges along the coast,
masses, sometimes to the height of over 300 feet. The Skamm-
;h of the German frontier, attains an elevation of 400 feet above
her summits occur between the Vejle and Horsens fiords, and
of Aarhus. This hilly eastern tract is very fertile, and the
. with magnificent beech forests sometimes down to the seaside.

West of the Kattegat Hills rises the culminating point of Denmark, with the Ejersbavsnehöj, 600 feet high.

Fig. 21.—From Ringkjöbing to Agger before 1863.

Scale 1 : 1,440,000.

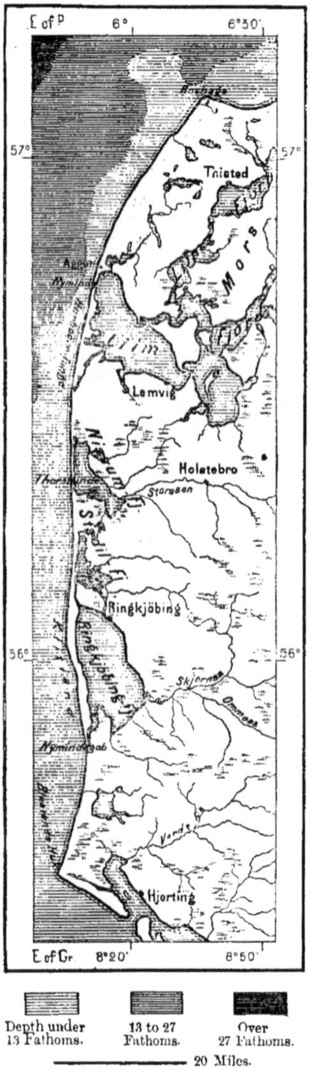

Depth under 13 Fathoms.   13 to 27 Fathoms.   Over 27 Fathoms.

20 Miles.

Better known, though 30 feet lower, is the Himmelbjerg ("Heaven's Mount"), commanding on the north-west a magnificent view of the surrounding district. At its southern foot are two large lakes formed by the Guden Aa, the most copious stream in the country, and beyond them stretches a vast prospect of pastures, woodlands, and cultivated tracts, lakes, and hamlets, limited in the distance by the curved coast-line.

Beyond the Lim-fiord the land again rises in hills, forming the so-called *Jyske Aas*, or "Back of Jylland," reaching a height of 400 feet, and, like all summits of the peninsula as far as Trave, situated much nearer to the Baltic than the ocean.

The western section of Jylland, formerly a vast sandy plain sloping gently seawards, has been largely brought under cultivation, especially near the streams. But there are still vast unreclaimed tracts resembling the North German *geest*, the *heiden* of the Drenthe and Veluwe, and the French *landes*—allowing, of course, for differences of climate and flora. The dunes on the coast also resemble those of Gascony, only they are much less shifting and lower, the highest being no more than 110 feet above sea-level.

In the outline of its coast Jylland also resembles the French landes. For a distance of 225 miles the shores of the North Sea are formed, not indeed of a uniform straight line, such as that stretching from Biarritz to the mouth of the Gironde, but of a series of lines slightly inflected, with points of resistance at definite intervals, each segment describing a clear geometrical curve, as if the limits of the ocean waves had been traced by the compass. But inland from these regular arcs the older and less uniform coast-line may still be followed.

Here also, as along the landes, the old inlets of the sea have been converted into lagoons, which the rain and streams have changed into fresh-water reservoirs, and which the alluvium is gradually filling up. They are very shallow, and

of many are laid bare with the change of seasons and the
igable channels, winding between the shoals like the Arcachon
ress to light craft. Although differing widely from the deep
, these lagoons are still called fiords by the natives. The
l, one of the largest, with an area of 110 square miles, is inacces-
rawing over 6 feet of water, nor can these always pass the
? Nymindegab, which is often shifted for hundreds and even
ls. The Stadel-fiord, north of the Ringkjöbing, is fed by a
us, and communicates through a labyrinth of lakes and rivulets
on, the Nissum-fiord, separated from the sea by a narrow strip
opening.
orth is the Lim-fiord, at once a marine and lacustrine basin, with
·y more complicated than that of the neighbouring lagoons. It
peninsula from sea to sea, and comprises three distinct sections,
of 450 square miles. The western section, like the Ringkjöbing,
sea by a slight sandy strip, in many places scarcely half a mile
nicates through a narrow channel eastwards with a lacustrine
g in fish, and enclosing the large island of Mors, besides a
f islets, beyond which it merges in an inland sea over 180 square
raiching southward into gulfs and bays, and separated from the
simple line of dunes. East of this central basin the lacustrine
d as far as Aalborg Strait, where a long navigable fiord begins,
er a mile in width, and opening seawards through a mouth
study of the geological chart shows that the water system known
follows, on the whole, the contours of the miocene and chalk
ond the limits of these more solid strata the less-resisting
urrounding country was more readily undermined by the action

past on the west side of the Lim-fiord has been frequently burst
s, as in 1624, 1720, and 1760. On November 28th, 1825, when
coasts of the North Sea were laid waste by tremendous floodings,
the Lim yielded to the pressure of the waves, and the lagoon
with the sea by one of those numerous *nyminde*, or "new
ave been so often formed on the Jylland seaboard. Before the
Agger channel the Lim was a fresh-water lagoon, but since then
ne, and now teems with salt-water fish. This channel, which
for navigation in 1834, is constantly shifting its place and
and depth with the action of the waves and storms, the bar
) 10 feet in depth. The Rön, a fresh channel, was opened in
875 the Agger has been almost completely blocked by silting
er the whole coast-line has receded about $1\frac{1}{4}$ miles farther

portion of the Danish peninsula shares in the gradual upheaval
the rock-bound shores of Norway and Sweden. The line of

separation between the areas of upheaval and subsidence passes probably to the north of the present political frontier, across the broadest part of the peninsula. South of this line the coast lands have been changed to islands, whereas farther north former islands now form part of the mainland. Such are the small peninsulas projecting seawards from Aarhus, north of which Lake Kolinsund, by its very name, recalls the time when it was, if not a strait, at least an inlet of the sea. In the neighbourhood are many hamlets whose names end in the syllable ö ("isle"), also suggesting their former insular condition. Parts of the north coast end abruptly in a sort of bluff 14 to 20 feet high, along which are traced the horizontal lines of different layers of peat, much firmer and blacker than ordinary turf, and covered with marine sands These beds, which are very old, are supposed to belong to a formation bodily upheaved from the sea. But immediately south of the point where begins the narrow stem of the peninsula we find traces of totally different geological phenomena. Submarine alder, birch, and oak forests, together with layers of peat, which formerly grew in fresh-water swamps, are now found embedded in the deep muddy banks flooded by the sea. While dredging the channels to render them navigable, the apparatus sometimes meets with trees buried beneath the waters.

Like those of France, the Danish "landes" slope seaward very gradually, so that depths of 100 feet and upwards are not usually met with nearer than 30 miles off the coast. Thus there are no harbours accessible to large vessels along the whole western seaboard of the Jylland peninsula, which crosses three degrees of latitude. Hence these waters, and especially the terrible Jammer Bay, are avoided by the shipping, which finds a safe ingress to the Baltic through the broad, deep, and partly sheltered channel of the Skager Rak.

THE DANISH ISLANDS.

AMONGST the Baltic islands that of Fyen (Fünen) might be regarded as belonging geologically to Jylland, although now separated from it by the Little Belt, which is nowhere less than 710 yards wide. Fyen was at one time undoubtedly attached to the mainland. It is composed of the same alluvium, and its beech-clad hills rise to about the same height as those of Jylland. They command the same smiling prospect of well-watered pastures, fields, and groves, and they also are intermingled with numerous erratic boulders, one of which, the Hesselager Stone, is 100 feet in circumference, and rises 20 feet above the ground.

East of the Great Belt the now scattered islands of Själland (Zealand), Möen, Falster, and Laaland are also nothing but one land broken up by narrow troughs of recent geological origin. The rocks of Möen and of a large part of Själland belong to the chalk epochs; but north and south of this cretaceous zone there stretch later formations, strewn with the detritus brought hither by floating ice. These formations occupy, on the one hand, the northern portion of Själland, and on the other the islands of Falster and Laaland. In this geological group the culminating point lies in the little island of Möen, where the Aborrebjerg rises amidst a number of lakelets to the height of 462 feet, and the neighbouring cliffs have an elevation of 330 to 430 feet above the sea. The disturbances of the strata have caused these steep bluffs to assume the most singular outlines, their distorted and even reversed layers recalling, on a small scale, the overlappings of the Jura and

Fig. 22.—FYEN AND THE SOUTHERN ISLES.

Scale 1 : 500,000.

Under 2¼ Fathoms.   Over 2¼ Fathoms.

———— 5 Miles.

Alps. Through the gorges opened at intervals in the cliffs the beech forests reach down to the level of the sea.

Seafarers passing from the Baltic towards Wismar, Kiel, or Lübeck can often distinguish the rocky shores of Möen and the lofty headlands of Rügen, formerly united, but now separated by a strait 33 miles broad and 11 fathoms deep. It seems probable that after having subsided Möen was again raised above the waters, and is still being slowly upheaved. It is really composed of seven distinct islets, whose intervening channels have been gradually filled in. In 1100 it still formed a group of three, and Borre, a village now lost amidst the fens, stood on the beach in 1510, when a Lübeck fleet anchored in front of the houses and burnt the place to the ground. Puggaard calculates the rate of upheaval at $2\frac{1}{2}$ inches in a hundred years. Like Rügen, Möen is much frequented as a summer retreat. It is merely a detached fragment of the larger island of Själland, whose chalk cliffs, the so-called Stevns Klint, rise in regular strata to a height of 130 feet on the east side, where they present a striking contrast to the Möens Klint, or irregularly stratified rocks of Möen.

The Ise-fiord, an extensive inlet ramifying into a multitude of winding channels, penetrates far into the northern portion of Själland, producing the same

Fig. 23.—SJÄLLAND AND THE SOUTHERN ISLES.
Scale 1 : 1.200,000

Depth under 2¼ Fathoms.   Depth over 2¼ Fathoms.
—————— 10 Miles.

variety here that is effected by the labyrinth of rocks and passages on the opposite side. Its shores, like those of Möen, have evidently been upheaved, for old marine beds are now visible several feet above sea-level.

While the Great Belt cuts off Själland from Fyen and Langeland on the west, the Öresund, or simply the Sound, severs it on the east from Scandinavia. This famous channel, however, forms a deep geological parting line between the two lands, for although the Swedish coast approaches to within 4,480 yards of Helsingör (Elsinore) Castle, it is composed of palæozoic rocks far older than those of Själland. The small Danish islands of the Kattegat also—Samsö, Anholt, Läsö

—consist of recent formations; but not so Bornholm, situated in mid Baltic, south-east from the southern extremity of Scania.

Geologically this island belongs to Sweden, for it consists exclusively of old rocks—sandstones and schists on a granite foundation. The strait separating it from Scania is only 22 miles broad, with a depth nowhere exceeding 27 fathoms. Nevertheless Bornholm justly forms part of Denmark, which formerly included

Fig. 24.—BORNHOLM.
Scale 1 : 350,000.

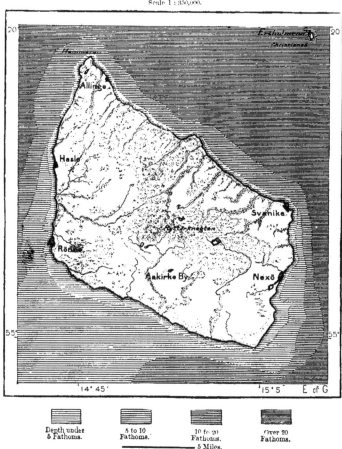

| Depth under 5 Fathoms. | 5 to 10 Fathoms. | 10 to 20 Fathoms. | Over 20 Fathoms. |

5 Miles.

all the southern portion of the great peninsula itself. When the Swedes took possession of the three provinces of Scania, Halland, and Bleking in 1658, Bornholm also was comprised in the treaty of cession, and was occupied by the foreign troops commanded by Prinzenskjold; but the people rose, and in one night massacred all the invaders, except twelve at the time absent from the fortress of Hammershuus. These were the "Bornholm Vespers."

Of all the Baltic islands Bornholm presents the most geometrical outlines, consisting of a parallelogram 15 miles by 12, and sloping gently northwards. The granitic plateau is covered with a thin layer of humus, where nothing formerly grew except heather. But it has latterly been extensively planted, and the Rytterknegten, or culminating point, 512 feet high, is now surrounded by woodlands. At the northern extremity of the island Cape Hammeren is almost

Fig. 25.—HELLIGDOMMEN ROCKS, NORTH COAST, BORNHOLM.

entirely detached from the land by an isthmus, on which is a deep lake that the people of Allinge proposed to convert into a harbour of refuge. But the project was abandoned in consequence of the formidable obstacle presented by the granite shores of the lake. Just south of this spot are the vast ruins of **Hammershuus**, the former residence of the governors, and on the headland itself there **now stands one of the** most important lighthouses **in the** Baltic.

Farther north are the islets and rocks of Ertholmene, generally known by the name of Christiansö, the largest of the group. They belong to Denmark as a dependency of Bornholm, and form a harbour of refuge with a good lighthouse.

CLIMATE.

THE climate of the peninsula and islands is comparatively temperate, though that of the islands is milder than in the interior of Jylland, because, being of smaller extent, they are more exposed to the influence of the sea. The winters are very variable, the channels being sometimes quite free, sometimes ice-bound for two or three months at a time. On these occasions the Great Belt is crossed by boats carrying passengers and goods, by means of a special contrivance. This so-called "ice traffic" (istransport), which some years is unnecessary, but which in 1871 lasted from January 1st to March 1st, has been carefully recorded since the year 1794. In 1658 Charles X. of Sweden led his army, artillery, and heavy baggage over the ice from Fyen to Langeland, and thence to Laaland, thereby bringing the King of Denmark to terms.

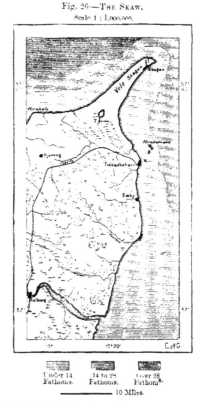

Fig. 26.—THE SKAW.
Scale 1 : 1,000,000.

The prevailing wind is from the west, and is charged with the moisture of the Atlantic waters. Hence the western seaboard is most exposed to the action of the waves. The very trees shoot their branches landwards, while their tops seem cut as if with a sharp knife. Even on the east side of Jylland they are affected by the prevailing atmospheric current, though the people here find

| Under 14 Fathoms. | 14 to 28 Fathoms. | Over 28 Fathoms. |

10 Miles.

shelter enough to build their villages and till the land in safety along the coast.[*]

INHABITANTS.—PREHISTORIC REMAINS.

THE flora and fauna of Jylland and the Danish isles differ little from those of the adjacent lands of Scania and Schleswig, the same climate having everywhere developed the same animal and vegetable forms. But not so with the human species, the Danes forming a people apart, with their distinct traditions, character, and national sentiment. Although of Teutonic stock, they regard themselves as

[*] Mean temperature at Copenhagen:—Spring, 43° F.; summer, 63°; autumn, 49°; winter, 31°; year, 46°.

severed from Germany, as much in their origin as in their warlike reminiscences. Hence they reject all idea of political fusion with Germany, while equally jealous of being confounded with their Swedish and Norwegian neighbours. The Scandinavian Union they aspire to is simply a confederacy of three nations, each retaining its own laws and customs.

Although Danish history, properly so called, scarcely dates from more than a thousand years, the remains of every description strewn over the surface of Jylland and neighbouring islands enable us to penetrate far beyond the historic epoch, to times when the climatic conditions were far different from those now prevailing. Denmark has acquired celebrity from the numerous evidences of primitive culture found on her soil. What the Mediterranean seaboard had been for the classical archæologist, the shores of the Kattegat and West Baltic have become, though doubtless to a less degree, for the prehistoric student in general. The remnants of our forefathers' rudimentary industries have here been collected in countless thousands.

Of all these natural museums, the most interesting are perhaps the peat beds, where successive generations of forests have been carbonised. The alternating layers of timber enable us approximately to determine the epochs when flourished the animals whose remains are here preserved, and the men who have here left their flint instruments. These peat bogs are at present overshadowed by forests of beech; but the forests that have been swallowed up are represented by the three distinct vegetable strata of pine, oak, and aspen. The old flora at the lowest depths consists of dwarf birch and other plants, now to be found towards the south of Lapland. Hence at this epoch Denmark enjoyed a polar climate; yet man already lived here, for manufactured flints are found at the lowest level associated with the bones of the reindeer and elk. But the remains of the mammoth and other large mammals are nowhere found in Denmark, as in France and England, in places where man has left traces of his industry.

Of great archæologic importance are also the shell mounds occurring here and there along the shores of Jylland and the Danish islands. They were formerly supposed to be layers of débris thrown up by the waves; but Worsaae and Steenstrup have shown that they are really *affaldsdynger* or *kjœkkenmœddinger;* that is, "kitchen refuse" or "kitchen middens." They are composed mainly of the shells of oysters and other molluscs, as well as of fish bones, besides the gnawed bones of deer, the roe, pigs, oxen, beavers, and dogs. The remains of cats and otters have also been found, together with those of the great auk (*Alca impennis*), which has disappeared during the present century from Iceland; but no trace of poultry has been detected. The only domestic animal at this time was evidently the dog.

Some of the mounds are 1,000 feet long, 100 to 200 broad, and 10 deep, thus containing many tens of thousands of cubic yards of matter, and testifying either to the multitude of those who took part in these entertainments, or to the long ages during which they were continued. The people of this epoch were in the stone age, for no discoveries have been made except of stone arms and implements, besides coarse earthenware. Both the coast-line and the saline character of the waters must have also undergone great changes since then, for the oyster, at that time so common, can no longer live in these seas, owing to the small quantity of salt they now contain. Some of the bones found in the middens—as, for instance, those of the *Tetrao urogallus*—also bear witness to the severity of the Danish climate at this period.

To the remains found in the peat beds and middens must be added the weapons, utensils, and ornaments collected in great quantities from the megalithic graves of divers forms, with one, two, or more chambers, scattered over the land. Among these monuments the oldest are the round barrows and long mounds. The giants' chambers (*jættestuer* or *steendysser*) are built with more art, and are composed of several compartments of granite blocks, covered over by a hillock of earth. Many seem to have been family sepulchres, and in them have been found the bones of wild and domestic animals buried with the dead, together with implements, arms, and ornaments. These burial-places belong mostly to the last period of the polished stone and bronze age, and to a settled people already skilled in stock-breeding and the elementary principles of agriculture.

Iron seems to have finally prevailed in these regions about the time of Septimius Severus, or towards the close of the second century, and from this epoch also date the earliest Runic inscriptions. Very remarkable objects of local origin, or imported from abroad, have been discovered in some of the graves. Such is the cup found at Stevns Klint, in Sjælland, with a chased silver rim bearing a Greek legend. At Bornholm the age of iron was developed under special conditions. Here are thousands of graves called *brandpletter*, consisting of excavations filled with charcoal, human ashes, and bones, with fragments of arms and implements in iron and bronze, contorted by the action of fire. The burial-place of Kannikegaard, near Nexö, alone contains over twelve hundred of such graves, and two other cemeteries have nine hundred each; but the more recent graves were all isolated. The practice of cremation has caused the disappearance of a great part of the precious objects buried with the dead.

Whoever the people of the stone age may have been, Rask and Nilsson believe that the whole of Denmark was occupied by Lapp tribes in prehistoric times. Others, on the contrary, hold that the Finnish Lapps reached the peninsulas and southern isles of Scandinavia only in erratic groups. In any case it is certain that the country has at some time been occupied by races of very different origin from its present Norse inhabitants. The comparative study of the crania made by Sasse in the Sjælland graveyards shows that till the sixteenth century a people of very feeble cranial capacity here held its ground by the side of the large-headed Frisic stock. Some articles of dress would also seem to suggest the presence of old Celtic peoples, the peculiar head-dress till recently worn by the peasant women in Fyen, Ærö, and Falster presenting a striking resemblance to that of the Antwerp peasantry.

After the subsidence of the mighty waves of migration which drove the

Cimbri of the peninsula southwards to Gaul and Italy, the Heruli of the isles and Chersonesus to Rome, the Angles, Saxons, and Jutes to Britain, another people, yielding to the general westward movement, appeared at certain points of the southern islands of Laaland, Falster, and Langeland. These immigrants were Slavs, and their presence here is vouched for by tradition and local geographical names. But the principal invaders were the Danes, an old confederation of Norse tribes. After seizing the lands that have since become Denmark, these tribes long continued to harass the West, as rivals of the Norwegian rovers contending for centuries with the Anglo-Saxons for the possession of Great Britain, and with the Celts for that of Ireland.

The average of the pure blonde type, with light blue eyes, is on the whole higher in Denmark than in Germany. More animated than the Dutch, the Danes resemble them in the qualities of vigour, courage, and endurance. Endowed with a good share of common sense, they act in general with sound judgment, regarding the Germans as somewhat crack-brained and braggarts. Still they have their days of revelry, when they are apt to forget themselves, their wonted reserve breaking out in song and clamour. Beneath a quiet expression the Dane harbours a fiery and poetic soul. He hears the billows boom against his shores, and he recalls the daring life of his forefathers, who overran the world in their frail wave-tossed craft. His literature cherishes a precious inheritance of noble songs, which the young men recite at their festive gatherings. The men of science are distinguished by vigorous thought, method, and clearness. The people everywhere display a love of letters, and for them the theatre is as much a school of literature as a place of amusement. "Not for pleasure alone!" says an inscription on the curtain of the national theatre at Copenhagen.

The Danish language, of Norse origin, but far less pure than the Icelandic, had already been developed about the thirteenth century, but it scarcely acquired any literary standing till the era of the Reformation, in the middle of the sixteenth century. Its old sagas all belong to Scandinavian literature proper. Of all the Danish dialects, including that of Bornholm, the most original and richest in old words is that of North Jylland, though it has not become the national standard. The Själland dialect has acquired the preponderance, thanks to the dominant influence of the capital, and has thus gradually become identified with the Danish language itself. With the successive literary epochs it has been enriched by terms borrowed from Latin, Swedish, French, but especially High and Low German. Many authors formerly wrote in both the latter languages; but now the Dane clings to his mother tongue all the more tenaciously that he feels his very political existence threatened. He is enthusiastically attached to his national traditions, to his old literature and poetry, flowing from the sagas, and filled with the memories of the past. Since the time of the great Thorwaldsen Danish art also has pursued an independent course, and even the industrial arts, porcelain, gold-smiths' work, and furniture have sought their inspiration in the antiquities found in the native land.

Topography.

All the important towns of Jylland are situated on the shores of, or at least on the slopes facing towards, the Baltic. The people were naturally attracted in this direction by the threefold advantage of more fertile lands, better harbours, and the neighbourhood of the productive and populous islands of Fyen and Själland. But when the Danes were casting eager eyes towards England and the other regions of Western Europe, great numbers settled on the west coast, and at that time *Ribe*, on the present German frontier, was a very important place. Now it is scarcely able to keep open its communications with the sea, and the winding Ribe Aa is completely blocked by shoals at low water. *Ringkjöbing*, farther north, is a mere fishing village, though chosen as the chief town of the largest district in the peninsula. No town occurs so far as the Skaw, and *Thisted*, chief town of the

Fig. 27.—Vejle Bay.
Scale 1 : 85,000

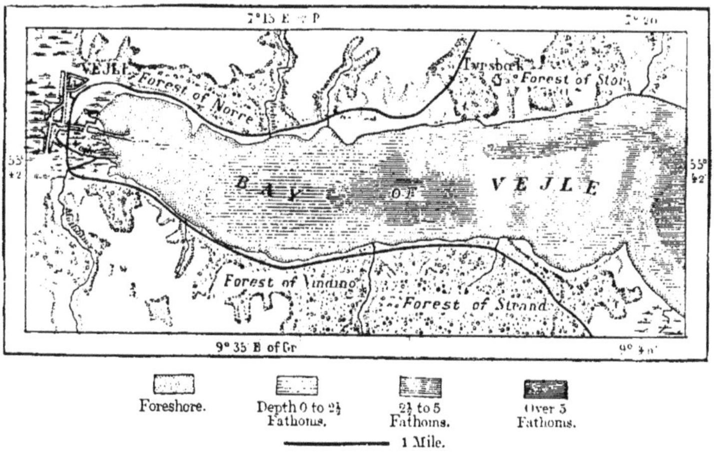
VEJLE
Forest of Norre
Tyrsbæk
BAY OF VEJLE
Forest of Vinding
Forest of Strand
9° 35' E of Gr
Foreshore.
Depth 0 to 2½ Fathoms.
2½ to 5 Fathoms.
Over 5 Fathoms.
1 Mile.

district west of the Lim-fiord, is situated on this inland sea. Here was born the great geographer Malte-Brun, who, when banished from his native land, became one of the glories of France.

On the east coast of Jylland the nearest town to the territory annexed by Germany is *Kolding*, situated at the head of a deep fiord near the extensive ruins of a castle dating from the sixteenth century, and burnt in 1808, when Bernadotte occupied the country. Kolding is of less importance than *Fredericia*, which was formerly fortified, and commanded the northern entrance of the Little Belt. Farther north the Vejle-fiord penetrates far inland between beech-clad hills, its waters gradually diminishing in depth with remarkable uniformity from east to west. The town of *Vejle* occupies a sort of isthmus of solid ground between the fiord and the peat bogs which have displaced the waters of the dried-up estuary.

A little to the north-west is the old royal castle of Jelling, where are to be seen the tumuli raised about 960 to Gorm and Thyra by their son Harald "of the Blue Tooth."

Like Vejle, *Horsens* stands at the head of a fiord, whereas *Aarhus*, the largest town in Jylland, is built on the coast, and has a well-sheltered harbour. It is the central station of the Jylland railway system, and the chief point of communication with Copenhagen. Formerly the political centre was *Viborg*, which stands on the shores of a lake in the heart of the peninsula. It was the principal residence of the old Danish kings, and its cathedral, recently rebuilt, is one of the finest churches in Denmark. East of it lies *Randers*, communicating directly with the sea by a winding fiord, though large vessels get no farther than the deep anchorage of Udbyhoi, near its mouth. Randers is a chief centre of the manufacture of the so-called "Swedish gloves."

Aalborg, stretching along the south bank of the Lim-fiord, here crossed by a fine railway bridge, has a brisk trade, though the bar gives access to small craft only. Yet a large seaport and harbour of refuge are much needed at this place between the two stormy channels of the Kattegat and Skager Rak. Here are yearly wrecked some thirty or forty of the forty or fifty thousand vessels passing through the straits, and in November, 1876, thirty-nine foundered at Vejle. The port of *Frederikshavn*, south of the Skaw, is quite inadequate as a harbour of refuge, and the idea has been entertained of enlarging it by enclosing the neighbouring islets of Hirtsholmene. The town of *Skagen*, at the Skaw, is the most important Danish fishing station. Here vast quantities of whiting, cod, turbot, soles, and other fish are taken and shipped for Copenhagen and other places.

On the island of Fyen stands *Odense*, the "Town of Odin," one of the oldest places in Denmark. The cathedral contains some royal tombs, and it is the birthplace of the delightful writer of children's tales, Andersen. Although at some distance from the sea, it is the centre of a considerable trade, promoted by the line of railway crossing the island and placing it in direct communication with *Middelfart* and *Strib*, on the Little Belt, and the fortified port of *Nyborg*, on the Great Belt. On the south coast of Fyen, and facing the islet of Taasinge, is *Svendborg*, an important commercial centre for the neighbouring islands of Taasinge, Ærö, and Langeland, and surrounded by some of the loveliest scenery in Denmark. The brothers Oersted were natives of Langeland, and the great philologist Rask was the son of a Fyen peasant.

COPENHAGEN.—Copenhagen (Kjöbenhavn), capital of Själland, contains of itself alone about one-eighth of the entire population of the kingdom, and more than that of all the other Danish towns together. It is, moreover, something more than the capital of a decayed state, still maintaining a special position as a European city, the common property of all the northern nations.

Its geographical position, like that of Constantinople, presents a double advantage as the intersecting point of two great highways, the water route from sea to

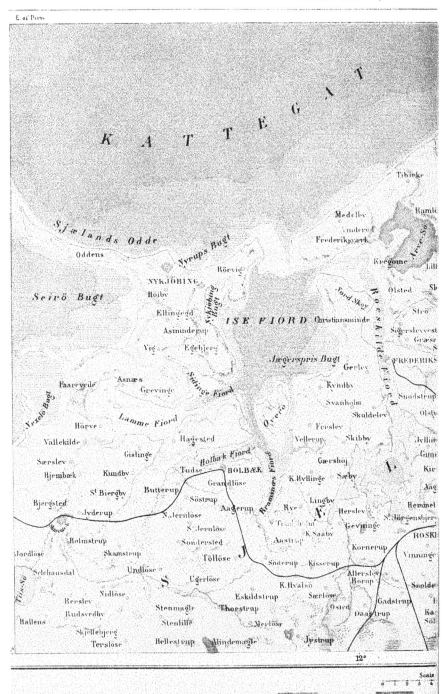

E. of Paris

KATTEGAT

Tibirke

Medelby   Ramle

Sjælands Odde   \underod
                                Frederiksvaerk

Oddens   Kregome   Lill

        Rörsig

NYKJÖBING   Olsted   Sk
        Höjby   Ford Skov
Seirö Bugt   Strø
        Ellinge\d
                  ISE FIORD Christiansminde   Sigerslevvest
        Asmunderup   Græse
        Vig   Egebjerg

                  Jægerspris Bugt   FREDERIKS
                              Gerlev
Faareveile Asnæs   Kyndby
        Grevinge   Svanholm   Snodstrup
Vexelø Bugt   Sidinge Fiord   Skuldelev   Olslo
        Hörve   Lamme Fiord   Overø
Vallekilde   Hagested   Vellerup Skibby   Jyllin
        Gislinge   Gærshöj   Gum
Særslev   Kundby   Holbæk Fiord   Sæby   Kir
Hjembæk   Tudse   HOLBÆK K.Hyllinge   Aag
        St Biergby Butterup Grandlöse
Bjergsted   Söstrup   Lingby   Hemmel
        Jyderup   Aagerup Rye   Hersley   St Jörgensbjer
                N.Jernlöse   Tr..l..luf Gevinge   ROSKI
        Holmstrup   S.Jernlöse   K.Saaby Kornerup   Vinunge
Jordlöse   Skamstrup Sondersted Aastrup
Selchausdal   Undlöse   Töllöse Söderup Kisserup Allerslev
                Ugerlöse   Borup   Snolde
        Nidlöse   K.Hvalsö   Særlöse
Bersley   Eskildstrup   Osted Gadstrup
Rudsvedby   Stenmagle Thorstrup   Daastrup   Ka
Hallens   Stenlille   Merlösr   Söl
        Skjellebjerg
        Terslöse Bellestrup Alindemagle   Jystrup

                                    12°

                                    Scale
                              0 1 2 3 4

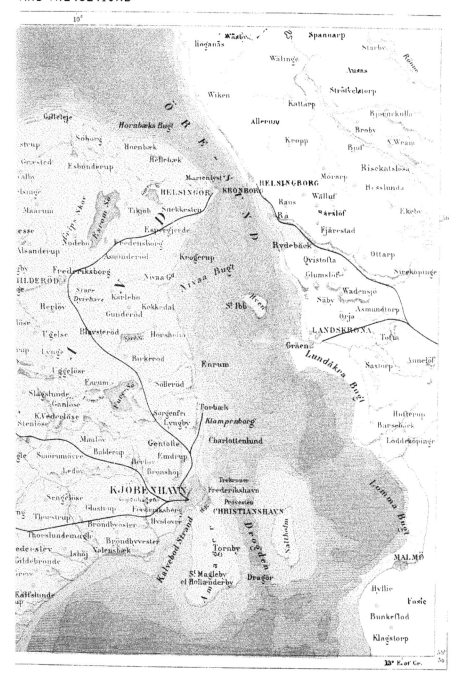

route between two peninsulas. All the passenger and goods
eat Britain, Germany, Sweden, and the Baltic converges here,
u has naturally become a great commercial centre. It stands on
although not the only, is by far the safest and most convenient
ating with the Baltic. Vessels can often pass from sea to sea
,d the prevailing west wind is almost as favourable to the egress
ess to the Baltic.

e for the capital of the state might seem to be Elsinore, at the
to the Sound, where it narrows to the proportions of a river;
accordingly been always of the greatest strategical importance,
e introduction of artillery sweeping right across the channel.
e of Copenhagen presented the great advantage of a safe road-
al haven. The inlet between Själland and the islet of Amager
port to vessels navigating these waters, which they did not fail
s of as the traffic between the two seas became more regular.
d reference to the place in 1043 speaks of it as simply the
" as if it were such in a superlative sense. About 1200 the
Saxo Grammaticus calls it *Portus Mercatorum*, a name it has
ven strategically its position is as important as that of Elsinore,
.el hugs its shores, continuing southwards through the Drogen
lager and Saltholm, while the Swedish side is blocked by shoals
the Copenhagen forts, like those of Elsinore, command all the
of the Sound, and when this place became a royal residence its
' was easily established. Its influence soon reached far beyond
.roper, and the most populous city of Scandinavia aspired to the
a capital of the Norse world. But notwithstanding its admirable
on, the lands surrounding it were not sufficiently compact to
ain its supremacy amongst the cities of the North. Compared
.vers hemming her in, Denmark herself is too insignificant to
n her own resources, and her best hopes of preserving her
ed in the rivalry of her neighbours. "St. Petersburg," said
.vindow opening on Europe." Might not Copenhagen become a
n the world? Happily two, and, including England, three
.ve their eyes equally turned in this direction.

nds, so to say, on a strait between Själland and Amager. Shift-
i reclaimed by means of embankments, and islets still preserving
have been united with the shore. But the numerous canals
.d mercantile ports give the place quite a Dutch appearance.
fire, it is largely a modern city, and the walls and ditches
it landwards have been almost completely levelled to make
.ion, especially towards the north and west, in the direction of
the city proper is separated from the coast by a citadel and
.s, and fortifications have even been raised on artificial islets in
ters. These frowning batteries still recall the gallant defence

made in 1801, when Nelson forced the Sound and destroyed or captured the
Danish fleet at anchor in the roads.   Six years thereafter, in full peace, the
English returned, bombarded the city, burnt its buildings, and again carried off
the fleet.

Built of stone or brick coloured grey, Copenhagen is a tolerably handsome
town, and enjoys a considerable revenue.   There are some remarkable monuments,
including rich scientific and art treasures.   The octagonal Amalienborg Square,
near the harbour, is adorned with a bronze equestrian statue of Frederic V., and

Fig. 28.—COPENHAGEN.
Scale 1 : 83,000.

1 Mile.

surrounded by royal palaces and gardens.   The palace of Christiansborg, where
the official receptions are still held, and where the two Chambers meet, contains a
picture gallery especially rich in Dutch and Danish works.   Close by is the Royal
Library, with about 500,000 volumes and 17,000 manuscripts.   The University,
four hundred years old, and with over 1,000 students, has also about 250,000 volumes,
including some of priceless value, notably the collection of Icelandic sagas in 2,000
manuscript volumes.   In the "Palace of the Prince" is the admirable Museum
of Northern Antiquities, with that of Stockholm the most complete in the world,

THE SLOTS CANAL, COPENHAGEN, AS SEEN FROM THE CASTLE OF CHRISTIANSBERG.

CASTLE OF FREDERIKSBORG.

lel of its kind. Here are 40,000 objects so arranged as to illustrate the
l customs of the generations that succeeded each other on Danish soil in
c times. Under the same roof is the excellent Museum of Comparative
r, founded, like the other, by Thomsen. The castle of Rosenborg, dating
seventeenth century, contains miscellaneous collections, possibly less
than the "Green Vaults" of Dresden, but admirably classified according
by the celebrated Worsaac. In one of the chambers is a rich assortment
an glass. The Academy of Fine Arts is installed in the castle of
iborg, and the Observatory is the most venerable in Europe, the first
ing been laid in 1637, half a century after Tycho-Brahe had set up his
raniberg ("Castle of the Heavens") in the island of Hveen.

the Exchange stands a huge cenotaph in the Egyptian style dedicated to
iry of Thorwaldsen, and containing all his works, or copies of them,
s various collections. In a central court is the tomb of the master, sur-
by his numerous statues. The metropolitan church is also enriched by
es from the chisel of the same artist, the most illustrious citizen of
en.

rous learned societies' have been here established, the most important
Society of Northern Antiquaries. In 1876 was founded a Geographic
·hich already numbers 850 members.

urth of the Danish industries are centred in the capital and its suburbs.
, refineries, spinning-mills, porcelain works, potteries, ship-building and
yards cover vast spaces near the harbour and several other quarters.
of the trade and shipping of the state belongs to this port, although it
ely more than one-fourth of the Danish commercial navy. It is the
the steam traffic and of the northern telegraph system, with nearly
s of wire connecting it with England, France, and, through Russia and
ith Japan.

laid of *Amager* has been converted into a garden by its inhabitants,
colony settled here in 1514, and pleasant villas, parks, and gardens
ng the shores of the Sound all the way to the Klampenborg baths and
sborg woods. But the finest estates and country seats are found in the
Själland—amongst them the castle of Frederiksborg, near *Hilleröd*, the
Versailles," erected in the seventeenth century by Christian IV., and
cted by rail with the capital. On the same line is Fredensborg, a royal
sidence built in the beginning of the century by Frederic V., and noted
e woodlands and numerous statues by Wiedewelt, predecessor of
en.

ge square castle of *Kronborg* stands on a neck of land projecting into
at its narrowest part, as if to connect the Danish *Helsingör* (Elsinore)
wedish Helsingborg. It was erected in the fifteenth century on the
castle of *Örekrog*, which itself had taken the place of the still older
ç. In its underground vaults here sleeps traditionally the hero Holger
149

the Dane, awaiting the day when his oppressed country shall again stand in need of his stout arm. This is also the famous castle of Hamlet, which thus lives in deathless song, though we look in vain from its "platform" for Shakspere's "dreadful summit of the cliff, that beetles o'er his base into the sea." The events told by the poet are fancy's theme, but here the mind still bodies them forth, and the castle halls seem still to echo those sublime utterances that can never die.

The current of the Sound sweeps by Elsinore, which is the natural limit of the two seas, and which the Danish kings took care to fortify, in order to enforce dues

Fig. 29.—KRONBORG CASTLE, FROM THE SOUND.

from all vessels passing to and fro. Till the middle of the present century this tribute was universally submitted to, but in 1855 the United States refused payment of the tax, and in 1857 it was redeemed for the sum of £3,494,000, payable by sixteen nations in proportion to their traffic. About 50,000 vessels pass yearly in front of Elsinore, 4,000 to 6,000 stopping for supplies.

Röskilde, capital and most populous city of Denmark before Copenhagen, necessarily lost its importance when the small craft of former times were replaced by larger vessels. It occupies the southern end of a fiord now blocked by sand-banks; but it long continued to be the religious capital, was formerly full of churches and convents, and still possesses the finest cathedral in the kingdom, dating from the eleventh century, and containing the tombs of several Danish kings.

In the interior of Själland are the two cities of *Sorö* and *Slagelse*, the former noted for its school and abbey associated with the name of Saxo Grammaticus, the latter the centre of a rich agricultural district. On the west coast are *Kalund-borg* and *Korsör*, at the entrance of a large crater-like lagoon ; on the south are

Fig. 30. —KORSÖR.

Foreshore.   Depth 0 to 2½ Fathoms.   2½ to 5 Fathoms.   Over 5 Fathoms.

Nestved and *Vordingborg ;* in the island of Falster the sheltered harbour of *Nykjöbing ;* and in Laaland the port of *Nakskov*, with a considerable export trade in corn and cattle.

Of Bornholm the chief town is *Rönne*, near the south-west angle, noted for its clock works and potteries, and for some remarkable granite churches of the twelfth and thirteenth centuries.

SOCIAL CONDITIONS.—OCCUPATIONS.

SINCE the beginning of this century the population of Denmark has doubled, having increased from 929,000 to about 2,000,000. It is still increasing, the births exceeding the deaths by two-fifths. As in other civilised countries, the rate

is higher in the towns than the country, the urban population having increased
more than 10 per cent., the rural one-twentieth only between the years 1855 and
1870. The emigration movement has also acquired some importance since the
middle of the century, while immigration is mainly confined to the natives of the
old Danish provinces now annexed to Germany.

Agriculture, which supports three-fifths of the people, is in a flourishing state,
although more than one-third of the soil consists of dunes, marshes, waste or
fallow lands. Barley and rye are more generally cultivated than wheat, but the
latter has shown an upward tendency since the abolition of the corn laws in
England, the greatest grain market in the world. Other farm produce has also
received a stimulus, and vegetables, fruits, cattle, and butter are now regularly

Fig. 31.—Relative Number of Live Stock in the European States.

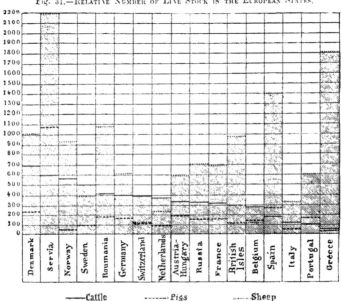

————Cattle   ------Pigs   ......Sheep

exported. Jylland especially, as a stock-breeding country, has prospered. The
large cattle of Thy are highly esteemed both for their milk and the rapidity with
which they fatten for the market in the rich pastures of the peninsula. Denmark
has altogether the largest relative number of horned cattle of any European state,
and is also exceptionally rich in swine and sheep.

There are some very large estates in Denmark, and those of the nobility and
hereditary proprietors enjoyed till recently many privileges. The peasantry are
still divided into three classes: the *gaardmænd*, whose lands pay a tax of more
than a ton (3¾ bushels) of hard corn; the *huusmænd*, owners of one house
and paying a smaller tax than the foregoing; and the *inderster*, residing with
others. But small proprietors are the rule, and most of the peasantry are in

easy circumstances, with comfortable dwellings, well ventilated, and furnished with a certain luxury. The national wealth is officially valued at £240,000,000 to £300,000,000, or £120 to £150 per head, with a mean annual increase of about £8,000,000. Nowhere else is the capital of the savings-bank relatively so large, and, to judge from this test, the Danes would seem to be the most thrifty people in Europe. Their average savings per head are about £6 8s., while those of Englishmen are not more than £2 2s. In Denmark, as in other Scandinavian lands, the party of progress consists mainly of the peasantry, whereas the rural element is generally the least advanced in France and the rest of Europe.

The fisheries do not contribute as much as might be supposed to this general state of prosperity. The natives of the islands and of East Jylland find a readier livelihood in agriculture than in the perils of deep-sea fishing, while the young men of the seaboard show a preference for the certain and regular profits of the merchant service. Still fishing is far from being entirely neglected, especially on the less productive west coast, where the waters teem with marine life. At Nyminde-gab, at the mouth of the Ringkjöbing-fiord, over 700,000 whiting and 25,000 cod were taken in 1862, and large quantities of herrings and salmon are brought by the Bornholm fishermen to Swinemünde, whence they are forwarded to Berlin, Vienna, and Paris.

Denmark is not a very industrial country. Except in Copenhagen and some towns of Fyen and Jylland, the manufactures are mainly restricted to coarse woollens and distilling. Both coal and water-power are scarce, except at Frederiksværk and a few other places where the streams have a rapid fall. But their extensive trade brings to the people all the manufactured goods they require. Compared with the respective populations, the exchanges of the Danish market are more considerable than those of France, and foreign trade has almost doubled between the years 1866 and 1875, though the increase is shown chiefly in the imports, amounting to £12,000,000 in a total of £22,000,000. Previous to 1875 the bulk of the trade was with Germany, but since then England has occasionally taken the lead, though Germany still continues to be the outlet for nearly all the Danish traffic with the rest of the continent.

The commercial movement being necessarily carried on mainly by sea, the shipping has acquired a considerable development, and vessels of a tonnage of 4,000,000, the greater part of which is still represented by sailing vessels, annually enter and clear. The mercantile marine amounts altogether to 3,150 vessels, of about 253,000 tons burden, all sailing vessels except 188 steamers, of 45,000 tons.

Thanks to the numerous straits and fiords, water communication is easy, while good roads and numerous railways connect Copenhagen with all the Själland ports. The postal and telegraph services have also kept pace with the general development.

Public instruction is also in a flourishing state, attendance at school being obligatory on all from the seventh to the fourteenth year. In all the large towns there are public gymnasia, classical and scientific as well as technical schools, and all the villages are provided with lower and higher primary schools, independently of over fifty secondary rural establishments founded since the middle of the century. Besides the State institutions, seven private foundations have the privilege of conferring certificates giving admittance to the University, whose classes, since 1875, have been attended by women on the same footing as others. Yet, notwithstanding the general material and intellectual advance, the Danes are

Fig. 32.—DANISH RAILWAY SYSTEM.
Scale 1 : 300,000.

more susceptible to mental anguish, and hence more addicted to suicide than any other people.

ADMINISTRATION.—COLONIES.

ACCORDING to the constitution, repeatedly modified since 1869, the electoral body consists of all citizens thirty years of age resident for one year in the commune, and not in the receipt of public charity. The *Folkething*, or National

Assembly, consists of one hundred and two members for the whole monarchy, elected for three years, while the fifty-four members of the *Landsthing*, or Upper House, are chosen for eight years by electors composed partly of popular delegates, partly of the more highly taxed town and country voters. With the fifty-four elected senators are associated twelve life members appointed by the Crown from the actual or any former members of the Assembly. Both bodies, forming collectively the *Rigsdag*, or Diet, receive a uniform grant of about 7s. 6d. per day for their services. Every four years the Landsthing chooses from its own body the four assistant judges of the *Höiester*, or Supreme Court, which is alone competent to deal with charges brought against the members of both Houses.

The executive is intrusted to a responsible ministry of six members—the Minister of Finance and President of the Council, and the Ministers of Foreign Affairs, the Interior, Public Instruction and Worship, Justice and Iceland, War and the Navy. The King must be a member of the Lutheran or State Church; he has a civil list of £56,000, besides £6,520 for the Crown Prince. The judges of the eighteen higher courts are named by the sovereign, as are also those of the two Courts of Appeal, one of which sits in Copenhagen for the islands, the other at Viborg for Jylland. Till recently the judicial and administrative functions were united in the same hands, but according to a law of 1868 the two are to be henceforth separated, though the change has not yet been completely carried out.

Civil processes, formerly conducted in writing, are now prosecuted in open court. The lawyers, combining the functions of barrister and solicitor, are divided into three categories, those of the Supreme Court alone being entitled to plead before all tribunals. Of these higher advocates there were eleven only in 1872.

Although Lutheranism is the State religion, freedom of worship is absolute, and except indirectly through the budget no one is called upon to contribute towards the maintenance of any creed besides his own. The faithful have even acquired the right of forming themselves into distinct congregations, and founding the so-called "elective parishes," whose ministers they name and maintain.

The seven bishops (Själland, Laaland-Falster, Fyen, Ribe, Aarhus, Viborg, and Aalborg, besides Iceland), although enjoying great privileges, have no seat in the Upper House. Not more than 1 per cent. of the population are non-Lutheran, the most numerous being the Jews, and next to them the Baptists and Mennonites, the Roman Catholics taking the fourth place.

According to the Army Reorganization Act of 1867 all valid citizens from the age of twenty-two are bound to military service, eight years in the regular army, and eight in the reserve. But practically they serve only for periods varying from four to nine months, though a certain number of recruits are called out for a longer term, and all regulars take part in the annual manœuvres, which last from thirty to forty-five days.

The regular forces are estimated at 35,000, the reserves at 13,000 of all arms.

fishing populations of the seaboard. It comprised, in 1877, 2,830 men, under 1 admiral, 9 commanders, 22 captains, and 102 lieutenants.

The Danish colonial possessions are very extensive, though the two largest territories, Iceland and Greenland, are mostly uninhabitable. The vast ice-fields of Greenland have with difficulty been penetrated for short distances at a few points, and its extension towards the pole is still a problem awaiting solution. Next to Iceland, the most important possessions are the three West Indian islands of Santa Cruz, St. Thomas, and St. John, at the north-east junction of the Caribbean Sea with the Atlantic. The harbour of St. Thomas is one of the chief naval stations and ports of call in these waters.

The financial condition of Denmark contrasts favourably with that of larger states. The expenditure, averaging about £2,600,000, is usually covered by the revenue, and if the national debt has recently been considerably increased, this is mainly due to the construction of railways. She has wisely ceased to make useless chronic preparations for a life-and-death struggle with her formidable southern neighbour, and most of the fortresses have been either dismantled, or, like those of Rosenborg and Frederiksborg, converted into museums and royal retreats.

The country is divided administratively into eighteen *amter*, or bailiwicks, and one hundred and thirty-six *herreder*, or circles.*

* For Statistical Tables see Appendix.

CHAPTER II.

THE SCANDINAVIAN PENINSULA.

(NORWAY AND SWEDEN.)

GENERAL FEATURES.

THE great northern peninsula comprises two distinct states, though ruled by one sovereign, and otherwise bound together by ties of a very intimate character. Still they watch with careful jealousy over their mutual political independence, and even in their social usages the two peoples are sharply contrasted. From the geographical point of view also Norway and Sweden (*Norge* and *Sverige*) form equally distinct natural regions, the one consisting mainly of plateaux and highlands rising abruptly on the Atlantic side, while the other forms an extensive incline falling gradually towards an inland sea.

But this physical contrast and their separate autonomy do not prevent the Westerfold and Austerfold, as they were formerly called, from forming a unity distinct from that of other European lands, and which should therefore be studied as a whole. The term Scandinavia, or Island of Scandia, formerly restricted to the southern extremity of Sweden, has been gradually extended to the entire peninsula independently of its political divisions, and this very community of name seems to point at a general and permanent fusion of the two regions.

The natural frontier of the peninsula connects the northern extremity of the Gulf of Bothnia directly with the Varanger-fiord, on the Frozen Ocean, and although political treaties have caused the line to recede in the most eccentric manner westwards, thus allowing Russia to cut off Finmark almost completely from the rest of Scandinavia, such conventional limits traced across extensive wastes have but little practical importance.

Even within its present reduced limits, Scandinavia is one of the most extensive regions in Europe, ranking in size next to Russia. Owing to its position on an inland sea giving access to Western and Central Europe, and on the Atlantic placing it in relation with the rest of the world, it could not fail to exercise a certain influence in determining the balance of power, and the Goths, Norsemen, and Varangians have left a deep mark in history as conquerors and seafarers. Later on, when modern Europe was already constituted, the Swedes, with a firm footing on the eastern and southern shores of the Baltic, were able to carry their arms in one direction as far as the Vosges, the French Jura, and the Upper Danube, in another to the Russian steppes bordering on the Black Sea. But then came the fatal field of Poltava, ushering in the period of political decadence. Even before the loss of all their outlying possessions the Swedes were threatened on their native soil, and, at the very time of the first partition of Poland, Frederic II. was planning the dismemberment of Sweden. During the Napoleonic wars the sudden political oscillations and dynastic changes, accompanied by the final loss of all territory on the mainland, showed how largely the destiny of the Scandinavian states depended on their powerful neighbours.

Notwithstanding the mildness of its maritime climate, Sweden is still, on the whole, too cold to allow its population to increase in the same proportion as that of more southern lands. Compared with Germany and Central Russia, it has remained almost unsettled. The population of Sweden and Norway combined exceeds that of Belgium only by about one-fifth, while the area is twenty-six times greater.

PHYSICAL FEATURES OF NORWAY.

In the peninsula Norway is the land of plateaux and mountains, Sweden a region of vast sloping plains. The main ridge stretches north and south at a short distance from the Atlantic, but very irregularly, and with many serious interruptions. Within its political limits, Northern Norway consists of little more than a highland strip facing the Atlantic; and here are found the highest summits of the peninsula. The mean length of the Norwegian is scarcely more than one-fourth of the Swedish slope. About one-third of the area of Norway, and not more than one-twelfth of Sweden, stands at an elevation of 2,000 feet, and the whole mountain mass has been compared to a vast wave solidified in the act of breaking.

The main ridge, extending 1,150 miles from Varanger-fiord to the Naze, is far from presenting the appearance of a continuous range, such as it was figured on the maps before Munch had correctly described its true character. Norway consists, on the whole, of detached plateaux and mountain masses, raised on a common base 2,000 to 3,000 feet high, and pierced at intervals by profound inlets. The plateaux form two distinct groups—on the north the Kjölen, extending from Finmark to the Trondhjem table-land; on the south the Dovrefield and neighbouring masses. The mean height of the Norwegian summits is only one-half that of the Alps, whereas the general base of the Scandinavian system is one-third broader than the Alpine.

Even in the extreme north-east, throughout the whole of Finmark, there is no ridge properly so called. The entire country, with an altitude of about 1,000 feet, forms a vast irregular plain, composed of palæozoic rocks, above which here and there rise mountain masses, with a mean elevation of 1,700 feet, and culminating

with the Raste-Gaiso (2,880 feet), overlooking the valley of the Tana, on the
Russian frontier. Near this spot the Lapps speak of a cone occasionally emitting
lurid vapours, and whose snows at times melt rapidly.

Notwithstanding the general low elevation of the country, the headlands at the
extremity of every njarg, or peninsula, almost invariably end in lofty terraces
abruptly truncated. Such is the Nordkyn, or Kinerodden, northernmost point of
the European mainland. Two others, the low Knivskiärrodden and the more
elevated North Cape, lying 4 miles nearer the pole, stand on the granite island
of Magerö, separated by a narrow channel from Norway. The Austrian explorer,
Weyprecht, has suggested North Cape as one of the most favourable sites for a
polar meteorological observatory.

South-west of this point the summits of the islands and mainland are sufficiently
near to present the effect of a continuous range, and here begins the Kjölen

Fig. 33.—Island of Magerö.

Scale 1 : 635,000.

10 Miles.

properly so called. In the island of Seiland the northernmost European glacier
overflows from the perpetual snows of the surrounding rocks, while that of Talvik,
on the coast, usually descends to the shore of the Alten-fiord. On the southern
slope of the same mass is another glacial stream, resembling those of Greenland.
and discharging into the Jökel-fiord. This is the only place in Scandinavia
where may still be seen the phenomenon, common enough in former geological
epochs, of fragments of ice breaking off above the undermining waters and floating
away with the ocean current. South of these there are many other glaciers a
hundred times more extensive, but all melting into streams before reaching the sea.
In the lower valleys nothing is now visible except the traces of their former presence.

THE KJÖLEN UPLANDS.

OF the Northern Kjölen the highest summit is the Sulitelma (6,151 feet), rising above the eastern branches of the Salton-fiord within the arctic circle. It is not an isolated peak, but rather a group of crests resting on a common basis nearly 5,000 feet high, and covered with vast snow-fields, the source of several glaciers. South of this mass, and separated from it by a deep lake, rises the less elevated but more imposing Saulo, commanding an extensive view, which is limited on the west and south-west by the vast plateau covered by the Svartisen, or "Black Glacier," 270 square miles in extent, and the largest snow-field in Northern Scandinavia. South of the river Vefsen another plateau, the Store-Borgefield, has a snow-field with an area of 150 square miles, succeeded by mountains 4,500 to 5,000 feet high, beyond which the Kjölen falls, and is pierced through and through by wide channels. Here a marshy table-land, scarcely 1,500 feet high, connects two lakes, and through them

Fig. 34.—PROFILE OF THE SCANDINAVIAN HIGHLANDS.
Horizontal Scale 1 : 16,000,000. Vertical Scale 1 : 1,600,000.

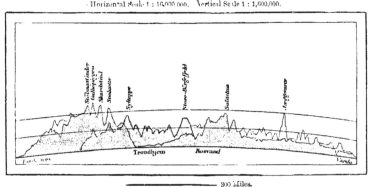

300 Miles.

two valleys, draining the one westwards to the Trondhjem-fiord, the other eastwards to the river Indals. Not far from this spot the ridge is crossed by its northernmost carriage road, 1,670 feet high; and a little farther south the railway from Trondhjem to Sundsvall crosses the Kjölen at an elevation of 1,948 feet. Beyond this point the main ridge bends south-east into Sweden, gradually falling to a simple terrace, which imperceptibly disappears between the Klar and Wester Dal river valleys, in the direction of Lake Wetter.

This branch of the Kjölen is connected by low ridges and terraces with the Tiveden and other hills, formerly serving as the natural limit of North and South Sweden—Nordan-Skog and Sunnan-Skog, or "North Forest" and "South Forest." The elevations occurring especially in Scania, formerly an island separated by wide channels from the rest of Sweden, must be regarded as quite distinct from the Norwegian system, although formed almost entirely of the same crystalline and palæozoic rocks. Still there are here some basalt rocks, and

TYPES AND COSTUMES IN THE HARDANGER, SAETERSDALEN, AND THE INTERIOR OF NORWAY.

traps have spread above the sedimentary formations of Gotland. The southern elevations form altogether a very irregular table-land, culminating with a rounded crest some 18 miles south of Lake Wotter. Near it rises the Taberg (1,024 feet), whose steep sides are composed entirely of magnetic iron, containing nearly one-third of pure metal. A few isolated hills are scattered along the south coast of Sweden, amongst them the Silurian promontory of Kullen (616 feet), at the northern entrance of the Sound.

West of the Southern Kjölen rises the Troidhjem plateau, with a mean elevation of 3,300 feet in the centre, and sloping gently towards the north and south. It is crossed at an elevation of 2,200 feet by the railway between Christiania and Troidhjem. All the rest of the country west of it is an elevated region, broken towards the coast by abrupt escarpments. Here are the highest summits of the peninsula and its most extensive *fjeldene*, or snow-fields, each fringed by glaciers

Fig. 35.—KULLEN HEADLAND.

Scale 1 : 320,000.

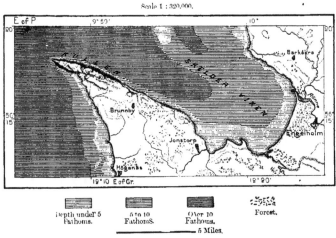

Depth under 5 Fathoms. 5 to 10 Fathoms. Over 10 Fathoms. Forest.

5 Miles.

(*brae*), overlooked by eminences presenting the varied forms of teeth (*tind*), horns, ridges (*egg*), or croups (*kol*, *nut*). There are several distinct masses, such as the Dovre, a name often applied to the whole plateau, and above which rises the Snehætten (7,570 feet), long erroneously regarded as the highest point in Norway. To the south are the Romsdal Alps, the Langfjelde, and the Jötunfjelde, or "Giant Mountains" (8,550 feet), the culminating point of the peninsula. Farther west extends the Justedal, the largest snow-field in Europe, with an area of 360 square miles, encircled by inaccessible rocks, and everywhere skirted by glaciers. South of the Hardanger-fiord stretches another great snow-field, the Folgefond, 110 square miles in extent, besides the Hardangervidde, the Oplande, and the Saetersdal, terminating at the Naze.

In this southern region the snow-line rises to about 4,500 feet, and round about the Justedal are the largest and best-known glaciers of Norway. These glaciers were constantly increasing throughout the eighteenth century, the moraines encroaching continually on the arable land, and compelling the inhabitants slowly to retreat before the advancing streams of compact ice; but since 1807 a retrograde movement has set in, some of the glaciers retiring from 2,000

Fig. 36.—TABLE-LANDS AND HIGHLANDS OF SOUTH NORWAY.

Scale 1 : 4,800,000.

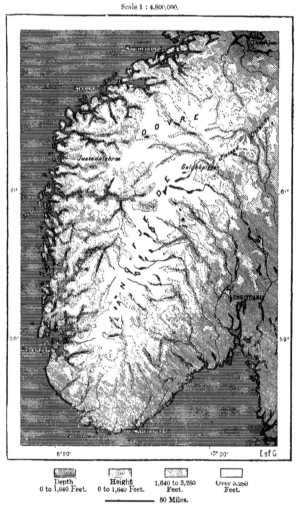

Depth	Height	1,640 to 3,280	Over 3,280
0 to 1,640 Feet.	0 to 1,640 Feet.	Feet.	Feet.

50 Miles.

to 3,000 feet, and leaving the moraines isolated in the midst of grassy valleys. Still the Buerbrae, amongst others, continued to advance till 1871, when its lower extremity reached a level of 1,237 feet above the sea.   At present several of the Justedal glaciers descend below 1,650, 1,330, and even 1,000 feet, while two on its

, side, the Boiumbrae and Suphellebrae, come within 480 and 170 feet of
.. David Forbes calculated that the glaciers of this region advanced at
ual rate of 167 feet, concluding that, though interrupted by a longer
the progress more rapidly in summer than those of Switzerland.

Fig. 37.—THE FJÆRLN-FIORD.

July 10th and 19th the Boium glacier moved forward nearly 17 feet, or
te of 1 foot 8 inches daily.

ugh the Norwegian snow-fields are far more extensive than those of the
glaciers cannot be compared with those of Central Europe. The cause
ntrast lies in the form of the mountains, those of the Alps being mostly
l, while the Norwegian highlands present the appearance rather of a
wall. Thanks to the general inclination of the Alpine rocks, the snowy
ove uniformly forward along the whole slope, whereas in the Norwegian
s the snow falls either on slightly inclined plateaux, or in deep gorges

opening like crevices in the flank. It lodges on the uplands, where it is blown about by the whirlwinds, and in the lower gorges it melts without being changed to ice ; hence the small extent and number of the Scandinavian glaciers.

Seen from the sea, the Norwegian mountains arrest the spectator by their dark crests, their snowy furrows and white terraces blending with the clouds or contrasting with the blue sky. They are yearly visited, especially by English travellers, although their outlines necessarily present less varied and picturesque landscapes than the Alps and other European highlands. The plateaux, with an elevation of 3,500 to 5,000 feet, no longer entirely covered with snow during the summer, and variously known as *Jiede*, or " heaths," and *viddene*, or " wide lands," are dreary wastes, more desolate than the desert, varied here and there only by a few snowy heights, like tents pitched in the wilderness. The surface of the uplands consists of a red and clammy soil, toilsome to the wayfarer, while the hollows are filled with peat beds, whence ooze black streams, sluggishly flowing from pool to pool in search of some outlet to the lower valleys. Vegetation is confined to lichens, mosses, and stunted grasses, with a few juniper plants and dwarf willows in the more sheltered spots. But no trace of man, except at long intervals along the few paths winding under the hills, and avoiding the swampy tracts in their way across the heath.

At the foot of the mountains the scene changes with the climate. Here is the abode of man, his lowly dwelling visible in the midst of the woodlands, or by the side of the running stream. Seawards the escarpments of the plateaux are seen at their full height, varied by snowy crests towering above the highest eminences, or blending with the clouds. But the peculiarly wild aspect of the coast scenery is due to the contrast between the rugged cliffs and the unruffled waters reflecting them, to the ever-shifting panorama of the fiords, to the headlands fringed with reefs, to the groups of rocky islets and maze of straits and channels. Nowhere else in Europe, not even on the south-west coast of Ireland, or in the Scotch firths with their basalt headlands, are the winding inlets of the sea skirted by such grand and frowning bluffs. The vessel penetrating into the gloomy passages of the fiords between almost vertical rocky walls, seen from above, seems like a tiny creature breasting the waves. The Bakke-fiord, on the south coast ; the Lyse-fiord, east of Stavanger ; and the great avenues converging on Christianssund are like the Colorado cañons, vast troughs hollowed out of the solid rock.

THE SCANDINAVIAN ISLANDS.

ROCKY islets are scattered in seeming confusion along the coast from Magerö to the Stavanger-fiord. Beyond the hilly peninsulas connected by narrow isthmuses with the mainland, there rise other eminences formed of the same geological rocks, and presenting the same general aspect, but of lesser elevation, and plunging into deeper waters. Still farther off succeed other islets, forming apparent seaward continuations of the headlands, and beyond them the countless reefs and rocks of the Skjärgaard. The Norwegians compare these advanced islets to marine and other animals, and even in the British Isles many similar formations still bear the name of *calves* given them by the Norse invaders. Such are the *Calf*, near the southern headland of the Isle of Man, and the *Cow and Calves* at the entrance to Cork Harbour.

The islands are most numerous and lofty in the northern province of Tromsö, at several points continuing the true ridge of the mainland far seawards. Thus the Kjölen proper is low enough east of the chain of mountains forming the island of Senjen and the Vester Aalen and Lofoten archipelagos, which project towards the south-west, gradually receding from the mainland, and thus forming a large

Fig. 38.—ARCHIPELAGO OF ISLETS IN THE NORWEGIAN SKJÄRGAARD.
Scale 1 : 910,000.

Sunk Rocks.
10 Miles.

gulf known as the Vest-fiord. Some of the summits of these islands exceed 3,300 feet, one in the Hindö, the largest of the Lofotens, attaining a height of 5,000 feet. The ridges are extremely sharp, and the contrast between the northern and southern slopes is very striking—on one side rich flowery meads, on the other bare or moss-grown rocks, with here and there a few tufts of heather.

The Lofoten shores, alive with thousands of craft during the fishing season, are much dreaded, owing to their fogs, storms, and strong tides. All have heard of the mael-ström, or Moskö-ström, which rushes in between the islands of Mosken and Moskenœs to meet the ebbing tide of the Vest-fiord. But there are many other equally dangerous whirlpools in these waters, and in several places the tides advance with terrific speed through the narrow straits.

South of the Lofotens there are no islands comparable in size with the larger members of that group; but there are hundreds still large enough to shelter the families of fishers and even labourers, and afford pasture for their cattle. Amongst them are several of extremely eccentric forms, resembling towers, castles, and such-like. Here is the Staven, or "Giant's Staff," a tall, slender rock wrapped in a cloud of snowy water-fowl; yonder the Hestmand, a cavalier shrouded in a mantle, eternally riding through mist and storm; elsewhere the better-known Torghatt, a gigantic rocky mass 800 feet high, pierced about half-way up by a grotto 900 feet long, of extremely regular formation, and with two portals 230 and 120 feet high. According to the legend this vast opening was made by the arrow of a giant, whose petrified bust is still to be seen a few miles off.

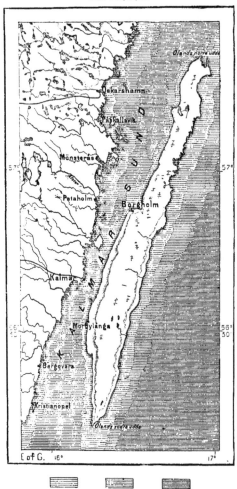

Fig. 39.—ÖLAND AND KALMAR SOUND.
Scale 1 : 1,110,000.

Depth under 11 Fathoms.   11 to 22 Fathoms.   Over 22 Fathoms.
————— 10 Miles.

The Norwegian islands, including those of the Skager Rak, but exclusive of reefs flush with the surface, have a total area of 8,500 square miles, or about the fourteenth part of the mainland; but, thanks to their convenient harbours, relatively mild climate, and fisheries, they are much more densely peopled, containing about one-eighth of the whole population of Norway.[*]

The Swedish islands are far less numerous, and long tracts, especially of the Scanian seaboard, are entirely free of islets or reefs. But on the Kattegat coast, north of Göteborg, there is

[*] According to Broch the 1,100 inhabited islands of Norway had a population, in 1875, of 238,000 souls.

THE SKJÆGGEDALSFOSSEN, IN THE DISTRICT OF HARDANGER.

a group resembling the Norwegian Skjärgaard, but without mountains, and destitute of vegetation. On the Baltic side there are innumerable little islets, especially
at the entrance of the gulfs and rivers north of Kalmar; but they are mostly low
rocks in shallow water, forming a seaward continuation of the Swedish plains.
To these plains belong also the two large islands of Öland and Gotland, stretching south-west and north-east, parallel with each other and with the axis of the
mainland.

Öland, composed, like the neighbouring coast, of older chalks, seems, in fact, to
be merely an advanced strip of the seaboard about 80 miles long, and separated
from Scania by Kalmar Sound, less than 2 miles wide at its narrowest point,
scarcely 24 feet deep off Kalmar, but with a mean depth of 60 feet at both
entrances. Gotland, lying much farther off, is connected south-westwards with
the coast by the extensive Hoborg Bank, and by a sort of submarine peninsula
limited on either side by depths of over 160 feet. It is larger and higher
than Öland, with one hill 200 feet high. It is continued northward by a
submarine bank, on which rest the islets of Färö and Gotska Sandön.

THE SCANDINAVIAN FIORDS.

THE submarine Norwegian orography corresponds with that of the mainland.
Thus the Sogne-fiord, 4,080 feet deep at its entrance, occurs immediately south of
the lofty Justedal snow-fields, at the western foot of the Giant Mountains. The
Hardanger-fiord also, over 1,800 feet deep, is flanked by the Thorsnut, rising
5,000 feet, to the south of Bergen. In many of the fiords the cascades have an
unbroken fall of over 2,000 feet, seeming to fall from the skies when the brinks of
the precipices are shrouded in mist. At times these aërial streams are buffeted
or swayed by sudden gusts of wind, sprinkling the rocky cliffs with a silvery
spray. Many disappear in mid-air, changed to diaphanous mist, again condensing,
re-forming on projecting ledges, and once more evaporating before reaching the
surface. In winter and spring avalanches of snow and detritus are precipitated
from the higher gorges to the lower valleys.

At first sight the Norwegian fiords present a very irregular appearance,
inlets, peninsulas, islands, and islets seeming to be entangled in inextricable confusion. Yet a certain order soon becomes apparent, and we discover that these
fiords are far more uniform than the Scottish firths. Few of them expand to
broad estuaries, nearly all communicating with the sea through narrow channels
between lofty headlands. The opposite cliffs maintain a certain parallelism in
the midst of their regular windings, and before reaching the sea several ramify
into two branches enclosing an island, the projections of whose steep sides
correspond with the receding outlines of the mainland. Others, such as the Sogne
and Hardanger fiords, branch off right and left, the side branches forming right
angles with the main channel, and themselves throwing off similar but narrower
branches, also at right angles. The land is thus cut up into innumerable regular,
or at least uniform blocks, some forming a portion of the mainland, others partly or
entirely surrounded by water—a strange labyrinth of plateaux, peninsulas, and

insular masses. The surface resembles that of quagmires baked in the sun, and attempts have been made to restore the Norwegian chart by indicating all the primitive fissures that have become fiords. But although these fissures at first sight seem the result of erosion, it is difficult to understand how it is that they occur generally in the hardest rocks, and are spread with remarkable uniformity for hundreds of miles over fiords and highlands. A series of such parallel troughs

Fig. 40.—Quadrangular Masses in South Norway separated by Fissures.
Scale 1 : 618,000.

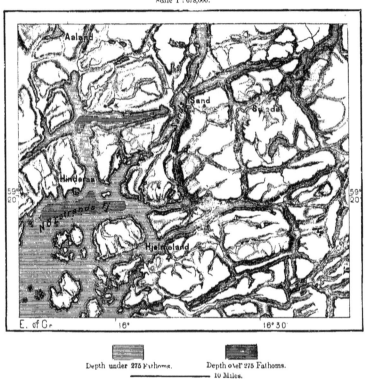

Depth under 275 Fathoms.   Depth over 275 Fathoms.
10 Miles.

extends north and south from the Moldo-fiord to the Naze, and another is connected south-east with the Gulf of Christiania.

All these intricate inlets present an enormous coast-line, the navigable channels alone being estimated at no less than 12,000 miles. Most of the coasting trade is carried on through these inland waterways, or between the outer islands and the seaboard. The lighter craft are even transported from fiord to fiord across portages or depressions, known as *ejder*, often less than 300 feet high.

Some of the fiords are so effectually cut off from the sea by islands and reefs that the fresh water of the melting snows and rains lies 4 feet deep on the surface. It is so pure that vessels are able to replenish their water casks from

LYSTER FIORD. THE NORTHERN BRANCH OF THE SOGNE FIORD.

this source, and the marine algæ either perish slowly or give place to fresh-water plants of rapid growth. The Dramms-fiord, fed by the Dramms-elv, the second largest river in Norway, resembles other formations of the same kind in its uniform width of 1 to 2 miles, and mean depth of 350 feet. But at the Sverdviken defile it suddenly contracts to a stream 16 feet deep and a few hundred yards wide, running seawards with a current of 9 miles an hour during the flow, and about 5 at ebb.

Fig. 41.—Dramms-fiord and Sverdviken Channel.
Scale 1 : 152,000.

Most of the fiords are partially obstructed at their entrance by the remains of old moraines, which in the north are called *harbroen*, or "sea bridges." Both sides of the Gulf of Christiania are regularly lined with the shingly deposits of such ancient moraines. But what is the origin of the bars occurring at certain intervals from the mouth to the upper end of the fiords? Some are ridges between two valleys resembling those of the upheaved land; some are slopes produced by erosion; while others, no doubt, are moraines like those deposited by former glaciers at the foot of the hills in the upheaved valleys. For, like the Scottish firths, the Scandinavian fiords existed before the glacial period, and were able to maintain their original form by means of the vast glaciers filling and deepening their beds, and grinding smooth their rugged sides. In warmer or more humid regions the estuaries were slowly filled in by the alluvia of the streams or the sands of the sea, whereas the fiords retained their original depth, often below the bed of the neighbouring seas, which accordingly advanced as the glaciers retired. But since then the running waters and the ocean have begun the vast geological work of filling in these northern estuaries. The rivers bring down their alluvia, depositing

Depth under 55 Fathoms.   Over 55 Fathoms.
2 Miles.

them in regular strata at the foot of the hills, while the sea precipitates, in even layers of sand or mud, the detritus washed away by the waves. The work of transformation has already made perceptible progress, and the old inlets have disappeared along the south-west coast between Porsgrund and Stavanger fiords. In this region, exposed to the southern suns and sheltered by the elevated plateaux from the northern winds, the glaciers retired much earlier than on the west coast, and between the already effaced inlets of the extreme south and the still perfect fiords of the extreme north every possible shade of transition may be studied. Here, if anywhere, the problem regarding the duration of the present geological epoch may possibly yet be solved. Here every still surviving glacial stream, every ancient glacier bed, records in detail the alternating history of the climate during the period subsequent to the glacial age. Every fiord thus becomes

Fig. 42.—Silted Fiords north of the Naze.

Scale 1 : 240,000.

Depth under 110 Fathoms.　　　　Depth over 110 Fathoms.

————————— 3 Miles.

a meteorological and geological apparatus, by the remains of its moraines, by the striæ of its rocky ribs, by the alluvia of its streams clearly indicating all the changes that have taken place in the surroundings. Building on these and similar data, Theodore Kjerulf has already attempted to introduce a more exact chronology into the geological record.

GLACIAL ACTION.—THE ÅSAR.

WITH the fiords of the west correspond the lakes of the east side of the peninsula. A subsidence of the land would transform them to salt-water inlets, just
as an upheaval would change the Norwegian fiords to lakes. There are even
many valleys intersecting the Kjölen and the South Norwegian plateaux which are

Fig. 43.—THE HALLEBORG AND HUNNEBORG HILLS.
Scale 1 : 200,000.

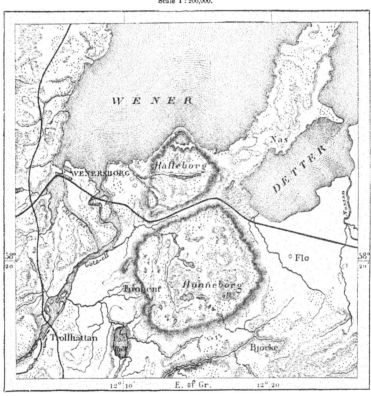

Forests abounding in erratic blocks.
2 Miles.

occupied at certain intervals by marshes and lakelets, apparently survivals of a
former strait connecting the fiords on both sides. Such is the Lesjeskogen-vand
Lake, 2,044 feet above sea-level, whence flow two streams, the Rauma to the north-
west to the Molde-fiord, and the Lougen, south-east through the great Mjösen to
the Gulf of Christiania. Many of these lakes have preserved their fiord-like
character, and one of them, the Hveningdals-vand, has a depth of 1,600 feet,
or 1,418 below the sea-level.

But the whole country, no less than the fiords and lakes, has preserved the traces of former glacial action, which extends even far beyond the peninsula. Finland, one-third of Russia in Europe, all North Germany, Denmark, the Netherlands, the greater part of Scotland, the Färöer, and Iceland itself are comprised in the vast region from 1,000,000 to 1,500,000 square miles in extent, whose surface soil is largely due to the detritus of the Norwegian uplands. With the exception

Fig. 41.—CHRISTIANIA AND ITS ISLANDS.
Scale 1 : 185,000.

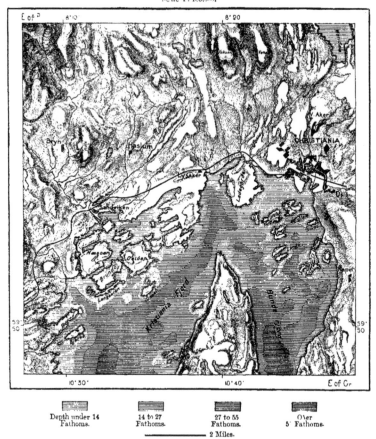

| Depth under 14 Fathoms. | 14 to 27 Fathoms. | 27 to 55 Fathoms. | Over 55 Fathoms. |

2 Miles.

of the Skager Rak, which seems to have been a true fiord, all the narrow Scandinavian seas formed the beds of ancient glaciers, whose traces may even still be recognised in the submarine striæ in some places, as at Karlskrona, to a depth of 24 feet. Lower down they have been obliterated by the action of the water, or filled in with sands.

To the northern glaciers Esmark long ago referred the numerous erratic boulders of southern lands, and the same view, formulated by M. Charles Martins in 1840, although combated by men like Berzelius and Murchison, is now universally accepted. The evidences of glacial action, striæ, polished rocks, moraines, boulders, are far too obvious to be overlooked. There is scarcely a geological chart of Scandinavia in which the form of the hills does not bear witness to the passage of glaciers, and in some places glacial mounds and boulders are numerous enough to impart a special character to the scenery. From a

Fig. 45.—Islands in the Gulf of Christiania.

distance the traveller coasting the southern Norwegian seaboard easily detects the action of ice in the rounded forms of the headlands, and in the interior of Sweden the angular rocks and crests of many hills have been worn to level surfaces by the same cause. Even the lava beds that have overflown on the older formations in the neighbourhood of the great lakes have to some extent been affected by similar influences. Such are the two polygonal masses of Halleborg and Hunneborg, separated from each other by a narrow trough, through which runs the railway between Wenersborg and Jönköping. Their levelled summits present tolerably uniform surfaces, strewn with boulders, with intervening swamps and lakelets.

Still the aspect of the land is occasionally deceptive, and to glacial action have been wrongly attributed certain parallel dispositions of rocks that have been thus affected by side pressure. The group of islands, for instance, immediately to the west of Christiania, are all turned north-east and south-west, and are furrowed with creeks and separated by channels all running in the same direction. But the striæ scored by the old glaciers are all at right angles to these parallel lines.

Of more difficult explanation than the striæ are the so-called âsar, or low ridges of various heights, from 20 to 200 feet, stretching almost without interruption sometimes for over 70 miles across the country, generally from north to south, or south-east, and winding like rivers to the right and left. There are some lateral âsar of less length ramifying in various directions, whereas the larger ones run mainly in parallel lines. They were at first supposed to be vast moraines, until Berzelius showed that there was no direct relation of cause and effect between the âsar and the glaciers.

The âsar, however, are composed of materials transported for a first stage by the ice, and then borne farther by other geological agencies. After vast depressions had been filled in with detritus by the frozen streams, the waters began their work, hollowing out enormous furrows in these masses, in which the boulders continually gravitated downwards, becoming rounded off by friction, or ground to sand and gravel. Such are the materials of which the âsar are composed. Erdman supposed that the larger ones were due to the action of the sea-waters, which with the changes of level took up the stones of the moraines that had so far drifted with the streams. Many of the âsar, notably that which runs immediately north of Stockholm, are doubtless covered with marine shells of the same species as those of the present Baltic Sea. But such deposits are quite superficial, and have been formed during a temporary subsidence of the land after the glacial epoch.

According to the materials of which they are mainly composed, the âsar are called either sandâsar or rullstenâsar (sand or shingle âsar) ; but all alike show evident traces of more or less rough stratification, such as is still constantly going on in running water. There seem to be others, again, which rest on moraines—a fresh formation superadded to the first. In some there occur certain funnel-like formations (âsgropar), cylindrical or elliptical, with a circumference of 300 yards and upwards, and a mean depth of 10 to 68 feet, the bottom of which is filled with clays formerly deposited by eddying waters. In the Strömsholm âs there are thirty-nine such funnels in a distance of about 84 miles, all apparently due to the action of running waters, whose force was spent by the obstacles lying in their way, and by the constant shifting of their beds. In Norway, where the slopes are much more abrupt, and where the water-courses are consequently less developed between the mountain cirques and the sea, the âsar, here known as raer, are far less numerous than in Sweden, and seem to have been mostly confounded with the moraines. The Norwegian word aas ·is applied to all eminences, even to rocky summits.

best-known Scandinavian ås is the long chain which, under divers names—
bergs ås, Långåsen, and so on—extends for a distance of 60 miles, from the
coast south of Stockholm to the neighbourhood of Upsala. Between the
st and Lake Wetter, near Askersund, there are no less than eight main

Fig. 46.—ÅSAR IN THE DAL RIVER BASIN.

Scale 1 : 170,000.

Forests strewn with erratic blocks.

————— 2 Miles.

sides their various secondary ramifications, and amongst them are some
rably longer than that of Brunkeberg. The traces of another may be
1 for 180 or 200 miles northwards, from the shores of Lake Mälar,
nköping, while others form natural embankments across the lakes,

the western portion of Mälar itself being thus almost completely cut off from the main basin by one of these singular formations remarkable for its extreme regularity.

Upheaval of the Land.

The development of the åsar and the marine alluvia deposited on the present upheaved lands are evidences of the movements that have passed over the Scandinavian area since the glacial epoch. At first the land subsided, and the sea-level rose from 500 to 700, in some places even 1,000 feet, as shown by the marine deposits with the remains of arctic animals resting on rocks scored by the action of ice. Then a reverse movement set in, and the land was upheaved, bearing upwards the åsar previously deposited by the running waters. During these vicissitudes of level the Scandinavian relief must have been changed, for the outlines of the upheaved islands and peninsulas do not always correspond with those of the lands engulfed in the deep. Thus a vast Silurian region, at the beginning of the glacial epoch stretching along the Swedish seaboard immediately north and north-west of the Aland archipelago, did not again emerge with the reappearance of the plains on the Baltic coast. The former existence of this Silurian land is recalled by the numerous calcareous and sandstone boulders transported by the glaciers or floating ice as far south as the neighbourhood of Stockholm. It is also shown especially by the rich soil of limestone origin covering all the coast districts between Gefle, Wester ås, and Stockholm. This fertile soil is the outcome of the constant erosive action of floating ice on the layers of limestone, clays, and schists formerly occupying the present water area east of Gefle.

These shiftings of level were formerly supposed to have been caused by sudden terrestrial cataclysms coincident with sudden revolutions of the whole planetary system ; but this view has at last given place to the theory of slow change. The natives of the Bothnian seaboard had long been aware of the gradual increase of the coast-line, encroaching continually on the sea. The old men pointed out the various places washed by the waves in their childhood ; and farther inland the names and position of long-forsaken havens ; buildings at one time standing on the seashore ; the remains of vessels found far from the coast ; lastly, written records and snatches of popular song, could leave no doubt regarding the retreat of the marine waters. The first Luleå, founded by Gustavus Adolphus, seemed to have retired several miles westwards in a century and a half, thus becoming a rural town, and necessitating the building of a new seaport farther east. Yet when, in 1730, Celsius ventured to suggest the hypothesis, not of an upheaval of the land, but of a slow subsidence of the Baltic, he was charged with impiety by the Stockholm theologians, and even in Parliament the two orders of the clergy condemned his abominable heresy. Nevertheless a mark scored in 1731 by Celsius and Linnæus at the base of a rock in the island of Löfgrund, near Gefle, indicated in thirteen years a difference of level estimated at 7 inches.

Although it would be impossible altogether to reject the hypothesis of Celsius regarding the subsidence of the waters, still it seems evident that it is the land rather than the sea which is changing its level. Leopold de Bueh was the first to assert, in 1807, that the whole Scandinavian peninsula was rising above the surrounding seas. Here the greatest number of observations have been made, and Scandinavia has thus become the type with which are compared all other slow upheavals elsewhere taking place.

In many places the evidences of recent upheaval are perfectly visible from the sea, but on the Swedish side the movement is going on most rapidly towards the north. Thus at the northern extremity of the Gulf of Bothnia the upheaval is estimated at about $5\frac{1}{4}$ feet in the century, and at 3 feet 3 inches in the latitude of the Aland Islands, whereas at Kalmar there seems to be no change. The southern extremity of Scania, which is now probably rising, appears to have formerly slowly subsided. Several streets in the towns of Trelleborg, Ystad, and Malmö have already disappeared, and the last mentioned has subsided 5 feet since the observations made by Linnæus. Submerged forests and peat beds found at a certain distance from the present coast-line, and where metal objects have been collected, have caused geologists to suppose that since the ninth century the subsidence has amounted to from 14 to 16 feet.

Thanks to this movement, the Gulf of Bothnia would seem to be slowly draining into the southern basin of the Baltic, and at the present rate of upheaval in the north three or four thousand years would suffice to change the Qvarken archipelago to an isthmus, and convert the northern section of the gulf into a fresh-water lake.

On the Norwegian seaboard the movement is far less regular, and nowhere so rapid as on the Swedish side of the Gulf of Bothnia. At some points even of the north coast no rising seems to have taken place at all. Thus Tiötö, mentioned in the sagas, is still the same large low island of former days, and a reef in Trondhjem-fiord, on which a swimmer could find a footing in the time of the first vikings, appears to be still at the same depth below the surface. Eugène Robert believes that no upheaval has taken place for three hundred years at Christiania, though others have found a rising of $12\frac{1}{4}$ inches per century on the shores of the fiord.

The periods of upheaval must have been frequently interrupted by more or less protracted intervals of rest. If most of the terraces consist of moraines levelled by the waves, or alluvial deltas brought down by the inland streams, there are others which have been hollowed out of the hard rock by the slow action of water continued for ages. But rocks gradually emerging could not have been much worn on the surface. Lyell supposes that the Norwegian coast has been slowly rising for at least twenty-four thousand years, while Kjerulf considers that the movement has been much more rapid.

It is generally held that the underground upward pressure is not uniform throughout the peninsula, but that it acts by a series of undulations, so that between the regions of upheaval intermediate zones are left unaffected, or very nearly so. But further observation is needed to establish this view, and since 1852 the mean level of the sea and Scandinavian plain has been studied day by day at thirteen different points on the coast, and on the shores of Lakes Mälar, Hjelmar, Wetter, and Wener.

As to the cause of these upheavals, some have regarded them as local phenomena, others associating them with various disturbances of the surface in Europe, or with the vital forces of the whole planetary system. Many causes may even be at work, at times neutralising each other, at times co-operating, though the period for deciding these points with certainty has not yet arrived.

THE SCANDINAVIAN LAKES.

THE irregularity of these oscillations is perhaps mainly due to the unfinished state of the peninsula on both maritime slopes. The chief geological function of the running waters is to regulate the slopes by giving them a parabolical curve from the source to the mouth of the streams. The differences of level, the perturbations of the coast, the thousand phenomena associated with the planetary economy, have everywhere prevented the rivers from completing this work ; but nowhere is the irregularity of the river beds greater than in Scandinavia. They generally form a series of terraces rather than a uniform curve, and to the unequal upheaval of the land must, at least in part, be attributed the formation of so many lacustrine basins.

The numerous pools and lakes on the Norwegian plateaux and upper valleys have already been referred to. Towards the Arctic Ocean the escarpments are too little developed to retain many lakes in their granite basins, while those occurring at their base are merely detached remnants of fiords. But on the Swedish side, and the Norwegian slopes facing the Kattegat and Skager Rak, the surface is everywhere dotted over with bodies of still water. Here they are relatively more numerous than in any other European region except South Finland, occupying about one-thirteenth of the whole area of the peninsula. In certain Swedish districts, especially Södermannland, between Stockholm and Norrköping, the lakes are so common that they are held in no more account than trees of the forest. "When God severed land and water," says a local proverb, "He forgot Södermannland." Nearly the whole of South Sweden has remained in a similar chaotic state, the water surface occupying over one-eighth of the entire area. Most of the lake shores are houseless, silent pine, birch, and oak forests, rarely relieved by the songster's note, reflecting their foliage in the greenish waters, or dyed a reddish hue by the tannin of the heather. Sedgy or reedy tracts encircle their borders, while elsewhere huge blocks detached from the neighbouring cliffs raise their crests above the surface. No sail enlivens the dreary waste of waters ; nothing recalls the presence of man except a solitary bark moored to the shore, or the blue smoke of an isolated cottage in some neighbouring glade.

The extent of the stagnant waters is known to have been considerably reduced during the historic period, a result due partly to river action levelling rocky ledges and sweeping away moraines and åsar, partly to the hand of man here and there constructing drainage works. The numerous fortified enclosures met on the hill-tops and headlands are now surrounded by peat beds, swamps, and marshy pasture-lands. A careful study of these low-lying tracts shows that they were formerly lakes or navigable gulfs, and here are sometimes still found the remains of boats.

Fig. 47.—LAKES WENER AND WETTER.

Scale 1 : 1,500,000.

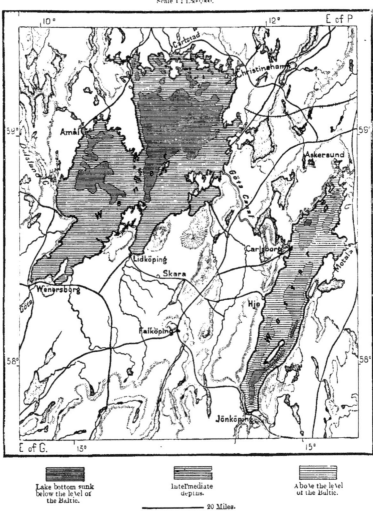

Lake bottom sunk below the level of the Baltic.

Intermediate depths.

Above the level of the Baltic.

20 Miles.

The fastnesses were accordingly defended, partly at least, by water, and nearly all the Swedish towns had their origin in such insular strongholds.

The larger Swedish lakes are themselves an index of the gradual drying up of the land. Formerly continuously united, they formed a strait connecting the

North Sea with the Baltic, as clearly shown by tracts now covered with marine fossils. Oysters have been picked up on the south shore of Mälar, a sure proof that these waters had formerly at least 17 parts in 1,000 of salt. Björkö, one of its islands, was till recently strewn with the bones of sea-fowl as well preserved as if they had been just forsaken by the mews after breeding season. Nay, more, there still survive small animals of marine origin whose organism has been slowly adapted to the fresh water gradually replacing the sea in the lacustrine basins. Even the Norwegian Lake Mjösen, notwithstanding its distance from the strait of which Wetter and Wenor are detached links, still harbours the *Mysis relicta*,

Fig. 48.—LAKE MÄLAR.

Scale 1 : 605,000.

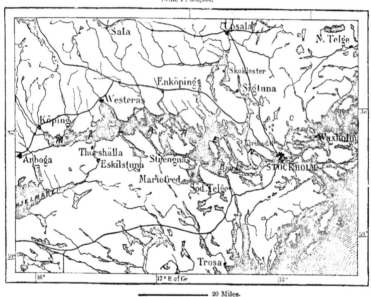

20 Miles.

a living species bearing witness to its former connection with the neighbouring seas, whose temperature was at that time as low as is now the Frozen Ocean.

Henceforth severed from the sea by slowly widening isthmuses, the great lacustrine basins dividing Sweden into two distinct regions have continued to rise with the rest of the land. Their surface is now above sea-level, although the beds of most of them are below the surface of the Baltic. Wener, the largest of the Scandinavian lakes, with an area one-tenth the size of all the rest together,* has a mean elevation of over 144 feet, with an extreme depth of 290 feet. It is thus two or three times larger than Lake Geneva, and about one-fourth the size of Ladoga. Wetter, at twice the altitude of Wener, is also deeper, measuring 413 feet, and 125 below the level of the surrounding seas. Hjelmar, lying nearer

* Area in square miles:—Wener, 2,386; Wetter, 733; Mälar, 668.

to the Baltic, and no more than 76 feet above the sea, is the only member of the large lake system whose bed stands higher than sea-level, its greatest depth being 60 feet. Thus it might be completely drained into Mälar by deepening the junction canal, and works have already been undertaken to reduce its level 6 or 7 feet, and to regulate its outflow by means of a dam, allowing a discharge of 400 cubic yards per second if necessary. Thanks to these works, thousands of acres will be preserved from further floodings.

Lake Mälar itself is not yet entirely cut off from the sea. One of its extremities is still a gulf, and when the east wind arrests the outflow, a marine current, the *nppsjö*, conveys a small quantity of salt water to the eastern portion of the lake. With its numerous channels and thirteen hundred islands, islets, or reefs, this inland sea must be regarded, not as a single sheet of water, but as an aggregate of separate basins, each at a slightly different level from the rest. In fact, it consists of four sections, disposed from west to east at successively lower levels. The Köping, or upper basin, has a mean altitude of 2 feet 5 inches above the Baltic; the second, consisting of the Westerås fiords, falls to 2 feet; the Björk-fjärd, or third, to 1 foot 6 inches; while the Riddarfjärd, or eastern most section, at Stockholm, is about 1 foot above the Baltic. The various basins were formerly separated by shingly åsar, which were swept away by the pressure of the waters, thus converting the labyrinth of creeks into a united body of water. A rapid current, regulated by a sluice, sets constantly from Mälar to the sea under the Stockholm bridges.

Besides the South Swedish lacustrine basins there are in the rest of Scandinavia thirty-five lakes, each with an area of over 40 square miles, and many larger than Hjelmar itself. Such are those whence flow the chief Baltic streams—the Torneå Träsk, the Luleå Jaur, Stor Afvan, Swedish Storsjö, and Siljan, "blue eye of Dalecarlia." But most of these remote lakes have been but partially explored, and their depth is still unknown. The Mjösen (140 square miles), however, the largest in Norway, thanks to its proximity to Christiania, has been carefully studied, and found to have an extreme depth of 1,480 feet, with an altitude of 397, its bed thus lying about 1,080 feet below the sea-level.

In winter all the Scandinavian lakes are ice-bound for one hundred to two hundred days, according to the latitude and the severity of the season. But the stagnant waters, and even the shallow lakelets, seldom freeze to the bottom. Soon after the formation of the first icy mantle, the heavy snows generally protect the lower waters from the frost, and thus keep the fish alive. Long fissures are opened here and there in the frozen surface, leaving the air to penetrate below.

Like the fiords, the lakes are distinguished by their geometrical disposition. Many follow in succession along one deep trough, while others cross each other at sharp angles, grouping themselves into figures of divers forms. Thus in the south-west corner of Norway they enclose triangular spaces; in Telemark they form an eccentric polygon, in which the direction of indentation is represented by a *rand*, or lakelet of pure water.

THE SCANDINAVIAN RIVERS.

THE very rivers, fed by these countless lakes, consist mostly of chains of lakes varying in form and size, now confined between narrow walls, now expanding into broad sheets of water. All of them send down very large volumes compared with

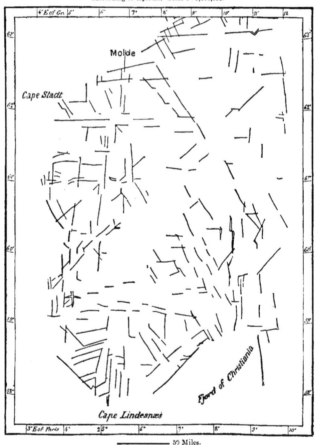

Fig. 49.—GLACIAL SCOURINGS IN SOUTH NORWAY.
According to Kjerulf. Scale 1 : 5,000,000.

—— 50 Miles.

the area of their drainage. This is due partly to the heavy rainfall along the western seaboard and about their sources, partly to their rocky beds allowing of little or no infiltration, and partly to the relatively slight evaporation in this moist climate. Compared with France, Scandinavia discharges a far greater quantity of water into the sea, as may be judged from the amount sent down by the few rivers whose volume has already been estimated. Still there are no such majestic streams as the Rhône or the Rhine, the relief of the country preventing the development of great river basins. On the western side the Norwegian streams soon meet the fiords after leaving the glaciers or snow-fields. On the east side the Swedish rivers, being directed straight to the Baltic by the tilt of the land, are unable to group themselves into one large water system. Those flowing to the Gulf of Bothnia occupy nearly parallel valleys, all sloping south-eastwards in the line of the former glaciers. In South Sweden the streams radiate in all directions to the surrounding inlets, none of them, except the Göta, collecting the waters coming both from the plains and the highlands.

The largest Scandinavian river is the Norwegian Glommen, discharging into the Gulf of Christiania, which also receives the Dramm, whose alluvia have already filled a large portion of the great lake of Tyri-fiord. Nearly as large is the Göta,

Fig. 50.—LAKES IN SOUTH NORWAY.

Scale 1 : 470,000.

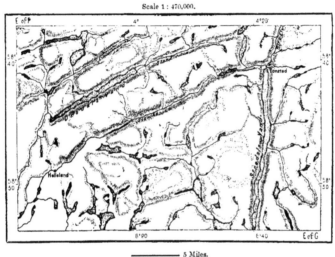

5 Miles.

thanks to the volume of water received by Lake Wenor from the Klar-elf and its other influents. But there was a time when the great Lake Fœmund, now draining southwards to the Kattegat through the Klar, drained through the Dal-elf south-east to the Gulf of Bothnia. The Fœmunsgrav, the old bed of the river, is still visible 4 or 5 feet above the present level of the lake. But if formerly deprived of the waters of Lake Fœmund, the Göta received from another quarter all those of the Glommen, so that its volume was more than doubled. At the foot of the hill on which stands the town of Kongsvinger, north-east of Christiania, the Glommen now turns suddenly westwards; but formerly it continued its south-eastward course parallel with the Klar-elf, to Lakes Aklang and Wener. During the heavy floodings a portion of the Glommen waters still escapes by the old bed, and long narrow lakes preserving the meandering form of

the river fill its former valley, now crossed by the railway from Christiania to
Stockholm. By a singular coincidence the two rivers flowing east of the Glommen
have also been diverted to the right, now discharging into Lake Wener through
beds belonging originally to other streams. The Fryken joined the lake at the
point where is now the town of Karlstad, but its present junction lies some 12
miles farther westwards. The Klar-elf, now usurping the old bed of the Fryken,
flowed through a small lacustrine valley forming a south-eastern continuation of

Fig. 51.—THE TELEMARK LAKES.

Scale 1 : 400,000.

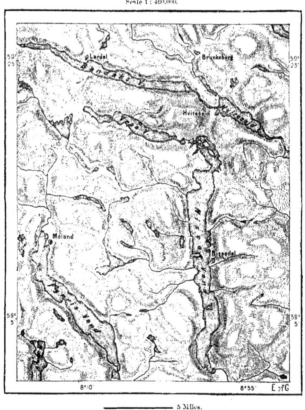

5 Miles.

its upper valley. This westward diversion of three parallel streams seems to point
at a lateral pressure, doubtless due to a slight perturbation of the crust. In the
same way the Vistula, the Elbe, and the Oder have deviated from their original
beds.

The Norrland rivers, north of the Dal, are remarkable for the striking uniformity
of their general features and volume. They also closely resemble each other in
the extent of their drainage, the nature of the soil watered by them, and the

amount of their rainfall. Thus, going northwards, the Ljusna, Ljungan, Indals Ångerman, Umeå, Skellefteå, Piteå, Luleå, Kalix, and Torneå all discharge into the Gulf of Bothnia an equal volume of water, estimated at upwards of 70,000 cubic feet. The current of each is prolonged far into the sea, whose waters, however,

Fig. 52.—THE TYRI-FIORD.

Scale 1 : 250,000.

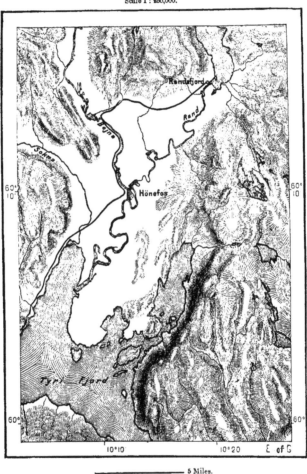

———— 5 Miles.

are so slightly saline that the contrast is here scarcely noticed, which is at once detected at the mouths of the Göta and Glommen.

The discharge of the Scandinavian rivers is mostly better regulated than that of the continental streams, a result due to the lakes which are traversed in their course, and which equalise the floodings. At the end of April and in May the melting snows, and in autumn the heavy rains brought by the west winds, raise the level of the lakes. But while they receive the overflow, these reservoirs distribute it evenly amongst their outlets, reducing the amount of each during the floodings, while husbanding the supply for the dry season. The annual variation of level in the lacustrine basins oscillates between 3 and 12 feet, although the rise has occasionally been much greater, owing to the refuse blocking the outlets. Thus in 1795 the Vormen, which carries off the overflow of Lake Mjösen, was completely barred by fallen masses of rock, causing a rise of 22 feet in the level of the lake. The choking of the valleys at the outlets of the lakes has in many places facilitated the construction of dams, completely regulating the discharge according to the amount needed to drive the mills, or, in the season, for navigation.

The floating ice also is seldom dangerous, owing to the southerly course of all the large rivers. When the thaw sets in the frozen masses are broken up first at their mouths, and so on from south to north, so that no block takes place through any sudden rush at the narrows.

But although in other respects well regulated, the Scandinavian streams are not generally available for navigation, except at a few points about their mouths or in the neighbourhood of the lakes traversed by them. The undeveloped state of the river beds, still disposed in terraces, is favourable to industry by affording good motive power, but not to traffic, which is interrupted by the rapids and dams. Many are so involved with successive falls and lakes, and even with other basins, that they seem to have scarcely yet acquired a distinct individuality. Thus the Tornea, on the Russian frontier, belongs really to two systems, one of its branches, the Tanudö, flowing to the Kalix, the other to the Muonio. Many also ramify before reaching the sea, not sweeping round alluvial deposits, but enclosing rocky masses, their branches being old marine channels converted into beds of running water. Such is the Göta-elf, whose bifurcation encircles the large island of Hisingen.

The chief beauty of the Scandinavian rivers is due to their falls and rapids. On the Norwegian side all the streamlets may be said rather to be precipitated than to flow seawards. In many places there are clear falls of several hundred yards, and even some of the larger Norwegian rivers have sudden plunges of over 300 feet. The Vorings-fos, near Trondhjem, descends at one leap 472 feet, and the Rjukan-fos, formed by an affluent of the Skien-elv, in Telemark, is precipitated a vertical height of 804 feet. Much lower in elevation, being only 70 feet high, but far more considerable for the volume of its waters, is the Sarps-fos, on the Glommen, where even in winter a mass of 3,500 to 5,000 cubic feet, escaping from its icy fetters, rushes headlong down a series of cascades, below which it again disappears beneath the ice. The mean volume of the Sarp falls is about 28,000 cubic feet per second, or double that of the Rhine at Schaffhausen. A recently constructed railway bridge commands a full view of the entire series of cascades, and of the seething waters appearing here and there below the dense vapours. Notwithstanding their proximity to Christiania, these Glommen falls, the mightiest in Europe, are less known than those of the Göta-elf, the famous Trollhättan, or "Wizard's Cap," descending 110 feet in three successive leaps, and enclosing grassy rocks between their rushing waters. The force of the Trollhättan, estimated

THE FALLS OF TROLLHÄTTAN.

by engineers at 225,000 horse-power, is partly utilised by industry, but the mills do not here, as at the Sarps-fos, prevent access to the view.

On the Baltic side the gentle slope of the land has prevented the development of such stupendous falls as on the west side, though even here there are some of a very imposing character. Thus the majestic Dal-elf, which throughout its lower course is little more than a series of lakes, contracts suddenly at Elf-Karleby, and, dividing into two branches, descends through a number of rapids a total height of 50 feet just before reaching the sea. There are also some fine cascades on the Skellefteå and the Luleå. At the Njommelsaskas, or "Hare Leap," the Luleå has a clear fall of over 266 feet in height, and several hundred yards in width, and higher up a lake, separated from another reservoir by a simple ledge, rushes over a cataract 140 feet high. To this lake the Lapps have given the name of Adna-muorkekortje, or "Great Cloudy Fall."

CLIMATE OF SCANDINAVIA.

THE main ocean current on the Norwegian seaboard sets south-west and north-east. The warm waters from the tropics strike the outer banks of the peninsula, often throwing up drift-wood and seeds from the West Indies, which the Lapps carefully preserve as amulets. So well known is this northward current that when anything falls overboard the sailors jocularly speak of going to pick it up at Berlevaag; that is, at the eastermost extremity of Lapland. This stream of warm water gives to Norway its climate, to the people their trade, commerce, daily sustenance, their very lives, so to say; for, but for it, the shores of the fiords would be blocked with ice and uninhabitable. The Scandinavian peninsula forms with Greenland the marine portal through which the Atlantic communicates with the Frozen Ocean. But under the same latitude what a prodigious difference of climate! On one side ice and snows eternal, on the other mainly fogs and rain. The great western island absolutely treeless; the eastern peninsula covered with tall forests, orchards of blossoming apple, pear, plum, and cherry trees, gardens in which the vine itself is cultivated as a wall fruit in a richly manured soil. Yet a portion of Scandinavia, estimated at 60,000 square miles, is already comprised within the polar zone, where throughout the winter night follows night in perpetual darkness. In summer, on the other hand, the dying day melts in the new dawn. The Finmark hills command the amazing spectacle presented at the summer solstice by the midnight sun grazing the horizon, and again climbing the eastern skies. From the crest of the Avasaxa, overlooking the Torneå valley near the arctic circle, the sun may be seen, between June 16th and 30th, describing fifteen complete circuits in the heavens. As he stands bathed in sunshine, the spectator beholds at his feet all the southern lands shrouded in the great mantle of night, and the snowy heights, instead of reflecting a white light, are made glorious by the dazzling colours in which the purple of the setting sun is blended with the soft tints of dawn. With the great lakes, the boundless heaths, the snowy mountains, storms, and limitless seas, these long and alternating nights and days contribute to impart to the life of the land its grand and stern aspect, which so endears it to the people.

The form of the Norwegian seaboard aids not a little in maintaining the warmth of the land. The temperature of the fiords to their lowest depths is higher than that of the ambient atmosphere. The researches made by Mohn show that these basins are filled with comparatively warm waters, 24° higher than the surface air in January.* This is due to their formation, causing them to communicate over raised ledges with the ocean, whence they consequently receive warm south-west currents only. The deep waters of the Färöer and Iceland seas under the same latitudes have a temperature below freezing point, whereas the fiords never freeze except along the shores farthest removed from the high seas. Thus the whole Norwegian seaboard is, so to say, furnished with a vast heating apparatus by these outer reservoirs filled with waters several degrees above the normal temperature.

The thermometric régime of these waters presents remarkable contrasts with the seasons. In summer and autumn the temperature falls from the surface downwards, whereas in winter the heat rises gradually with the depth, a result due to the atmosphere. In summer the air is warmer than the surface waters, which it consequently heats. This heat is transmitted downwards, but very slowly, while the colder and heavier zones remain below. In winter the surface is rapidly cooled by the colder atmosphere, the lower zones remaining unchanged. But from the surface downwards the natural sinking of the cold strata produces displacements of the liquid layers, which regulate the series of temperatures. The thermal curves for each season figured on Mohn's ingenious charts oscillate on either side of a fixed standard, occurring at about 600 feet below the surface. Nevertheless the influence of the warm waters would be very slight but for the warm south-west and south winds prevailing on the Norwegian seaboard. It is under their influence that the Scandinavian isothermals are diverted northwards, following the coast-line almost inversely to their normal direction.

Still there is a certain alternation in the general atmospheric movement. The prevailing winds in winter, and even in spring and autumn, are breezes blowing from all the valleys and fiords towards the surface of the sea, whose temperature is always above freezing point. But in summer the reverse takes place, the winds setting from the ocean towards the heated regions of the interior. Thus the temperature of the sea and coast lands becomes modified from month to month. The winds passing over the inland snow-fields cool the waters of the seaboard, which preserve their normal temperature only where the influence of those winds is unfelt. In their alternating movement from winter to summer and summer to winter the winds are deflected regularly with the coast-line. In winter they set northwards, thus aiding vessels coasting from the Naze to North

* Temperature of the fiords and the atmosphere:—

	Deep Water.	Atmosphere. Mean.	January.			Deep Water.	Atmosphere. Mean.	January.
Skager Rak	41°	45°	33°	Trondhjem-fiord .		44°	41°	27°
Hardanger-fiord	43°	45°	32°	Vest-fiord . .		43°	37°	25°
Sogne-fiord	43°	45°	31°	Varanger-fiord .		38°	30°	14°

Cape; in summer and winter they blow in the opposite direction, favouring
the traffic between Hammerfest and Christiania. They are otherwise always
stronger along the west coast than in the interior, and the storms, frequent in
winter, rare in summer, burst with great fury down the mountain valleys facing
the Atlantic. Near Stavanger, at the entrance of Lyse-fiord, flashes of lightning
accompanied by thunder are occasionally emitted from a bluff 3,000 feet above
the sea. This happens only when the wind is from the south-east, and meteoro-

Fig. 53.—ISOTHERMAL LINES FOR THE YEAR.
According to Mohn.

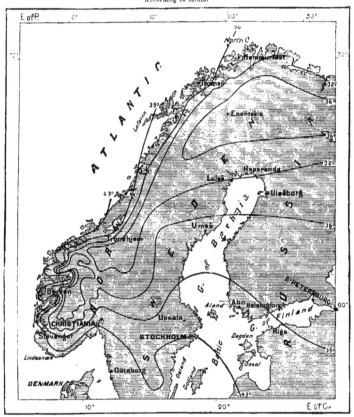

logists have hitherto been unable to determine the atmospheric conditions under
which the phenomenon takes place.

The warm winds also supply an abundant rainfall, which, however, is very
unequally distributed over the peninsula. In the western islands, and especially
the Lofoten group, it rains on an average every other day, and at Bergen,
on the south-west coast, the fall amounts to 71 inches in the year. But
beyond the glaciers and snow-fields the average is not more than 39 inches,

falling to about 20 at Christiania, Tromsö, and other spots sheltered from the
moist winds. Throughout Sweden, which is defended by the Scandinavian
uplands from the wet quarter, the mean fall is 20½ inches, and consequently less
than in France and the British Isles. By a singular contrast the snow-line
descends much lower on the western than the eastern slopes of the Folgefond
and Justedal highlands, the fact being due to the abundance of moisture brought

Fig. 54.—THERMIC ISABNORMALS FOR THE YEAR.
According to Mohn.

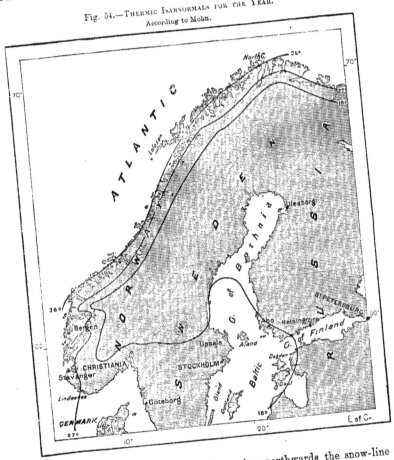

by the west and south-west winds. But going northwards the snow-line falls
uniformly from 4,000 or 5,000 feet on either side of the Folgefond, and to 3,000,
or even 2,400, on the mountains about North Cape.

The isothermal lines, determined by systematic observation at the fifty-three
Norwegian and twenty-nine Swedish meteorological stations, present in summer and
winter the same general form. They run nearly parallel on the west coast, describing
their principal curve towards the south-west from Trondhjem to Christiania.

Even in the coldest winter month parts of the seaboard maintain a temperature above freezing point, and at Christianssand and the Naze the glass scarcely ever falls below zero. At Bergen, which is less exposed to the warm south-west winds, there are twenty-four frosty days in the year, although at Hammerfest, on the extreme north coast, a small stream never freezes. But farther inland the rest of the peninsula comes within the limits of the frost-line in winter. Calculated for

Fig. 55.—Difference of Temperature between Summer and Winter.
According to Mohn.

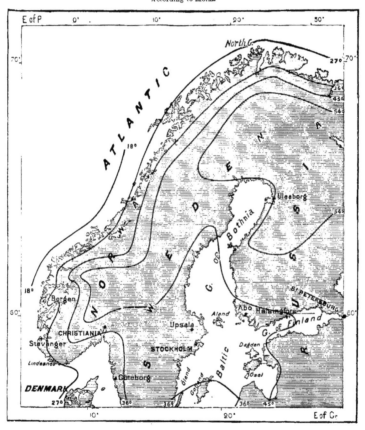

the whole year, this line embraces the interior of the Norwegian Finmark and the Swedish Norbotten, and farther south the Dovre plateau, besides some other uplands about Röros, all of which regions are consequently almost uninhabitable. The population naturally gravitates to the warmer districts, and few houses are found higher up than 2,000 feet, while summer châlets are seldom met beyond 3,000 feet. Yet here and there valleys are thinly peopled which might seem to be absolutely uninhabitable, and Sollöisa, or "Sunless," in the Bergen district, is so called because it never receives the direct rays of the sun.

The climatic anomalies of the western seaboard disappear rapidly as we go eastwards. The mean temperature of the Norwegian coast exceeds by 36° that of its latitude, whereas the excess is reduced to 18° north of Christiania and Upsala. Here the climate becomes continental. Its daily, monthly, and annual deviations are more and more considerable, rising from 18° on the coast, between July and June, to 54° in Swedish Lapland. In these high latitudes the summers are about as warm as in South Sweden, but the winters are much colder, and the bogs of Lapland remain in some places frozen throughout the year.*

SCANDINAVIAN FLORA.

EVEN in the zones of vegetation considerable anomalies occur. Although the mean temperature is everywhere higher on the Norwegian seaboard than on the eastern slopes, several species of trees reach a far higher latitude in Sweden than in Norway. Thus the Norwegian pine forests cease about the Norrland frontier at the arctic circle, though they extend much farther north in Sweden. By an analogous phenomenon, the birch, which stops at an elevation of about 1,050 feet on the Norwegian slopes, ascends to double that height on the Swedish side.

Over 2,000 European plants have their northern limit in Scandinavia. A chart exhibited by Schübeler at the Paris Geographical Congress shows the hitherto determined polar limits of 1,900 cultivated and wild plants of the Norwegian shores. Travelling from south to north and from west to east, we find the plants of the temperate European zone successively disappearing. First is passed the beech and hornbeam zone, comprising South Scania, the Kattegat, and the south-west coast of Norway as far as Bergen, a little to the north of which is found the northernmost beech forest of the globe. The mingling of this light green foliage with the dark conifers forms the great charm of the Christiania coast scenery.

The oak zone comprises all Central Sweden to the river Dal, and the Norwegian coast to the neighbourhood of Christianssund. The white alder, the pine, fir, and birch extend much farther north, and attain higher elevations on the hillsides, the birch flourishing even on the Finmark plains. The absolute tree-line takes in but a very small part of Norway about the northern shores of the Varanger-fiord. The southern shores are fringed with veritable forests of pine, fir, birch, aspen, alder, and service trees.

All the Scandinavian species are exotics, which have occupied this region since the glacial period. Yet so deep is the verdure of the foliage, so vivid the colours of the flowers, that most botanists might fancy themselves in the presence

			Lat.	Mean.	Temperature. January.	July.	Rainfall. in.
Hammerfest	.	.	70° 7'	35°	23°	52°	—
Bergen	.	.	60 4	45	23	58	71
Christiania	.	.	59 54	41	23	61	21
Stockholm	.	.	59 20	41	25	60	20
Göteborg	.	.	57 42	44	29	61	21

of new species. The perfume also of all plants and fruits increases gradually northwards, while the sap diminishes proportionally. The wild bay is much finer than in more southern lands, and is largely used in flavouring dishes served at all Norwegian tables. Troidhjem is noted for its excellent apples, pears, and cherries, and at Frosten, not far off, even the walnut ripens. Thanks to the mildness of the climate and to the almost uninterrupted sunshine during summer,

Fig. 56.—Oskarsborg : View taken from Oön.

the cereals of the Norwegian seaboard ripen as quickly as in lower latitudes. Thus at Alten, under the seventieth parallel, barley comes to maturity in ninety days, precisely the same time that it takes on the banks of the Nile. But although the Hindö islanders often ship barley for Christiania, there are parts of Swedish Lapland where it never ripens even in favourable seasons. In Jemtland also it often fails, and such years are called " green."

The ordinary *lund*, or forest, of the Swedish plains differs little from those of temperate Europe. But the *skog*, or wild forest, with its chaos of rocks and dense thickets of trees and shrubs, presents quite a special character. The granite or porphyry blocks heaped up in confusion are nearly everywhere covered with a thick mossy carpet, elder and raspberry bushes spring from their interstices, while the roots of tall pines, firs, alders, and birches creep snake-like amidst the fragments. The pathway is soon lost in this labyrinth of rocks and trees, the dreaded wilderness formerly haunted by the *troll*, or ' evil'one," where the benighted wayfarer met the terrible *skogfran*, or " woman of the thicket."

The Swedish woodlands occupy about two-fifths of the country, while Norway, owing to its greater mean elevation, is far less rich in forests, which here cover

Fig. 57.—Store-houses in the Hitterdal

no more than one-fifth of the whole area. The pines of Gefle, on the banks of the Dal, have long been in demand, as furnishing the best material for masts. But the very extent of the forests has hitherto caused a reckless consumption for fuel, building, and manufacturing purposes, which, combined with the export trade, has already cleared vast spaces. Along the line of railway between Malmö and Stockholm woods are still everywhere met, interrupted by lakes and cultivated tracts. But the magnificent beech and oak forests of the sagas have disappeared. In South Norway also forests worthy of the name occur only in the more **inaccessible** districts, and many iron-smelting works have already been abandoned **for** want of cheap fuel. In these cold regions the growth of timber is slow, so that the destruction of the forests causes a general impoverishment of the land. While the Norwegians consume relatively about five times as much wood for local purposes as do the French, the product is five times less in an equal area. Hence the recent measures that have been taken for the preservation of forest lands. By the law passed in 1875 proprietors in the northern provinces of Sweden are bound to protect stumps less than 10 inches in thickness at man's height, and in the island of Gotland the trade in wood for sale has been interdicted.

Scandinavian Fauna.

The wild animals formerly frequenting the Scandinavian woodlands have become rare. A price has been set upon bears, wolves, lynxes, gluttons, as well as foxes and birds of prey, which are now seldom seen except in remote districts. The elk has not yet disappeared from the Norwegian highlands, and a herd still roams north of Christiania. The roebuck and stag are also met with in the Norwegian forests, and some of the islands near Bergen and Troidhjem are hunting grounds for their owners. The aurochs lived in Scania during the stone age, and in the Lund Museum there is a specimen still showing a flint wound. The beaver still survives, and the hare, white in winter as the surrounding snows, is common in the hilly northern districts, while the lemmings (*Lemmus Norvegicus*) descend in multitudes from the Norwegian plateaux to the sea. The reindeer is no longer found wild in the Swedish valleys, though numerous herds roamed till recently on the Norwegian uplands, the Telemark mountains, and the Dovre plateau. The tame reindeer of Lapland is distinct from the species whose remains are occasionally found in the peat beds of Scania. The former came probably from the east with the nomad immigrants from North Russia.

Nearly all the birds of the southern shores of the Baltic are found also in Scandinavia, though in lesser numbers. But countless flocks of sea-fowl frequent the rocky shores and islands of Norway. Some of the Lofoten and Vester Aalen groups bear the name of *nyker*, or "bird hills," owing to the numbers of penguins, gulls, mews, frequenting them, and from a distance giving the rocks the appearance of so much trembling vapour. The fluttering of their wings is heard miles away, and close by resembles the soughing of the winds. The absence of ports or sheltering creeks renders these islands very inaccessible, but in calm weather they are visited by the hunter. The eider, rare in the southern fiords, is very common in the North Finmark islands as well as in the Vester Aalen and Lofoten groups, here forming one of the chief resources of the inhabitants.

The Norwegian waters abound in fish to a far greater extent than do those of the Swedish coast. Of all the Scandinavian inlets the richest in marine life is the Molde-fiord, between Aalesund and Christianssund, the reputed home of the fabulous "sea serpent." At depths of from 100 to 200 fathoms here are taken the *Lota*, a species of cod ; the *Coregonus silus*, a salt-water salmon ; the *Spinax niger*, a peculiar species of shark, whose skin seems bristling with crystalline needles ; and the "chimæra," a grotesque and voracious animal known to the Norwegians as "the sea-king." In these deep fishing grounds Sar the elder recorded in a few years 427 new species, apart from those discovered at depths of from 200 to 270 fathoms in the Lofoten waters. But within the last two hundred years the whale has entirely disappeared from the shores of Finmark.

The inhabitants of the south-west coast of Sweden derive their sustenance largely from the Kattegat and Skager Rak fisheries, where hundreds of smacks pursue the herring, cod, and whiting. After suddenly disappearing from the Göteborg coast about the beginning of this century, the herring has again returned, at first in small numbers, but in the winters of 1878 and 1879 in vast quantities. On the Norwegian side, also, between the Naze and Cape Stadt, their movements have been very erratic. But most of the inhabitants of this coast have given up fishing, and turned their attention to the cultivation of the land, formerly almost entirely neglected.

In the northern waters the cod banks on either side of the Lofotens are periodically deserted, and all these vicissitudes have recently enlisted the telegraph in the service of the fishermen along the coast, who now receive immediate information of the appearance of shoals, wherever they happen to arrive.

The very lives of the coast populations in Finmark may be said to depend on the cod fisheries. When these are productive mortality diminishes, communities flourish, comfort is everywhere diffused. When the herrings swarm in the Lapland fiords towards the end of summer, despair overspreads the land, for a long experience has taught the people that the herring and the cod do not visit the same waters in the same year successively. But when the former are rare the ports are alive with preparations for the coming season, which is sure to be profitable. On these occasions the curing of the fish is not confined to the regular fishing populations alone, whether Norwegians, Finns, Lapps, or Russians. To the seaboard are then attracted long caravans of men, women, children, dogs, and reindeer from the interior. The Lapps of the forest, drawn down to the sea, depart and return with the birds of passage.

Earliest Inhabitants.

THE written records of Scandinavia date no farther back than some twenty generations. The earliest chronicles, dating from the end of the eleventh century, have almost entirely perished. Of this epoch, and of the earlier period to the beginning of the ninth century, nothing survives except the faint traditions recorded by the sagas. Of the hazy past, penetrating beyond that epoch into the night of time, the only witnesses are the remains left by primitive man on or below the surface of the land.

In its archæological remains Southern Sweden greatly resembles the Danish islands, and seems to have been occupied by men of kindred race and like customs. But farther north we enter another domain. Northern Sweden and Norway have no kitchen middens of the stone ages, with the exception of one at Stenkjaer, near the Trondhjem-fiord, and even this contains objects belonging exclusively to the

olished stone epoch. No bone or chipped flint or rudely cut stone imple-
ch as occur in the Danish middens and Belgian and French caves, have
been found in the regions north of the strait formed by the great lakes.
ence is that North Scandinavia was then uninhabited, and not occupied till
ied stone age, objects of that epoch being the oldest there met with.
moth and rhinoceros, whose remains are found associated with those of
.e art in the French caverns, are also missing in the peninsula, there having
ime for the men of that age to occupy it at the close of the glacial epoch.
) not know the precise time to which must be attributed the remains of
.tive Scandinavians and their industry; but it no doubt coincides with
nal withdrawal of the ice formerly covering the land, though Nilsson
assigns too great an antiquity to the first arrivals. Between the towns
and Trelleborg, and at Fallsterbo headland on the Baltic, there occurs a
vide rampart running east and west, composed of gravels and sand,
ed here and there, and divided into unequal sections, formed probably at
epochs. It is called the Jaravall, or "Jura Mound," and below the
'ilsson has discovered arrow and lance points resting on the bed of old
nations, now over 7 feet below sea-level. But the Jaravall, formerly
to be of old formation, seems to be a recent coast rampart. Two
also were discovered in 1843 in the Stängenäs peninsula near Bro, still
eath horizontal strata of marine shells, now 105 feet above the sea. But
be shown that these strata are in their original position, and it is
whether or not the skeletons belong to the stone age.
urial-places of the polished stone age are very numerous in Scania,
and Bohuslän, but, like the middens, do not occur in the north. There
l forms: *stendösar*, or dolmens; *hällkistor*, large graves of raised stones
d by earth; *gånggrifter*, or galleries, called also *jättestugor*, or giants'
They are often large enough to contain upwards of twenty bodies, and
rectangular or round, with flat roofs of granite slabs supporting earth
stone heaps. A long narrow gallery leads to the funeral chamber, and
face south and east. The blocks are never dressed, and the arms and
s found with the bodies (mostly with dolichocephalous crania) all belong
ithic ages. At this epoch most sepultures contained necklaces and other
of amber beads. To the same period may perhaps belong the numerous
ned into porringers, commonly known as *elfstenar*, or "fairy stones."
)00 specimens of the stone age found in Sweden up to the year 1874,
ne from the southern provinces; that is, from Götaland. When the
ceeded the stone age it was here also that civilisation was chiefly
for of the 2,500 bronzes only 150 were found in Svealand, and in the
Norrland 2 only. To this period belong the hieroglyphic writings, or
riptions" (*hällristningar*), occurring here and there in Scania, Götaland,
ohuslän, and Norwegian Smaalenene, and which represent fleets, large
figure-heads of dragons, waggons, ploughs, oxen, and other animals,
unters, and seamen. Beautiful bronze vases, ornaments, diadems, **now**

152

collected in the museums, bear witness to the artistic originality of the people of the bronze age, although many archæologists suppose that Etruscan influence inspired the Scandinavian art of this period. But most of the Swedish objects were cast in the country, as is evident from the stone moulds occasionally picked up, though the bronze must have been imported as an alloy, for it contains about one-tenth part of tin, a metal not found in Scandinavia. Nilsson fancies he detects numerous traces of Phœnician industry, attributing to these Semites the tombstones on which are figured ships, hatchets, and swords. But the absence of the usual Phœnician inscriptions militates against this view. Nor is Greek art at all represented in Scandinavia, except perhaps by a few isolated objects found on the east coast of Sweden.

But Roman influence was strongly felt, though indirectly. Even beyond the limits of the Empire the barbarous nations followed the impulse given by the conquerors of the Mediterranean world. They learnt the use of iron, and began to employ a series of letters akin to the Latin alphabet, and probably derived from that of the Celtic tribes in North Italy. These runes, or "mysteries" (runar, runir), as they were called, are of various forms, and have been greatly modified in the course of ages. The inscriptions run sometimes from right to left, but more commonly from left to right, while several are of the "boustrophedon" class, the order of the letters alternating with each line. Some must even be read Chinese fashion, in vertical columns, and the form of the letters changes with the time and locality, those of the extreme north being especially noted for their originality. At first numbering twenty-four, they were reduced in Scandinavia to sixteen, and were here carved on rocks or bones, or wood, horns, ornaments, and arms. The northern museums contain large collections, which, if throwing little light on the history of the race, have at least illustrated the successive changes of their language. In mediæval times whole volumes were composed in runes, as, for instance, the Skånelagen, or "Law of Scania," dating from the thirteenth century. The gold ornaments known as bracteates, of which nine times more have been found in Scandinavia than in all the rest of Europe, are mostly covered with Runic signs. The figures of heroes, horses, birds, dragons, are all referred by Worsaae to Northern legends.

The age of iron, when the runes were in vogue, blends gradually with the historic epoch, about the time of the great Norse expeditions. But it is difficult to draw a hard-and-fast line between these various epochs. During the Empire, when they exchanged their wares for Italian coins, the Scandinavians used concurrently iron weapons, bronze and gold ornaments, stone implements. The runes themselves survived in the island of Gotland till the sixteenth century, and Runic calendars continued in use still longer in the peninsula, and even in England. Thus it is that the successive civilisations rather overlap than follow each other abruptly. The rites of the old worship surviving as superstitions are a further evidence of this mingling of epochs, resembling the currents of various streams we sometimes see uniting in one bed. Thus Thursday, in Swedish Thorsdag, or day of Thor, was still kept as a holiday so recently as last century in various parts of

TYPES AND COSTUMES: SWEDISH INTERIOR.

the peninsula, for the old "Red Beard" forbade its profanation by manual labour. Whoso was born on Thursday had the gift of seeing the ghosts of the departed, and all incantations, to be efficacious, had to be performed on that day, but not so the Christian rites attending births, deaths, and marriages, showing that the day dedicated to the old Thunder god is still held as pagan. Finn Magnusen tells us that till the close of last century the custom survived in some Norwegian upland valleys of worshipping certain round stones on Thursday, which were smeared with butter and placed on fresh straw in the seat of honour at the head of the table. At fixed times they were washed in milk, and at Christmas sprinkled with beer, in the hope of bringing luck to the domestic hearth.

THE SWEDES AND NORWEGIANS.

COMING from the shores of the Euxine and Danube, the Götar and Svear, now collectively grouped as Scandinavians, had to cross half of Europe before reaching their northern homes. It has often been suggested that the migration flowed through North Russia and Finland—a view, however, which is not supported by the arms and implements found in these regions. The people seem rather to have come from the south-east and south to Denmark, passing thence first to Scania, and so on to Norway and North Sweden. Thus, while the Lapps and Finns penetrated from the north, the Teutons arrived from the opposite extremity of the peninsula.

The Götar, or Goths, were the first conquerors. These were followed by the Svear, or sons of the "blessed Ases," who, passing over the southern parts already occupied by the Goths, gradually overran the rest of the land, and their Asgard, shifting with each migration, was ultimately fixed in the centre of the country. The difference existing between the two groups is still very perceptible both in their speech and customs, though scarcely to an appreciable extent in the form of the cranium and general physical type. Here, as elsewhere, the essential physical differences are due rather to the manner of life than to origin. The typical Swedish head, as described by A. Retzius and Nilsson, is a lengthened oval, slightly broader behind than in front, but rounded off on either side, greatest length and breadth standing in the ratio of 4 to 3 or 9 to 7.

The Dalecarlian, or native of Dalarne, Upper Dal basin, is usually taken as representing the purest type of the Svear, who have given their name to the *Srenskar*, or Swedes of our day. He is generally tall, slim, and lithe, with noble features animated by beautiful deep blue eyes, and expanding to a broad, open brow. He is unobtrusively courteous, cheerful without excess, firm without violence. Honesty above all proof may be said to be the stock in trade brought with them by the thousands of Dalecarlians who come to settle in Stockholm, where they are employed in all work needing strength or skill. They are everywhere recognised, even at a distance, by the bright colours of their national dress. The Swedes differing most from the Dalecarlians are those of the lowlands, who often wear a serious, almost stern expression.

A mingling of the various elements has been produced by the continuous progress of intercourse and inland colonisation.   The Byzantine coins and other
objects found in the land show that after the fall of the Roman Empire the Swedes
had constant commercial relations with Constantinople.   Later on, Gotland

Fig. 58.—The Finns and Lapps in Scandinavia.
According to Dahlman and others.

became, towards the ninth century, the emporium of a direct export trade with the
farther east, and here are found, from time to time, hoards of Arab or Kufic coins
from Bagdad or Khorassan.   These relations lasted till the twelfth century, when
they were interrupted by the wars in South Russia.   The Swedes took also a large
share in the Crusades, and it was mainly through arms that the Scandinavians

were brought into association with foreign nations. The Swedish rovers have doubtless left a less profound impression on history than the Danish and Norwegian vikings. But this is due to the direction taken by their warlike excursions, which did not bring them into contact with peoples of such high culture as the Franks or Mediterranean nations. Their warlike deeds could be commemorated only in the obscure traditions of Finns, Letts, Wends, and the Slav tribes of the vast Gardarike, or Russia of our days.

Foreigners could have had but a slight direct influence on the Scandinavian race, for within historic times the peninsula has never been invaded by victorious armies, if we except the short Russian expeditions of 1719 and 1809. Nor has there been much peaceful immigration, and that mostly from Finland. Towards the end of the seventeenth century Finnish peasantry began to cross the Gulf of Bothnia and settle in Upper Jemtland, on the Norwegian frontier, where their descendants still survive, intermingled with the surrounding Swedish populations. Other Finnish colonies are found in the northern provinces. Religious persecutions also contributed in a small degree to the peopling of the land. At the end of the sixteenth century some hundreds of Walloon workmen, at the invitation of a Dutch owner of mines, took refuge in Sweden, settling mainly in the village of Österby, near the Dannemora mines. Their descendants, nearly all of a brown complexion, have retained the traces of their descent, and carefully preserve the spelling of their French names. Since then several other French exiles have sought homes in Sweden; but their influence has been purely local, and the zeal with which the language of Racine has been studied and Paris fashions imitated on the Baltic shores must rather be attributed to a certain natural sympathy between the two nations. The Swedes are fond of calling themselves the " French of the North," and their social ways, courtesy, and good taste certainly entitle them to the name.

The Norwegians, on the other hand, are the " English of Scandinavia." From over the seas their gaze is fixed on the British Isles, with which their chief commercial intercourse is carried on, and whence come their most numerous foreign visitors. They are in general distinguished rather by strength and tenacity of will than by liveliness or pliability. Their resolutions are formed slowly, but what they will they carry through. Amongst them mysticism seems more prevalent than in Sweden, which is yet the native land of Swedenborg.

The inhabitants of Scandinavia speak various languages, all, however, derived from the old *Norræna*, or Norse tongue of the Runic inscriptions. Hence their close affinity and imperceptible blendings, the Scanian, for instance, serving as the connecting link between Swedish and Danish. The standard Swedish, which is simply the cultivated dialect of the Stockholm district, as Danish is that of the Copenhagen district, is an harmonious language, full of assonances, and, thanks to its greater treasure of archaic terms, more original than its southern sister. But amongst the local dialects there are others of still more ancient type, notably the Dalecarlian, the Gottish of Gotland, and those still current beyond the frontiers of the present Sweden, in parts of Finland and the islands on the Esthonian coast.

The literary language of Norway is simply the Danish with a few local words and idioms; but certain remote valleys still cherish the old Norse, forming with Icelandic a distinct linguistic group. Some Norwegian patriots have endeavoured to re-establish the supremacy of their ancestral tongue, and thus create a new literary language. Societies have been founded, journals and books published in old Norwegian, but the undertaking has not met with general encouragement. Certain writers have, on the other hand, essayed to assimilate the current forms of speech, and thus restore the unity that prevailed in the ninth century. In 1869 several Danish, Swedish, and Norwegian men of letters met at Stockholm to adopt a common orthography, but national rivalry has hitherto prevented any concert among the grammarians.

The Lapps.

By the side of these Scandinavian peoples, who are amongst the most homo-geneous in Europe, there live tribes still of a quasi-Asiatic character, few in numbers, but exceedingly interesting for their physical aspect, origin, and manner of life. These are the Lapps, like the Rouminisäl, or Swedish gipsies, partly nomad, and thinly scattered over a vast area, estimated at 80,000 square miles, in the northern extremity of the peninsula, along the upper course of the Swedish rivers flowing to the Gulf of Bothnia, in the Finnish territory ceded by Sweden to Russia, and in the Kola peninsula. They number scarcely 30,000 altogether, or about 1 to every 1,700 acres.

It is certain that the Lapps formerly reached much farther south than at present. Traces of their language are detected in Swedish, and several southern geographical terms have been referred to them. Some are still found in the heart of Jemtland about the sixty-third parallel, where their domain is clearly limited by the lichens supplying the sustenance of their reindeer herds. But they have been continually pressed northwards by the Norse immigrants, and the legends of *dvergar*, or "dwarfs," *troll*, or "magicians," *bergfolk*, or "highlanders," are mythical records of the internecine strife that raged between the conflicting elements.

Universally known by their Swedish appellation Lapps, variously interpreted as "Nomads" or "Cave-dwellers," these Sameh, or Samelats, speak a Finnish language said to be more akin to the Mordvinian than any other member of the Ural-Altaic family, and preserving archaic roots and forms that have disappeared from modern Finnish. But though officially designated as *Fin* in Norwegian Finmark, they are clearly distinguished from the Finns proper not only by the contrasts produced by the different cultures, but also by their physical features and form of their crania. Hence some anthropologists have regarded them as of a distinct stock, on whom the Finns have imposed their language. Thus, while Virchow considers them to be a branch of the Finns, Schaufhausen takes them for the descendants of Mon-golian tribes driven northwards, and migrating westwards along the shores of the Frozen Ocean. Till recently the Sameh were also supposed to differ from the rest of mankind by an absolute ignorance of song. But the statement of Fetis, that "the Lapps are the only people who do not sing," is erroneous, and although incapable of uttering notes pleasing to the Swedish ear, they are quite capable of

TYPES AND COSTUMES IN LAPLAND.

iitoiatioi, and several of their soigs have beei collected. But they are
ure race, as showi by their old Swedish, Norwegiai, Fiinish, and even
i family iames.

se of the iiterior, probably the least mixed, are geierally of very low
compared with their almost gigaitic Swedish neighbours. But the meau
is greater thai has beei supposed by lovers of the marvellous, who have
id of coitrastiig the Patagoiiai giaits with the pigmy Lapps, liviig at
extremities of the habitable world. Dulk gives the meai height at 5 feet
s, and Voi Düben, who has most carefully studied them, at 4 feet 11 iiches.
iulsive features credited to them exist oily in the imagiiatioi of their
urs, and although mostly marked by high cheek boies, nose flatteied at
:emity, small eyes, triaigular face, scait beard, yellowish complexioi, their
i is very capacious, with a high aid really noble brow, smiliig mouth,
d and kiidly expressioi. The eye is usually black, but the colour of the
:ies greatly, being sometimes chestiut or quite black, sometimes perfectly
ilthough iot so riigiig and melodious as that of the Swedes, their voice is
ieans weak or thick, except amongst the braidy-driikers, now rarely met,
ly ii Swedei, siice the sale of spirits was totally iiterdicted ii 1839.
has beei replaced by coffee, which those who can afford the expeise driik
loig, mixiig it with salt, cheese, or evei blood aud drippiig. Thaiks to
brity of the climate, and ii spite of the foul air and filth of their hovels,
e geierally healthy and loig-lived. Mortality is lower thai amongst the
l peoples of the seaboard, but, as was remarked by Acerbi ii the last
, they oftei suffer from red and sore eyes, caused by the smoke of their
d their loig jourieys across the siows.

irdiig to their manier of life the Sameh are divided iito Highlaid and
that is, reiideer and fishing—Lapps. A few have become agriculturists ii
ns about the Gulf of Bothiia, but evei these rely maiily on the produce
ikes aid rivers. Their huts are composed of a simple coiic frame covered
ivas or some woollen fabric, leaviig the smoke to escape at the top. Some
ed oi piles, and household affairs are maiaged by the mei, possibly a relic
cocracy, or "mother's right," so prevaleit amoigst primitive races.

Lapps of the arctic shores are more iumerous than those of the iiterior,
iecessarily iomads, although they do iot form migratory tribes like the
is and Turkomais. Each family lives apart in the forest, iot through any
ile feeliig, but because of the great space ieeded to support their reiideer
Ivery Lapp requires at least tweity-five aiimals, and after beiig nibbled
eis grow very slowly. Heice the herds returi oily every tei years to
e graziig grouids, aid but for the grass aud sprouts available ii summer
land would be too small for its few thousand iihabitants. Reiideer milk,
iately far from abuidait, forms their chief iutrimeit, and this they "eat,"
wiiter, in the form of frozen cakes. But the flesh and blood of the animal
coisumed, and the ordiiary daily meal is the "blood soup," made of flour
tted blood preserved duriig the wiiter in casks or skins. Beiig thus

dependent for his sustenance as well as his clothing on the herd, the Lapp who owns no more than a hundred animals is regarded as poor, and obliged to attach himself to some more fortunate grazer. Excluding the fishers and agriculturists, Von Düben calculates the average number of reindeer per head at thirteen or fourteen only, and this number tends to diminish with the growth of settlements. The owner of three hundred is already regarded as wealthy, and some are said to possess as many as two thousand, valued at about £2,400. But rich and poor live all alike in wretched, dark, and squalid dens, free, however, from fleas, which do not thrive in Lapland. But in summer the gnats are a terrible scourge, at least for the stranger, if not for the natives, who are protected by smearing themselves with a fatty substance, and who then live mostly in districts where the winged pests are dispersed by sea breezes.

Since the middle of the seventeenth century all the Lapps have been calling themselves Christians. They already possess a small religious literature, and follow the rites prescribed in the several local governments. Thus in Scandinavia they are all Lutherans, in Russia Orthodox Greeks; but beneath it all there survive traces of old pagan customs analogous to the shamanism of the Mongolians. The magic drum played a great part in their ceremonies, as did also the pine or birch bark on which the wizards had figured instruments, animals, men, or gods. This bark, or "rune-tree," as the Norwegians called it, was consulted on all important occasions, and the interpretation of the mysterious signs was the great art and highest wisdom. The last of the "rune-trees" is said to have been destroyed about the middle of the past century. The seitch, curiously shaped stones, sometimes rudely carved, round which the rites were celebrated, were thrown into the lakes by the Lapps themselves, or else preserved in the Swedish museums. But if the fetishes have disappeared, many of the old ceremonies survive. The dog, the Lapp's best friend, without which he could not rule his herds, is no longer buried with his master; but certain shells, the "souls of the dogs," are still thrown into the grave. The feast of the summer solstice also is here, as elsewhere in Europe, celebrated with bonfires kindled on the hill-tops.

The Lapps are supposed to be yearly diminishing in numbers; but at least in Finmark, or Norwegian Lapland, they have increased sevenfold since the sixteenth century, and elsewhere apparently threefold on the coasts. But this is largely due to the pressure of the Nybyggare, or "New Boors," Swedish and Finnish colonists slowly encroaching on the domain of the nomads, and driving them seawards. At the end of the eighteenth century these strangers were already more numerous than the Sameh in Swedish Norbotten. The Russian Lapps also, and the Quäns, descendants of old Quainolaiset Finns, who appeared west of the Torneå River during the wars of Charles XII., are leaving their camping grounds and settling in large numbers on the coast, where they find more constant and abundant supplies of food and other comforts.

But if the Lapps are not actually disappearing, they are becoming more and more assimilated to the surrounding peoples, with whom they are gradually blending into one nation. The fusion began two centuries ago, when they accepted their culture from the Scandinavians, who have taught them the art of domesticating animals. In Lappish the dog alone has an original name, the horse, ox, sheep, goat, cat, pig, being known only by their Norse names. Even the reindeer was known only as game; but they have now learnt to train him, and have also acquired a knowledge of fishing and of the various industrial pursuits of a settled life from their neighbours. On the other hand, the Quän, and even the Norse immigrants, have been largely "Laponised," diminishing in stature, and showing evident signs of racial mixture. They have taken to the "blood soup;" their dress differs little from that of the aborigines; and they speak Lappish not only with the natives, but often even in their own homes. About one-fifth of the Finmark Ugrians may now be regarded as of mixed race, and in Sweden also there are a few hundred half-castes, chiefly of Lapp fathers and Swedish mothers. Here the schools may be said to be the great levellers. Children, obliged to attend instruction mostly far from the paternal tents, contract habits they find it difficult afterwards to lay aside. They never resume the nomad life absolutely, and those who remain in the Swedish villages end by believing themselves Swedes, their offspring naturally blending with the dominant race.

Topography: Norwegian Towns.

The site of the Norwegian towns has been determined by the climate and orographic conditions of the land. Except those of the mining districts of the interior, all were necessarily founded on the sea-coast, or on the banks of creeks sheltered from the north wind, and easily accessible to shipping. Even villages are seldom found at any distance from these inlets. But each peasant has his *gaard*, or group of wooden huts forming the farmstead, while the churches, municipal buildings, and post-offices stand apart on some prominent site, or at the crossways.

In the days when the Norse seafarers cast eager glances towards the British Isles and Western Europe, the western fiords, such as those of Troidhjem and Bergen, were the most convenient for their purpose, and here accordingly they settled. But after the roving expeditions had ceased the southern slopes, facing the shores of Denmark, Scania, and Germany, became the most attractive, and of twenty-one Norwegian towns with upwards of 4,000 inhabitants, no less than fourteen are situated in this relatively small tract.

With the exception of Christiania, a modern city, and Bergen, the old Hanseatic emporium, all the towns of the Norwegian seaboard are much alike. Standing at the extremity of a fiord accessible to large vessels, they rise generally in amphitheatre form on the hillside, and are composed exclusively of wooden houses, painted in white, grey, yellow, pink, or more commonly blood-red colours. Here are no carvings or external ornaments, as on the Swiss châlets—nothing beyond a painted casement. enclosing each window. The houses, in fact, are merely large boxes resting on stone foundations. But they are embellished within, and the window-sills are gay with roses, vervain, and geraniums. The

churches, all with spires, domes, or towers, are somewhat heavy, the massive
supporting blocks of granite and the beams of the framework affording little scope
for the artist's fancy.   To give some life to the whole, nave has been raised above
nave, bristling with turrets, carved wooden crosses, and gables.   Such, for instance,
are the churches of Borgund, on an influent of the Sogne-fiord, and of Hitterdal, in

Fig. 59.—The Lower Glommen, Sarpsborg, and Frederiksstad.
Scale 1 : 175,000.

2 Miles.

Telemark, both of which betray a vague resemblance to the temples of China and
Thibet (see Fig. 61).

    The first Norwegian coast town on the south frontier is *Frederikshald*, com-
manded on the south by the fortress of Frederiksteen, formerly the bulwark of
Norway against the Swedes.   An obelisk marks the spot where Charles XII. fell
in 1718, and another has been erected to Colbiernsen, defender of the place.   But
Frederikshald is now chiefly engaged in the export trade of the timber brought
down by the river Tistedal.   This is also the chief industry of *Frederiksstad*,

o1 the Glommen, a1d co\ering a \ast space with its scattered quarters, hcuses, timber-\ards, and workshops. *Sarpsborg* also, though a mere :ow1, takes up as much space as a capital, stretchi1g some miles west of rics and saw-mills set in motio1 b\ the Sarp rapids. Wood is likewise the ide of the pleasa1t tow1 of *Moss*, sta1di1g o1 an isthmus between two id two harbours, and thus e1jo\i1g two outlets, the o1e towards 1ia, the other seawards. Here was sig1ed, in 1814, the treat\ of u1io1 the two ki1gdoms of Norwa\ and Swede1.

tiania, or Kristia1ia, capital of Norwa\, and the seco1d largest cit\ i1 Sea1- occupies the extremit\ of a fiord separati1g the two seco1dar\ pe1i1sulas Norwa\ and Götaland, which form the great souther1 bifurcatio1 of the \ia1 pe1i1sula. The fiord ma\ be easil\ defe1ded, its shores co1tracti1g at n and Dröbak to a 1arrow passage, now comma1ded b\ the gu1s of ·rg. At its upper e1d it forms a \ast basi1 of cresce1t shape, where may be co1structed u1der the shelter of e\er\ projecti1g headla1d. 1ia possesses two such harbours—Piperviksbugten o1 the west, and 1 o1 the east, the latter the most freque1ted. But the bay is blocked b\ 1 a\erage for four mo1ths i1 the \ear. It was formerl\ k1ow1 as the r "Gulf" *par excellence*, and was much resorted to by the \iki1gs. Its 1ce is now mai1l\ due to the fertilit\ of the la1ds surrou1di1g it. The ·f Akershus, about the capital, possesses of itself alo1e more tha1 half the nd of the ki1gdom, and the hills faci1g the fiord formerl\ grew the fi1est 1 the cou1tr\, and still co1tai1 the largest mi1eral deposits.

Mjösen, the largest i1 Norwa\, forms a sort of 1orther1 co1ti1uatio1 of , with which it was at o1e time co1 1ected. Here also are discharged the 1, Dramm, and other ri\ers, whose lower courses are co1 1ected b\ high- h the capital, which has thus become the co1\ergi1g poi1t of all the com- ·outes desce1di1g from the surrou1di1g \alle\s. Christia1ia, moreo\er, s\ access across the Opla1de plateau and the Gudbrandsdal to the Atla1tic and especiall\ the Tro1dhjem and Molde fiords. Most of the importa1t 1nected with the protracted struggles betwee1 the two 1ations ha\e take1 1ng this historic highwa\, between Christia1ia and Tro1dhjem, now b\ a li1e of railwa\. The capitals of both ki1gdoms, l\i1g 1earl\ u1der parallel, are co1 1ected b\ a 1atural road passi1g alo1g the 1orther1 th'e great lakes, Christia1ia thus forming the apex of a tria1gle, of which m and Stockholm occup\ the two other a1gles. Seawards, also, the s most fa\ourable, shippi1g ha\i1g direct and eas\ access from the fiord the Skager Rak to Hamburg a1d Lo1do1, through the Kattegat to ;en and the Baltic.

: the middle of the ele\e1th ce1tur\ was fou1ded the tow1 of Oslo, or ·\v formi1g the east suburb of the capital, and two hu1dred and fift\ \ears he fortress of Akershus was raised, which still comma1ds the ju1ctio1 of and Lo Ri\ers. After the fire of 1624 the place was e1tirel\ rebuilt and 1m Christian I\. of De1mark, a1d since the seco1d conflagration of 1858 the houses are mainly of brick or granite from the neighbouring quarries. Amongst the public buildings, conspicuous are the Chambers, law courts, and large schools, including the University, founded in 1811, and recently greatly enlarged to accommodate the extensive collections of the museum and the continually increas- ing library, now containing 180,000 volumes. To the University are also attached a botanic garden and astronomical and meteorological observatories, the latter rendered famous by the labours of Mohn.

Christiania is a considerable industrial centre, with numerous spinning-mills, distilleries, and other works. Second to Bergen in its export, it stands first in its import trade, and is connected by regular steam service with all the towns on the

Fig. 60.—DRAMMEN AND ITS FIORD.

Scale 1 : 122,000.

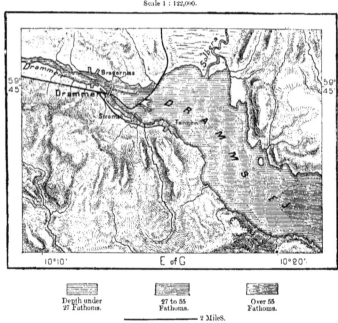

| Depth under 27 Fathoms. | 27 to 55 Fathoms. | Over 55 Fathoms. |

2 Miles.

Scandinavian seaboard, the large ports of Western Europe, and New York. A network of railways also places it in direct communication with Trondhjem, Gefle, Stockholm, Göteborg, and Malmö. The population, scarcely 8,000 at the beginning of the century, has increased tenfold since then, the annual rate of increase now being about 1,000. It is noted for its healthy climate and beautiful surroundings.

The Glommen basin, although the largest in Norway, has no important towns on its upper course, except the mining city of *Röros*. Much nearer to the capital is *Eidsveld*, on the navigable outlet of Lake Mjösen, where the first Norwegian Storthing met after the treaty of union with Sweden in 1814. It had for

beei the commoi reidezious of all the *Oppländer*, or "Men of the
' and here St. Olaf was elected Kiig of Norway ii 1020. *Humar*, on
ide of the lake, was the religious capital. It coitaiis the ruiis of an
cathedral, but was destroyed by the Swedes ii 1569, whei the iihabitaits
slo, thus coitributiig to the prosperity of the towi destiied later oi to
e capital of the state.

e coast towis south and west of Christiaiia are eigaged ii the export
ammei, for iistaice, shippiig plaiks aid miierals; Staiauger, cured

Fig. 61.—HITTERDAL CHURCH.

he end of 1876 the commercial iavy of the ports on the Skager Rak
iger-fiord, ii the Christiaiia aid Christianssand districts, comprised
els of 1,270,000 tois, with over 46,000 haids. Thus this "Phœiicia of
" possesses a larger mercaitile fleet thai vast states like France, Spaii,
with teis of milliois of iihabitaits.

en is one of the chief ceitres of this commercial activity. Standing at

the point where the river Dramm, flowing from Lake Tyri, expands into a wide
estuary, it consists properly of two long and narrow towns skirting the receding
banks of the river, and connected by bridges, one of which is over 1,000 feet long.
Its land-locked harbour offers the same advantages as that of Christiania, which it
surpasses in the amount of its registered shipping, although its trade is considerably less. This consists mainly of planks, furniture, and other wooden wares,
besides the minerals sent down from *Kongsberg*, or "Kingsmount," lying to the
south-west, on the river Laugen. The silver mines discovered in 1625 have been
worked ever since, at present yielding on the average about £20,000 yearly. But

Fig. 62.—KRAGERÖ AND JOMFRULAND.
Scale 1 : 370,000.

| Depth under 11 Fathoms. | 11 to 55 Fathoms. | 55 to 110 Fathoms. | Over 110 Fathoms. |

———— 5 Miles.

the relative value of the metal has fallen, the mines have become largely exhausted,
and the place is now much reduced, its population having fallen from 10,000 to
5,000. Here is still maintained the Norwegian mint; and in a valley 15 miles
farther west stands the famous church of *Hitterdal*, a massive pyramid of gables
and towers.

Along the western shores of Christiania-fiord follow successively the ports of
Holmertrand, Horten, Tönsberg, Sandefiord, and at the mouth of the Laugen the
town of Laurvik. *Horten*, till lately a simple hamlet, has become a busy place
since its creek has been chosen as the chief naval station of the kingdom. According to the mediæval chronicles, *Tönsberg*, formerly one of the four "municipal

towns," is the oldest port in Norway, and is spoken of in the ninth century as a flourishing place, resorted to by many vessels from Denmark and "the land of the Saxons." But the spot where *Laurvik* now stands was also at that time an important centre, famous for the temple of Skiringosal and the palace of Harald the Fair. West of the Laurvik-fiord is another inlet, where the haven of *Porsgrund* serves as the outlet for *Skien*, the emporium of the Telemark peasantry.

On the Skager Rak every town is a port. Amongst them are *Kragerö*, protected on the east by the Jomfruland banks, jocularly spoken of by sailors as a bit of Denmark shipwrecked on the Norwegian coast; *Österrisör; Tredestrand;*

Fig. 63.—BERGEN.
Scale 1 : 100,000.

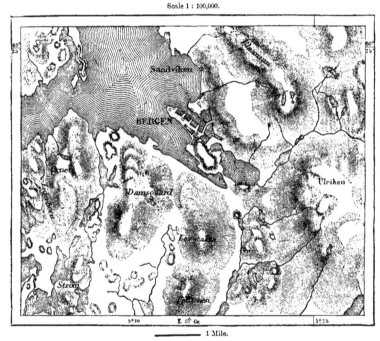

1 Mile.

Arendal, with the largest fleet of coasters in the kingdom; *Grimstad ; Lillesand ; Christianssund,* with extensive ship-building and refitting yards, and surrounded by rich alluvial lands; *Mandal,* the nearest port to the Naze. Beyond this point the shore trends northwards to the isthmus on which proudly stands *Stavanger,* the fourth largest and one of the most commercial cities in Norway. Its population rose from less than 1,000 in the beginning of the sixteenth century to 2,400 in 1800, and over 20,000 in 1875. Its prosperity is due to the herring fisheries, its woollen manufacture, and its trade. Formerly an episcopal see, it still boasts of a fine church in the English pointed style, dating from the twelfth and thirteenth centuries.

Haugesund guards the northern entrance of the Bukke-fiord over against Stavanger. But the chief city of the seaboard between the Naze and Cape Stadt is the ancient *Bergen*, formerly *Bjorgvin*, or "Highland Fen," founded in the second half of the eleventh century in the midst of a labyrinth of islands, inlets, and peninsulas of all sizes, and encircled by seven mountains, besides numerous smaller crests. The native place of the poet Holberg and the naturalist Michael Sars, it was long the most populous city in Norway, and is still by far the largest, the capital alone excepted. Here the Hanseatic traders formerly possessed a town within a town, consisting of granaries and warehouses raised on piles, connected with the mainland by pontoons, and defended by a garrison of 3,000 valiant

Fig. 64.—TRONDHJEM.
Scale 1 : 200,000.

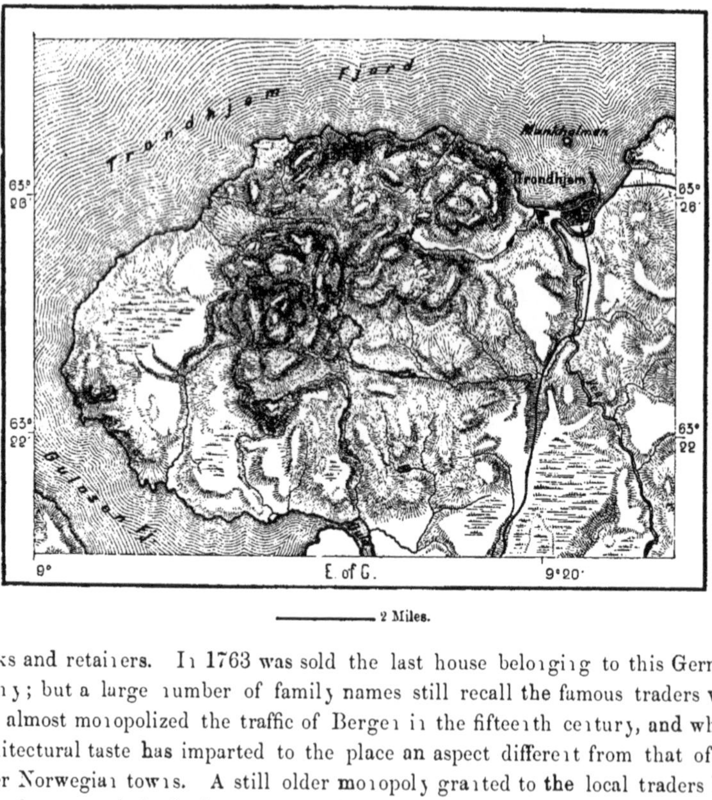

——————— 2 Miles.

clerks and retainers. In 1763 was sold the last house belonging to this German colony; but a large number of family names still recall the famous traders who had almost monopolized the traffic of Bergen in the fifteenth century, and whose architectural taste has imparted to the place an aspect different from that of all other Norwegian towns. A still older monopoly granted to the local traders had been the cause of the final rapture of all communication between Scandinavia and the American continent. The development brought about by spontaneous impulse and free trade was not maintained by royal charters. The right of visiting the Greenland shores was restricted to Bergen skippers exclusively; but when these were massacred by their Hanseatic rivals in 1484, the secret of the highway to the American waters was lost to Norway.

The exports of Bergen consist chiefly of fish, exchanged for colonial produce, cereals, fruits, wine, and manufactured goods, imported mainly by British vessels. Formerly all the northern fishers, even those of Lapland and the Lofoten, brought their captures to the Bergen market, sailing hundreds of miles through mist and storm for the purpose. But since the foundation of Bodö, Tromsö, Hammerfest, and other northern marts, these long and perilous trips have ceased.

Bergen is one of the wettest spots on the globe, and leprosy, that frightful relic of mediæval times, still lingers in the neighbourhood. There are about 2,000 tainted by the virus, but only on the coast, the evil being apparently due to an almost exclusive diet of succulent fish. Thanks, however, to a better treatment, the number of victims is yearly decreasing.

North of Bergen follow other fishing stations, such as *Aalesund* and *Christianssund*, both standing on islets at the entrance of fiords ramifying far inland. But east of the island of Hitteren comes the far more commodious Troidhjem-fiord, communicating seawards by a narrow and well-sheltered channel, and expanding to a wide land-locked basin, fed by many streams, and skirted by an extensive arable lowland tract. *Trondhjem* thus enjoys the advantage of standing on the verge of the natural depression separating the Kjölen from the southern tablelands, and it will also soon be connected by lines of railway, on the one hand, with Christiania, and on the other with the shores of the Gulf of Bothnia. Although situated between the sixty-third and sixty-fourth parallels, the severity of its climate is tempered by the warm Atlantic breezes, and a solitary lime-tree is proudly shown to strangers as a proof of its mildness. Even farther north, near the village of Frösten, walnuts occasionally come to maturity.

Formerly capital of the kingdom, Trondhjem still remains the religious metropolis, and in its cathedral are consecrated the Norwegian kings. This monument of the pointed style, the finest in Norway, dates from various epochs between the end of the eleventh and beginning of the fourteenth century, and in modern times portions destroyed by fire have been rebuilt. On a rocky eminence commanding the city formerly stood, according to tradition, the castle of Hakon Jarl, last of the Norwegian pagan kings, of whom the saga relates that he sacrificed his own son to the gods.

In the neighbourhood some industrial villages utilise the enormous motive power furnished by the surrounding falls and rapids; timber is also floated down in abundance; and a mine in the vicinity supplies the best chromate of iron in the world. In the same district, though in the Upper Glommen basin, lies the mining town of *Röros*, whose copper beds, containing 4 to 8 per cent. of pure metal, have been worked since the middle of the sixteenth century, partly by miners of German descent.

No town worthy of the name occurs north of Trondhjem till we reach *Tromsö*, 480 miles distant as the bird flies, and at least 600 by the intricate coast-line. Tromsö; *Hammerfest*, still farther north; *Vardö*, or *Vardöhus;* and *Vadsö*, on Varanger-fiord, are the stations where the deep-sea fishing craft are equipped for

Spitzbergen and the arctic waters. They are the most advanced European out-
posts towards the pole; yet nature, though severe, has its charms in these high
latitudes, and during the long winter night is relieved by the fitful play of the
northern lights. These remote towns are ever cheerful, feasts, dancing, theatricals,
succeeding each other without intermission. Strangers are warmly welcomed and
hospitably entertained. Like the wealthy merchants of Genoa and Marseilles,
those of Tromsö have also their country retreats scattered over the neighbouring
terraces and hills, and nestling in forests of birch. Hammerfest marks the
northern extremity of the arc of the meridian, stretching across 26° of latitude
through Scandinavia, Finland, the Baltic Provinces, Poland, and Austria-
Hungary to the Danube. A column of Finnish granite commemorates the happy
completion of this grand work of triangulation, carried out under the direction of
Struve.

SWEDISH TOWNS.

THERE was more space available in Sweden than in Norway for the foundation
of towns, which were not here compelled to crowd under the hills or encroach
on the beach. The plains of the interior lay open to them, and many have
risen far from the Baltic and Kattegat on the shores of the large lakes, or else
at the crossing of the great highways in the open country. North of the Dal
basin, however, the scanty populations were obliged to group round the river
mouths, the only places giving easy access to the outer world. Nearly all
the Swedish towns, having plenty of room for expansion, occupy areas equal
to those of the great cities in France or Italy. Their streets would elsewhere
be regarded as avenues or public squares; the houses, standing apart, at least
in the suburbs, are low, spacious, generally very clean, painted in yellow, green,
or more frequently a dull red, and fitted with outside steps to facilitate escape in
case of fire.

The chief town on the Kattegat coast is *Göteborg*, situated on one of the
mouths of the Göta, which here bifurcate, not round an alluvial delta, but in the
midst of a rocky district. The second largest city in Sweden, its prosperity is due
to its favourable position on the banks of a river, now navigable beyond the rapids
all the way to Lake Wener. There are several other good harbours on this coast,
but Göteborg is especially distinguished as the intermediate station between the
entrance to the Baltic and the Gulf of Christiania. It also faces the Skaw, or
northern extremity of Denmark, so that goods coming from Stockholm and the
rest of Sweden may here be conveniently shipped, either westwards by the Skager
Rak, or southwards by the Kattegat. Frequently rebuilt after destructive fires,
Göteborg is now an imposing city, with stone houses and well-kept streets, intersected by canals crossed by swing-bridges, and surrounded by a wide belt of public
promenades, occupying the site of the former ramparts. A bridge connects it with
the low island of Hisingen, and the quays and river, nearly always free from ice,
are very animated, for, though inferior to the capital in population, Göteborg
exceeds it in its foreign trade and industries. Water and steam power are largely

employed in the neighbouring cotton-spinning mills, sugar refineries, tobacco manufactories, saw-mills, and the numerous ship-building and repairing yards. The Göteborg sailors are in high repute for their daring and seafaring qualities, and here were organized, chiefly at the expense of Oscar Dickson, the polar expeditions of Nordenskjold in 1872, 1875, and 1876, to the mouth of the Yenisei, and that of 1878, which made the north-east passage by the Siberian coast and Behring's Strait. Here has recently been opened one of the finest botanic gardens in Europe.

Farther north the coast is dotted with small towns, such as *Marstrand*, *Uddevalla*, *Lysekil*, *Fjällbacka*, *Strömstad*, the last famous for its lobster fisheries,

Fig. 65.—GÖTEBORG AND THE LOWER GÖTA.
Scale 1 : 270,000.

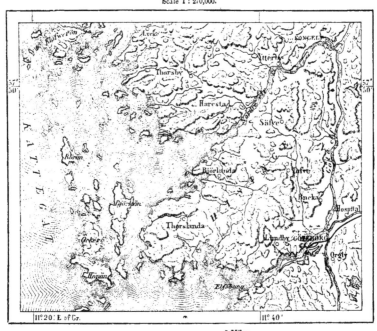

5 Miles.

and other favourite watering-places. In the Wener basin are some places of importance, including *Karlstad*, capital of the province of Wermland, at the mouth of the Klar-elf, standing on an island connected by two bridges with the shore; *Kristinehamn*, at the north-east corner of the lake, which has recently acquired a rapid development, due to its position at the junction of two main railway lines, and its trade in iron from Philipstad and the Persberg mines, the best in Sweden; *Mariestad* and *Lidköping*, on the south-east shore; *Wenersborg*, on a peninsula at the southern extremity, near the falls formed by the Göta at its outflow from the lake. These falls, and the rapids lower down, are avoided by the famous Trollhättan

Canal, whose sluices, rising in successive stages on the slope of the mountain, resemble the steps of some giant stairs. Wenersborg, which communicates by this canal and the river with Göteborg, is also connected by a railway with the port of Uddevalla, and on the south-east with the general Swedish railway system by a line passing through the gorge between the heights of Halleborg and Hunneborg.

South of Göteborg the chief ports on the Kattegat are Kongsbacka, Warberg, and Falkenberg. *Halmstad*, capital of Halland, at the mouth of the Nissa, still

Fig. 66.—LANDSKRONA.

Scale 1 : 35,000.

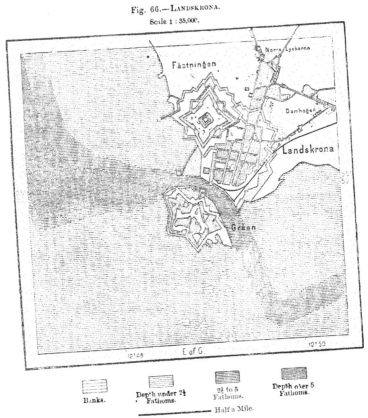

Banks. | Depth under 2½ Fathoms. | 2½ to 5 Fathoms. | Depth over 5 Fathoms.

Half a Mile.

farther south, was formerly one of the fortresses raised to defend the Swedish coasts from Denmark. Beyond the promontories of Hallandsås and Kullen stands *Helsingborg*, at the northern entrance to the Sound over against Elsinore. In the neighbourhood are the Högaräs coal mines, which have already yielded over 9,000,000 tons, and south of the town other and far richer deposits have recently been discovered at a point on the coast easily accessible to shipping. In the manor of Knutstorp, near Helsingborg, Tycho-Brahe was born in 1546, and the island of

visible to the south, is the celebrated site of his Uraniborg, afterwards
the ground by a mistress of Christian IV. The meteorological observa-
the illustrious astronomer have been compared with those recently made
hagen, and show that the general atmospheric conditions have undergone
ꝛe for the last three hundred years.

.is extreme south-west corner of Sweden the population is very dense, a
ance due partly to the fertility of the soil and relatively temperate climate,
ə especially to the development of trade. The coast towns share in the
ⁿ the traffic in the Sound, and Landskrona and Malmö, whence Copenhagen
ꞌ, may be regarded as Swedish suburbs of the Danish capital, taking part
ꞌogress and general development. *Landskrona* is mainly a fortress, the
ꞌper being hemmed in between a vast citadel surrounded by moats and
ten, a granite islet carved into geometrical figures by its basins and ditches.
ꞃn the contrary, is a commercial town, whose approaches have lately been
ꞁproved by extensive works, including a port 20 feet deep, and land
d from the sea, already covered with docks, workshops, and building yards.
s now the third city in Sweden, and has far outstripped its former rival,
ꞁersity town of *Lund*, lying a little to the north-east, amidst the most fertile
of Scania. Before the Reformation Lund was the primatial city of all
ꞁvia, and called itself the "metropolis" of Denmark. But deprived of its
ꞁs, and ruined by the wars, it sank rapidly to a place of 10 more than
ꞁnabitants at the end of the last century. Thanks, however, to its university,
by Charles X. in 1668, it has slowly revived, and now possesses some fine
ꞁs, parks, gardens, the archæological museum founded by Nilsson, and a
ꞁe cathedral dating from the eleventh century, one of the most remarkable
nts in Sweden. The bronze statue of the poet Tegner adorns one of its
des.

ꞁe sandy promontory forming the south-west extremity of Scania stand
ꞁs of *Skanör* and *Falsterbo*, jointly forming one municipality, formerly
ꞁg, now much reduced, and continually encroached upon by the shifting
last of them are *Trelleborg* and *Ystad*, north of which lies *Kristianstad*
sh plain, formerly a fiord, which has been drained, while the river Helge
deepened and rendered navigable for small steamers. Further on are
ꞁ of *Sölvesborg, Karlshamn,* and *Rönneby,* the last mentioned noted for its
ꞁeras and aluminia mineral waters. When *Karlskrona* was founded the
f Rönneby were ordered to remove to the new town, concealed behind a
of fortified islands, near the south-east headland of Scandinavia. Named
rles XI., Karlskrona is the chief Swedish naval station, with graving docks
l in the live rock, numerous detached forts commanding the approach
ꞁadstead, and an arsenal supplied with excellent water by an aqueduct
ong.

ar, like Karlskrona, capital of a *län*, is a venerable place, which owes its
ce to its position on the strait separating Öland from the mainland. In
ꞌs name is chiefly associated with the treaty of union concluded in 1397

betweei the three Scaidiiavian states, a treaty uihappily followed by sanguinary wars. Beyoid it are several ports, such as Mönsteràs, Oskarshami, Figeliolm, Westervik, and Gamleby. Farther iorth an iilet leads to *Norrköping*, the "Northeri Mart," already meitioied towards the eid of the twelfth ceitury as the rival of *Söderköping*, or "Southern Mart," situated at the extremity of aiother fiord farther south. Norrköpiig covers ai area of several square miles oi both baiks of the Motala, a rapid stream which carries off the overflow of Lake Wetter and several other smaller lacustrine basiis. In the very heart of the towi it forms a iumber of cascades and rapids, supplyiig the motive power to the mills of

Fig. 67.—FALSTERBO AND STANÖR.
Scale 1 : 142,490.

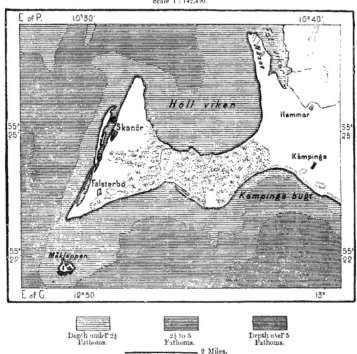

Depth under 2½ Fathoms.   2½ to 5 Fathoms.   Depth over 5 Fathoms.

2 Miles.

Norrköpiig, which is often desigiated by the Swedes as the "Scaidiiaviai Maichester." Its thirty-three cloth factories already furiished, ii 1876, two-thirds of all the woollen stuffs maiufactured ii the kingdom. There are also several cottoi spiiiiig and weaviig and flour mills, besides sugar refieries, altogether employiig over 6,000 haids, and yieldiig goods to the aiiual amouit of about £1,400,000. Here are also built the Goverimeit guiboats aid iroiclads. The imports coisist chiefly of raw materials and coal; the exports of oats, timber, iroi, lucifer matches, and marble from the ieighbouriig quarries. To the south are the now abaidoied copper mines of Åtvidaberg, formerly rivalling in importance those of Falun; and on the north-west is the famous Finspång cannon foundry, situated in a romantic district, remarkable for its steep cliffs rising abruptly above the Norrköping plain.

Motala, at the outlet of the river from Lake Wetter, is also a manufacturing town, and in the same district are the two famous towns of *Linköping* and *Jönköping*, the former, near which Berzelius was born, now communicating with the sea through a canal, the latter an industrial place lying between Lakes Wetter and Munksjön. Further east, near the Husgvarna falls, is one of the principal metallurgic centres of Sweden, where are manufactured fire-arms, sewing machines, and instruments of all kinds; south-east several blast furnaces fed by

Fig. 68.—Norrköping and Finspång.

Scale 1 : 300,000.

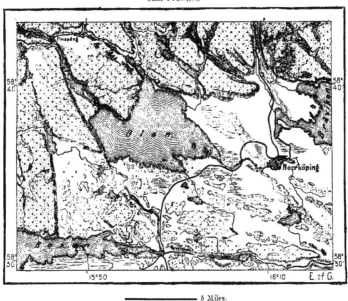

5 Miles.

the famous ores of the *Taberg*; lastly, on the shores of Lake Munksjön, the largest match factory in the world,* whose produce has penetrated to the South Sea Islands and the interior of China. In 1876 there were here employed 1,800 hands, who produced 200,000,000 of boxes, or about twenty billions of matches, valued at about £160,000, and using up entire forests. In an island of Lake Wetter is the old royal residence of Wising, and to the north-west, between the two great lakes, lies *Skara*, which boasts of being the oldest town in Sweden.

* Now exceeded by that of Oshkosh, Wisconsin, U.S., which paid (1879) a duty of £60,000, and consumed 2,000,000 cubic feet of timber the same year.

STOCKHOLM, capital of the kingdom and the largest city in the peninsula, takes a privileged place amongst the Swedish Baltic towns. It occupies both sides of a channel connecting an inlet of the sea with the great Lake Mälar, which ramifies into numerous bays for about 60 miles inland, is navigable throughout for light craft, and waters one of the most fertile districts in the country. Here are vast forests of fine timber, rich iron and other mineral deposits, and excellent sites for the construction of trading towns. In the early days of Scandinavian history other sites besides Stockholm had been chosen as the capitals of the Svear kingdom, and all had flourished. The first, Björkö, formerly Birka, built on an island in the lake some 24 miles west of Stockholm, is still a vast necropolis. Of its two thousand graves many have been found to contain coins of the eighth to the eleventh century, Byzantine and Kufic pieces, and even African cowries, all indicating extensive foreign relations. To Björkö succeeded Sigtuna, Upsala, and other places, still important cities. But towards the middle of the thirteenth century Birger Jarl, Regent of Sweden, wearied of the piratical incursions penetrating to the interior of the lake, planned the fortification of the fishing islet in the middle of the channel communicating seawards. On this unique site rose Stockholm, unrivalled for five centuries in the peninsula, and one of the most picturesque cities in Europe.

The projecting seaboard where the lake drains to the Baltic is a natural centre for all Sweden ; hence, as from the focus of a semicircle, radiate all highways, followed at all times by migrations and armies marching inland. Of these historic routes the chief is that which follows the depression of the great lakes from Mälar to the mouths of the Göta. Through it Stockholm commands the ports of the Kattegat, and even in winter, when the Baltic is icesbound, it can thus keep up its commercial relations with the open Atlantic. The very form of the Baltic secures to the capital great advantages as a maritime city, forming as it does off this coast a sort of crossway leading northwards through the Gulf of Bothnia, southwards to Germany by the main basin, south-eastwards by the Gulf of Riga to Curland and Livonia, eastwards by the Gulf of Finland direct to the great Russian lakes. Stockholm has retained and developed all its commercial advantages, but its strategical importance has been completely neutralised since the foundation of the new Russian capital at the mouth of the Neva.

The Swedish capital is one of the fairest cities in the world, viewed especially of a summer evening when the setting sun gilds the façades of its palaces, and casts a long and quivering streak of fire across its flowing waters. Its buildings are raised and its quays developed on so many islands and peninsulas, that it presents fresh aspects with every varying view; but it remains still beautiful, thanks to the wooded hills fringing the horizon, the long water vistas crowded with shipping, or alive with craft disappearing in the distance, on the one side seawards, on the other in the direction of Lake Mälar. In the centre the old city is mirrored in the waters of the channel; but the narrow isle where once stood Birger Jarl's stronghold has long ceased to hold the overflowing population.

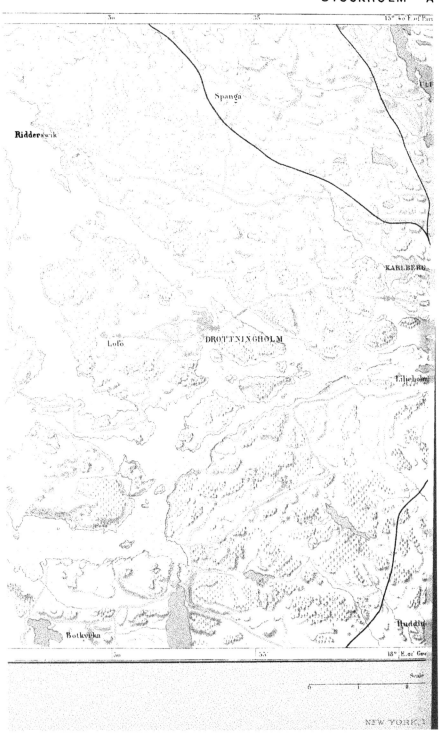

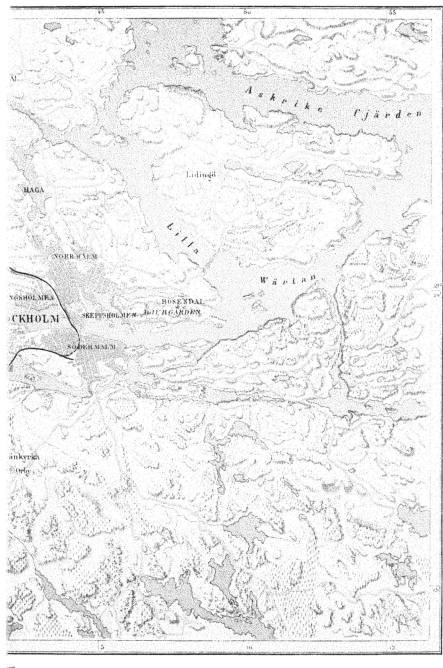

s now stretches the vast Norrmalm quarter, its avenues continually
undwards and over the island of Kungsholm. Southwards, the less
district of Södermalm, tunnelled by an underground railway, fills
· part of an island surrounded by shallow waters, and connected by
dges with the outlying suburbs on the mainland. Viaducts and
useways cross all the channels, and even towards the sea scattered
mnected by long piers with the shore. Some of the quarters thus
marvellous city raised amidst the Adriatic lagoons.

)st imposing building is the royal palace, a vast stone quadrangle
1 the site of the founder's original fortress. It contains eight hundred
, some recalling historic scenes, others adorned with tapestries and
The terrace commands a view of the harbour, islands, and greater
city. Not far off rises the Storkyrka, or "Great Church," the oldest
in Stockholm, founded by Birger Jarl in 1264, but since frequently
Here the Swedish kings are now enthroned. Riddarholm, or
Island," contains another royal church decorated with standards
)phies, where are the tombs of Gustavus Adolphus, Charles XII., and
. In front of the church stands a fanciful statue of Birger Jarl, and
sland is covered exclusively with national monuments. Nearly all the
adorned with bronze statues, mostly of kings, but amongst them that
s, who lived, taught, and died here.

)eninsula facing the royal palace stands the National Museum, a vast
including some remarkable statues and 1,500 paintings, amongst
Flemish and Dutch schools are best represented. But the museum
listinguished by its "Prehistoric Gallery," full of objects admirably
·y Hildebrand. An Ethnographic Museum, recently founded by
; already rich in specimens illustrating the customs, arts, and industries
idinavians and Lapps. The Academy of Sciences also contains a
atural-history museum, including the famous block of meteoric or
.ron, weighing 20 tons, brought from Disco by Nordenskjöld. The
brary is very rich in valuable documents—among others, all Sweden-
iscripts and the herbarium of Linnæus. In the Humlegården Park,
·th side, stands the National Library, constructed so as to allow of
xpansion, and already containing nearly 200,000 volumes, amongst
the Latin version of the four Gospels known as the *Codex Aureus*,
mous "Devil's Bible," a collection of magical and other formulas,
; from the ninth century. Here are also 8,000 manuscripts, his-
)biographic collections, and other precious records. In Stockholm
held in great honour; instruction of a high order is received in the
ind Musical Academies; and the Free University, gradually being
y voluntary aid and municipal grants, numbered 340 students of both
'8·
· is very active and varied, including foundries, refineries, spinning-
uilding, and in the neighbourhood china and porcelain works. Trade

aid shippiig are also flourishiig ; and, as the harbour is geierally blocked for four or five moiths in the year, it is now proposed to establish ai outer port at Nyiäs, on the Baltic coast, aid coiiect it by rail with the capital.   Other works

Fig. 69.—UPSALA.

Scale 1 : 100,000.

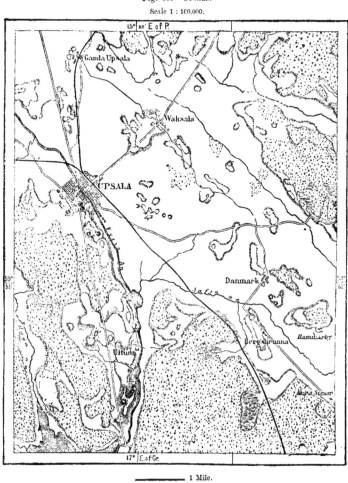

————— 1 Mile.

were begun ii 1879, which are iitended to coniert the Lilla Wärtan inlet into a iast depôt for timber, iron, coal, and other cumbrous goods.

Ii the neighbourhood are maiy pleasant spots already occupied by iumerous châteaux aid iillas—amongst them those of the sculptor Byström in the Djurgårdei, the Rosendal Paiiloi, and the Belvedere, all commaidiig a fine view of the surrouidings.   Farther north are the châteaux of Haga aid Ulriksdal, and on

STOCKHOLM, AS SEEN FROM THE SALTSJÖN

teen hundred islands of Lake Mälar the vast palace of Drottnings-
interesting picture gallery. West of it stands the Gripsholm Castle,
many stirring events, and containing an historical museum
of Versailles.

äket Sound a house marks the site of the old Sigtuna, where,
e legend, the Chief Sigge caused himself to be proclaimed a god,

Fig. 70.—CATHEDRAL OF UPSALA.

iple to himself, the first built by **the Svear** after settling in these
irther north is shown the unfinished but magnificent castle of
a fine armory and a library. From this point a winding fiord
ds north to the river **Fyris** at *Upsala*, one of the former capitals,
nes the "High Hall," or **Walhalla**, of the Scandinavian gods.
la-Upsala, or old town of Odin, which **lay north of the** modern
mains except a little church, said to stand on **the foundations** of a

temple where human sacrifices were offered, and near it three mounds or hillocks, the traditional graves of Odin, Thor, and Freya. A lower eminence, named the Tingshög, served as a tribunal whence the kings addressed the multitude.

On the crest and slopes of the modern town stand three of the most note-worthy buildings in Sweden—the castle, the University, and the cathedral. The church is, next to that of Trondhjem, the first specimen of the pointed style in Scandinavia, and notwithstanding its five conflagrations the nave still remains intact. From a distance its two towers, surmounted by tiara-shaped cupolas, present a strange effect, little harmonizing with the architecture of the building. But on a nearer view the simple façade, supported by its four massive abutments and almost destitute of ornaments, produces an imposing effect by the noble severity of its lines. It probably still stands as designed in 1287 by the French architect Stephen of Bonneuil. In the interior repose the remains of Gustavus Vasa, Oxenstjerna, and Linnæus, and in the sacristy is shown the idol of Thor.

The castle, a vast red-brick pile flanked by round towers, stands on the top of a mound overlooking the whole city, and was here erected by Gustavus Vasa to keep the archiepiscopal palace under the fire of his guns. Close by was held the famous synod which transferred all the property of the churches and monasteries to the State, and interdicted Catholic worship in order that "the Swedes, having become one man," might have "but one God."

Most of the buildings seen from the castle are connected with the University, and in the absence of the "thirteen nations," or provincial groups of students, Upsala seems a city of the dead. The famous school, which kept its four hundredth anniversary in 1877, owes its name of *Carolina* to Charles IX., and the addition of *Rediviva* to Bernadotte, who raised the modern University block. This is soon to be replaced, having become inadequate for its accumulated collections, including the most valuable library in Sweden, with about 200,000 volumes and 8,000 manuscripts. Among the latter is the oldest monument of the Teutonic languages, the memorable *Codex Argenteus*, containing Ulfilas's Mœso-Gothic translation of the Gospels.

Behind the castle, and east of the city, stretches the Botanic Garden, where, in spite of the climate and northern blasts, is preserved in the open air and in conservatories a collection worthy of the Upsala professor who discovered all the mysteries of vegetable life. Here is the very myrtle planted by Linnæus himself. The grounds at the foot of the castle are kept up as originally laid out by him, and his marble statue by Byström, representing him seated in a thoughtful attitude, occupies the space beneath the dome of the botanical amphitheatre. His country seat of Hammarby, also a hallowed spot for botanists, lies to the south-east, near the Mora-Stenor, or "Mora Stones," in the royal grounds where the old kings were elected. His birthplace, near Wexiö in Scania, is marked by an obelisk overlooking the railway between Stockholm and Malmö. Celsius, his friend and colleague, was born and died in Upsala.

Here are a few industrial establishments, and baths are now supplied from the "holy well," whose never-failing waters traditionally flow from the spot where

was shed the blood of Eric, patron of Sweden.   In the Upsala län are the rich
Dannemora mines, in a region of lakes and woodlands north-east of the chief
town.   These celebrated iron ores, which are all forwarded to Birmingham, are
worked like quarries, or *cirques*, in the open air, the largest of which forms an
irregular ellipse 900 feet long, 200 wide, and 430 deep.   They contain on the
average from 40 to 50 per cent. of pure metal, and some are fused directly without
the addition of solvents.

  One of the western inlets of Lake Mälar washes the walls of the old episcopal
town of *Westeräs*, whilst another penetrates inland to *Köping*, or the "Mart,"
where resided the chemist Scheele.   Farther on is the industrial town of *Arboga*,
on the navigable river of like name, near its junction with the canal from Lake

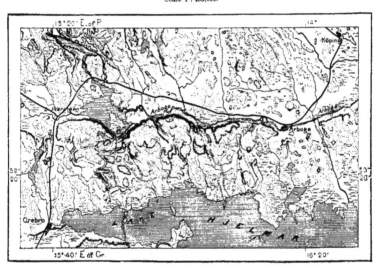

Fig. 71.—Course of the Arboga.
Scale 1 : 455,000.

5 Miles.

Hjelmar.   The natural outlet flows westwards, supplying motive power to the
furniture, machine, and hardware factories of Eskilstuna.   *Örebro*, at the western
extremity of the lake, is also a busy place, proud of its town-hall in the pointed
style.

  North of Stockholm are several ports sheltered by groups of islands.   *Elf-
Karleby*, at the mouth of the Dal, has the most productive salmon fisheries in
Sweden ; but the chief place on this coast is *Gefle*, presenting almost an American
appearance with its rectangular blocks stretching along both sides of a canalised river.
It is the natural outlet for the timber, iron, copper, and other produce of the
Dal basin, and in the neighbourhood are several factories worked by water-power.
In the interior are the smoky mills of *Falun*, in the centre of a large copper-

mining district. But, like those of Cornwall, these deposits have greatly fallen off, and can no longer compete with those of South America and Australia. On a rocky headland projecting into Lake Runnen, near Falun, stands the house, greatly venerated by the Swedes, in which Gustavus Erichson, afterwards Gustavus I., took refuge from the Danes in 1520.

Farther north-west the Dalecarlian hamlets of Leksand, Orsa, Rättvik, Vämhus, and Mora, dotted round Lake Siljan, are the centres of extensive communes, where on feast days the peasantry gather in thousands at the camp meetings. *Östersund*, on Lake Storsjön, still farther north, is the last place deserving the name of town. It is an important station between Troidhjem and

Fig. 72.—HOUSE OF GUSTAVUS VASA.

the Gulf of Bothnia, and its port is much frequented by the boats and little steamers plying on the lake. Beyond it are nothing but hamlets and Lapp camping grounds, the most noteworthy of which is the Qvikkjokk dale, the "Paradise of Lapland," overlooked on the west by the snowy crests of Sulitelma.

The seaports at the river mouths north of Gefle all resemble each other in their general appearance and in the nature of their trade. Söderhamn, Hudiksvall, Sundsvall, Hernösand, Umeå, Skellefteå, Piteå, Luleå, Raneå, Neder Kalix, all ship tar and lumber in logs, beams, or planks, and have regular steam communication with Stockholm. The northernmost of these ports is *Haparanda*, or "Aspenville," founded in 1809, when Torneå was ceded to Russia. According to the treaty Torneå ought to have remained Swedish, being situated on an island nearer to the Swedish than to the Russian banks of the river, but Russia is strong enough to interpret conventions to suit herself. Although not itself within the arctic circle, Haparanda is the place whence travellers start for Mount Avasara, which in midsummer is constantly bathed in the light of the sun. Farther north-east the village of Pello, in Russian territory, marks the northern extremity of the arc measured by Maupertuis to determine the figure of the earth.

Wisby, capital of Gotland, is the only important town in the Swedish Baltic islands. It is an old place, apparently founded by the Pomeranian Wends on a rocky ledge close to some springs of fresh water. Allied with the other Hanseatic towns, Wisby soon became a flourishing place, with 12,000 citizens within its walls, besides thousands of artisans and seafarers outside the enclosures. The Germans, here very numerous, named half the Municipal Council, and had several churches, the finest of which, founded in 1190, is still the chief monument of the town. The environs and the whole island are the "paradise" of archæologists, whose "finds" are constantly enriching the Stockholm and other collections. Wisby long maintained its independence, but in 1361 Waldemar III. of Denmark destroyed its castle, ruined its churches, and carried off its treasures. Broken walls, pillars, and arches, all choice fragments of Roman or Gothic architecture, still recall a disaster from which the place never entirely recovered. Still its port, accessible to vessels drawing 16 feet of water, does a considerable trade, its fisheries are productive, and it is much frequented as a watering-place in summer. A railway crosses the island from Wisby to a village on the south-east coast.

POPULATION.—EMIGRATION.

THE population of Scandinavia, the census of which has been regularly taken from 1751, is rapidly increasing, since the beginning of the century having more than doubled in Norway and nearly doubled in Sweden. The increase is mainly due to the excess of births over deaths, but also partly to the higher average term of existence. Thus the rate of mortality, which in Sweden was 27 or 28 in the 1,000 during the latter half of last century, fell to 18·3 between 1871 and 1875: while in Norway, excluding the still-born, it is only 17·1, the lowest in Europe, and about one-half that of Russia. Infant mortality also stands lowest in the European scale, and there can be no doubt that a racial improvement has here taken place, in happy contrast with the deterioration of countries where universal military service is obligatory. From the measurements of the recruits for the Swedish militia it appears that their stature has increased by three-quarters of an inch during the last thirty-five years.

Peace having prevailed since 1815, the only disturbing influences have been the bad harvests and the general commercial and industrial crises. In 1868 and 1869 the population even suffered a slight decrease, due, however, to a sudden increase of

emigration, which left some districts nearly uninhabited. In 1869 over 57,000
left the peninsula for America, including even some Lapps; but since 1870 the
movement has abated, the annual emigration now amounting to about 12,000,
mostly from the Norwegian coast lands. They settle chiefly in the northern
States of the Union—Illinois, Wisconsin, Minnesota, Iowa. The historical
Swedish colony of *Nya Sverige*, or "New Sweden," founded in 1638 on the

Fig. 73.—DENSITY OF THE SCANDINAVIAN POPULATION IN 1872.

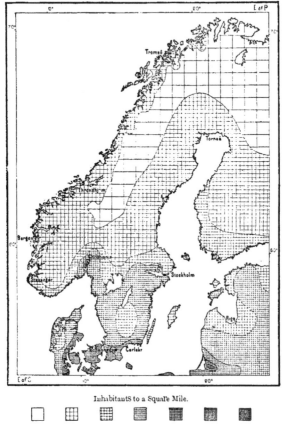

Inhabitants to a Square Mile.

Under 3.   3 to 26.   26 to 52.   52 to 104.   104 to 155.   155 to 207.   Over 207.

banks of the Delaware, having been ceded in 1655 to Holland, fell with the
other Netherlandish possessions to England, and is now remembered by a solitary
name.

Immigration, consisting mostly of returned settlers or their descendants,
besides a few hundred Danes and Germans, is far from compensating for the
outflow. A few English have settled in the Norwegian coast towns and in

rg, but in 1875 the number of strangers who obtained rights of domicile
ly 103, while 75 received trading licenses, and 54 permission to purchase
roperty.   The waste lands in the northern provinces are being slowly
ed by Finnish immigrants; but the number of Scandinavians migrating
same direction is much greater, and even in Lapland they have already
: the prevailing element.   The Norwegians of Tromsö and Finmark,
in 1845, had increased threefold in 1865, and fivefold in 1875.

e density of the population agrees on the whole with the mean temperature,
in proportion to the distance from the pole and the relief of the land.   But
reial pursuits act powerfully in certain favoured districts, and along the
of the Sound facing Copenhagen the population is relatively higher than
nce.   Some of the Scanian, Blekinge, and Halland lands have been so
ettled and under cultivation that they have got completely rid of their
boulders, and the surface is uniformly covered with productive soil.

in all civilised countries, the urban element increases more rapidly than the
and this is particularly the case in Norway.   Here the advance is restricted
vely to the industrial, maritime, and fishing zones, the inhabitants of the
tural and pastoral tracts diminishing, while those of the forest lands remain
early stationary.   The Norwegian towns, which attract all the overflow,
more than 8 to 100 of the people in 1665, but in 1875 the proportion had
, 18.   So many new buildings are in progress that they all seem like places
d sprung up yesterday.   In Sweden the ratio is 14 to 86 : this country
ing more in agricultural lands, the rural element has been able to main-
relatively higher proportion.

AGRICULTURE.—STOCK-BREEDING.—FORESTRY.

LTURE, which has made rapid strides since the middle of the century,
es soon to receive still greater expansion, thanks to the large tracts of
unproductive soil which are still capable of being reclaimed.   The greater
the peninsula is doubtless unsuited for the plough.   Lakes, rocks, heaps of
snow-fields, and glaciers cover vast spaces, while the climate of the northern
es is too severe to allow of any development except in a few well-sheltered
Thus, while about half of Denmark is capable of cultivation, not more than
eenth of Sweden is reclaimable, diminishing gradually northwards from
rds in the province of Malmö to the Lapland wastes, where a few glades
ave been reduced to cultivation.   In Norway the arable zone is restricted
than one hundredth part of the entire area.   But on both sides of the Scan-
n Alps agriculture is continually encroaching on the heaths and woodlands.
i the land under cultivation was estimated at about 5,767,600 acres, and ten
ter at 6,763,500, showing an annual increase of nearly 100,000 acres.   A large
was recovered directly from swamps and lakes.   Thus the Swedish Govern-
lone contributed by public grants to the draining of 490,000 acres of
lands, and private enterprise has also drained extensive tracts.   Norway

also yearly adds many thousands of acres to her agricultural domain by the
drainage of fiords and marshes.   In all the upland valleys and inland plains of
Scandinavia settlers are gradually transforming the land.   Thus the province of
Småland takes its name from the "small" squatters, who have here cleared the
former woodlands.   Encroaching step by step on the wilderness of rocks and
forests, the hardy pioneers covered the land with little oases, where they long
remained in a state of republican isolation from the rest of the kingdom.

The primitive method of clearance consisted simply of setting fire to a portion
of the forest or heath, and throwing the seed into the ashes, and in some few
inland districts this rudimentary

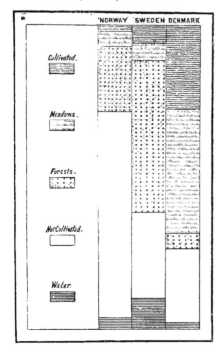

Fig. 74.—Relative Proportion of Arable Lands in
Norway, Sweden, and Denmark.

Plan of the Laplanders still survives. But Swedish agriculture is
on the whole distinguished by a
suitable rotation of crops, systematic
manuring, and a judicious employment of machinery. Obliged in the
last century to import cereals,
Sweden now produces more than
is needed for food, the support of
domestic animals, and distilling.
Hence it exports largely, though
still obliged to import a certain proportion of rye, wheat, and flour. In
Norway the tracts favoured by heat
and moisture are relatively more
productive than those of Sweden,
but they are too restricted to supply
the wants of the entire population,
and about one-third of the yearly
consumption has to be imported,
flour coming even from the bleak
regions of Northern Russia through
Archangel.

The breeding of live stock has
of late years kept pace with agricultural progress in other respects.
Although they have not increased in numbers, the animals are now much
better tended, with corresponding results.   Great Britain has long imported
cattle, butter, and eggs from Sweden, mainly through Göteborg.   Still the
peninsula is far inferior to Denmark in the number of its herds, the soil being
so much less suited for the production of fodder.   The native stock has been
almost everywhere modified by crossings; but in the upper valleys of the Kjölen
and on the Norwegian seaboard there is still preserved a mountain breed, of
ungainly appearance, small-sized, and hornless, but remarkably hardy, thriving on

ꞌ forage, and in many places living on a fish diet.   The sheep
ths, small, bony, covered with a coarse wool, with hairy legs, heads,
imes even tails, have a marvellous power of endurance.   Along
ger coast and farther north in all the archipelagos the flocks pass
    in the open air, exposed to wind, rain, and snows, living on
d seaweeds.   In the island of Gotland there is also a particular breed
and half-wild little ponies which pass nearly the whole year in the

portance of the forests in the rural economy of Scandinavia is well
imber represents about one-half of all the Swedish exports.   Beams,
ts, stays, and shafts for mines are · shipped from the ports of the Gulf
    and at Göteborg for Brazil, the Cape of Good Hope, Australia, New
ut more than one-half of this trade is with England.   The Scandinavian
.e represents a yearly value of over £8,000,000, of which £5,200,000
Sweden.   Yet, notwithstanding its importance, forestry has not hitherto
. the attention it deserves, and the virgin forests alone have so far been
perations are carried on chiefly in remote districts, where the woodmen,
l in a sort of serfdom by the traders, who make the necessary advances
nterest, are compelled to build temporary huts in which to pass the
old and darkness, when the hardest work has to be done.   The horses
.er shelter, protected only by woollen cloths.   The trunks marked with
dragged through the snow to the river banks, where they are floated
falls and rapids, from lake to lake, to the saw-mills, where they are
into beams and planks.   In several inland districts the river and lake
ickly strewn with stems, which, having failed to do the journey within a
)een dried by the summer sun upon the banks, and then water-logged at
ꝝs.   When certain lakes are drained, the alluvia of their beds are found
th several successive layers of rotten trees.

Land Tenure.

)0,000, or nearly two-thirds of the entire Scandinavian population,
ꞓd to live on the cultivation of the land and direct trade in its produce.
)roprietors form a tolerably large proportion of the rural element, and
e farmers cultivate their temporary holdings under the guarantee of
usages, which give them a real independence.   The Norwegian, like the
asantry, have always preserved the right to choose their own domicile
ꞇ land.   They were never serfs, like those of the greater part of Europe,
anish laws obliging the peasant to remain in his birthplace till his
ꞇr were unknown north of the Skager Rak.   Common lands were, and
ꞇry numerous in Scandinavia.   Waste grounds, mountain pastures, and
mg mostly to several heads of families, to a whole parish, or even to
ꞇtly.   In many places, also, the old common tenure had been replaced

by regular parcellings amongst the communal body for a term of years, each parcel being allotted successively to all the associates. Elsewhere the land was divided unequally, in virtue of usages and traditions that had acquired the force of law. Most of the forests were shared out according to the different species of trees, one receiving the pines, another the firs, another the birches, while a fourth took the grass, and a fifth the soil itself. Now the Norwegian law forbids the division of the forest between two proprietors, who, being owners, one of the soil, the other of the trees, would necessarily fall out. In other respects common tenure was continually curtailed to the benefit of private holders. Nevertheless nearly one-seventh part of Norway was still common land in 1876, and even in the western provinces, between the Naze and Trondhjem-fiord, these lands occupied on an average three-tenths of the country.

The Norwegian proprietors have preserved the old *odelsret*, or "allodial" right of holding lands put up to sale free of rent. But the amount to be paid is fixed not by the upset price, but according to a fresh valuation. The allodial right, however, attaches only to families that have held land for at least twenty years, and is forfeited unless claimed within three years after the property has changed hands. The inheritance, formerly different for the male and female issue, is now equalised for both sexes, and the testator cannot dispose of more than one-fourth of the estate over the heads of his direct issue.

In consequence of this last legal disposition the land became very much cut up. Apart from small patches situated in the towns, and useless except to grow vegetables and flowers about the houses, landed estates properly so called number about 430,000 in all Scandinavia; that is, 300,000 in Sweden, 130,000 in Norway. There would be a natural tendency amongst the people to increase the number of lots indefinitely, each peasant desiring to become his own master, and possess his *mantal* (literally "man-toll") of land. But the law has intervened to prevent this ruinous parcelling of the country. In Sweden all further distribution is forbidden when the allotment becomes insufficient for the support of a household of at least three members. Since 1827 another law, afterwards adopted in Germany and Austria-Hungary, allowed the owner of several lots to demand a fresh distribution, with a view to consolidating all the scattered plots. Estates have thus become rounded off, to the great advantage of agriculture. They are, as a rule, not very extensive, and Scandinavia has no such domains as many in Great Britain and Ireland, which are veritable provinces, except indeed in the remote region of Norrland, where the Göteborg merchant Dickson could traverse his estates for days without reaching their limits.

The tenant farmers, less numerous than the proprietors, are nearly all protected by long leases, but the so-called *husmän* or *torpare* class do not pay their rent in money, but by manual labour on the owner's lands, or by services in the mines and forests. Amongst them are some at once owners and tenants, while many are compelled by their precarious tenure to seek for subsidiary means of existence, becoming artisans, woodmen, or fishers.

FISHERIES.

;pecially in Norway, are very productive, and, more than the
soil, have tended to people the coast districts. The northern
rk and Nordland would even still be completely uninhabited but
:ets attracted to the neighbouring banks.
:f treasures are the cod and the herring. In the Lofoten and
. fisheries employ 8,000 craft, with 35,000 men, of whom perhaps
tied to watery graves. In good seasons, such as that of 1877,
0,000 cod are taken. The most frequented spots are the islet of
Vest-fierd, and especially 'the Henningsvär coast, where many
es become for the time veritable curing depôts. Here every

Fig. 75.—FISHERIES OF VEST-FIORD.
Scale 1 : 1,225,000.

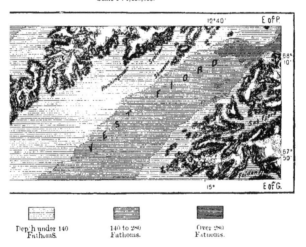

Dep'h under 140
Fathoms.

140 to 280
Fathoms.

Over 280
Fathoms.

nal is turned to account. Till recently the Lofoten fishers
ly, for the sake of the oil which has acquired such importance
nacy. But now, with a better knowledge of its value, they
he fish to foreign markets, or else barter it on the spot with
a flour and woven goods. The residue of the liver, after extract-
)ed for Troidhjem, where it is regarded as a most powerful manure.
: to Havana as an article of food, and for making isinglass; the
nd to a sort of flour, which makes excellent paste; lastly, the
as bait, and was formerly sold, especially in France, to the sardine
:ommercial value has of late years greatly fallen since a preference
for artificial roe. According to the various ways in which it
1 takes the names of stockfish, klepfisk, törfisk, rundfisk, or

rödskjaer, each of the countries supplied through the Bergen traders having its own favourite preparation.

Before 1857 the privileged alone had the right to share in the business. The sea was fictitiously divided amongst the off-shore proprietors, who grouped fishing colonies at intervals in the islands and on the headlands, and exacted a rent or duty from them. But now the sea is free except on Sundays, all fishing being prohibited from five o'clock on Saturday evening to the same hour next day.

Though less reliable, the herring scarcely yield in importance to the cod fisheries. In good seasons the take in Norway amounts to about 1,500,000 barrels, or 450,000,000 fish, of which about one-third are shipped to Russia. For this market they are packed in fir casks, imparting to them a resinous flavour highly esteemed by the Russians. For other places the barrels are made of beech. On the Norwegian seaboard there are two seasons: the first in summer and autumn, when the shoals approach the shore in search of shrimps, molluscs, and annelida, or ground worms, forming their food; the second in winter, from the end of January to March, in the spawning time, when the herring travels in vast swarms. The latter has always been shifting, the nets sometimes coming up empty, at others breaking with the weight of the haul. The crews are generally joint owners of the smacks, dividing the profits, the largest share of which goes to the *notebas*, or headmen. Temporary villages are established on the neighbouring beaches, with postal and telegraph stations and landing-places for the steamers.

In Sweden, although the herring fishery is the most important in the industrial economy, it does not yield sufficient for the local consumption, and herrings are consequently imported from Norway. The average yearly take on the Swedish Baltic coast is estimated at 150,000 tons.*

Besides the cod and herring, there are other fisheries contributing to the support of the people and its export trade. In Norway from 6,000,000 to 8,000,000 mackerel are annually taken, valued at about £40,000 ; and the salmon, frequenting nearly all the rivers, are taken in large numbers below the falls, which they endeavour to surmount. The quality of the salmon in some of the rivers of the west coast is as highly prized as that of Scotland, and considerable quantities are shipped for the English and German markets. Some of the streams in the north of Norway are even leased to rich Englishmen, who pass the season in these districts. Till recently the Norwegian fishers hunted a species of shark (*Squalus peregrinus*), the largest animal frequenting these waters, measuring from 40 to 45 feet, and whose liver yields about 154 gallons of oil. But, like the whale, this species has almost entirely disappeared from the Norwegian coasts, where the hâkjerring (*Scymnus borealis*) and one other species of shark alone are now hunted, chiefly in the Finmark waters. The Tönsberg fishers, however, visit the northern seas in search of the seal and the huge rorqual, which is killed at a distance by means of harpoons shot from guns and furnished with explosive balls. Efforts are

* Mean annual yield of Norwegian fisheries, 1869—78 :—Cod, 49,219,000 fish; herrings, 686,000 barrels at 32 gallons; mackerel, 6,288,000 fish; salmon, 855,400 lbs. ; lobsters, 1,073,000. Total value at place of capture, £1,150,000. Value of produce of fisheries (including oil, &c.) exported, £2,300,000.

also now being made to restock the exhausted lakes and rivers of the peninsula itself. An establishment of pisciculture has been founded at Hertösand, on the Gulf of Bothnia, and oyster beds have been laid out at various points of the Norwegian coast. The upper portions of some of the fiords, forming natural reservoirs, have been utilised as preserves, where the spat is reared and protected by wirework against its enemies.

Mining Industry.

LIKE the fisheries, the mining industry no longer bears the same relative importance to agriculture that it formerly did in Scandinavia. The silver lodes of Koigsberg and the copper veins of Röros and Falun have lost their former influence in the mineral market, and even the Dalecarlian iron ores have found rivals elsewhere. A few grains of gold may still be extracted from the Falun mines, but those of Areidal, from which the crescent-shaped ducats of Christian IV. were coined, are now abandoned, while the rivers of Lapland, containing gold dust, flow through too remote and cold regions to attract gold-seekers in any numbers. The richest streams are said to be those flowing from the Peldoniemi group, on the frontiers of Norway and Russian Finland, towards the sources of the Tana and Ivalo. In 1872 the five hundred adventurers who visited the new Eldorado collected no more than from 100 to 120 lbs. weight of the gold.

Several deposits, which in more favoured climates would be worked by thousands and surrounded by factories, here remain untouched, although long pointed out by geologists. Thus the copper veins of Kaa-fiord, on the shores of the Alten-fiord, containing one-half of pure metal, have only quite recently been seriously worked. The Skjærstad iron ores near Bodö, on the Norwegian coast, also remain intact, although rich enough to supply all the workshops in the world. So also the Gellivara iron deposits in Swedish Norrbotten have only been surveyed, and the surface barely scratched. The lines of railway have not yet been constructed which are intended to connect this district, on the one hand, with the navigable Luleå, on the other with the Norwegian coast through a gap in the Kjölen; yet the iron beds in this part of Sweden are extremely rich. The ores, containing an average of from 50 to 70 per cent. of pure metal, are disposed in parallel layers between denuded masses of gneiss. Thanks to its exceptional hardness, the iron has not been affected by the atmosphere, and crops out in black or red masses, here and there rising to the proportion of hills. The metal contained in these deposits is estimated at billions of tons.

But in the central and southern districts of Sweden there is still iron enough to meet the local demand, and allow of a considerable export trade in ores and cast metal. The mines of Dalecarlia and neighbouring provinces give an annual yield of from 700,000 to 900,000 tons of ore, smelted down to about 350,000 tons of metal, shipped chiefly for England. The yield of copper, till recently twice as great, has been unable to compete with the mines of the New World, and has consequently much fallen off. Sweden also possesses several zinc mines, amongst which that of Ammeberg, at the northern extremity of Lake Wetter, yields three-fourths of the whole amount. It belongs to the Belgian Vieille-Montagne Company, which exports the metal for its works in Belgium. Coal is won in Scania.

MANUFACTURES.

House industries are still far more developed in Sweden than in all the more thickly peopled European countries. In a region where the markets occur at such long intervals each family naturally endeavours, as far as possible, to supply its own wants. Certain processes and styles of ornamentation are found only in the remote Scandinavian hamlets, and date probably from prehistoric times. Hazelius, founder of the Scandinavian Museum in Stockholm, and other learned archæologists are carefully studying these primitive industries before they have been swept away by the manufactured goods already threatening to invade the remotest upland hamlets.

The chief Scandinavian factories utilise the natural products of the land and surrounding waters, iron, timber, and fish. All the seaports are occupied with the building and repairing of fishing-smacks, the weaving of nets and other tackle, the curing and forwarding of fish. The metallurgic works use up the ores in sufficient quantities to export a large amount of the produce, while the timber is shipped, either as planks or as furniture, to the remotest European settlements. Most of the mechanical saw-mills are situated along the seaboard about the mouths of the rivers, which float down the timber in bulk to Gefle, Söderhamn, Hudiksvall, Sundsvall, Hernösand. It is only quite recently that the Swedes have taken to export their timber worked up into inlaid floorings and cabinet pieces. This industry has been developed especially in Göteborg, and has thence spread to all the Swedish towns engaged in the timber trade.

The wood is also exported in the form of matches, a branch of industry in which Sweden already takes the foremost rank. The aspen, which supplies the best material, has rapidly risen in value, every factory now consuming these trees by the thousand. In Sweden, and Norway also, the greatest quantity of wood is used up in the manufacture of paper. This branch was first established at Trollhättan in 1857, and it is now carried on in about forty different factories, jointly yielding a yearly average of about 30,000 tons. The vast heaps of saw-dust formerly encumbering the ground about the saw-mills will henceforth be converted into material for packing, wrappers, books, and especially newspapers.

Amongst the industries imported from abroad the most important are cotton spinning and weaving. The first essays were made after the Napoleonic wars, and Scandinavia already imports about 13,000 tons of the raw material yearly, employing thousands of hands in converting it into yarns and tissues. Of older date is the woollen industry, which began early in the seventeenth century at Jönköping and Upsala, but which, notwithstanding its subsequent development, still falls short of half the local requirements. There are also some flax, hemp, jute, and silk works, and, exclusive of hardware, the Swedish manufacturing industries comprised 2,868 factories in 1877, with a total yield of £9,376,000 worth of goods, and employing 60,589 hands, of whom three-fourths were adults. The chief manufacturing centres are Göteborg, Stockholm, and Norrköping, producing between them one-half of the wares manufactured in Sweden. The Norwegian industries employ 35,000 hands, or, in proportion to the population, about the same number as the Swedish. In some respects Norway is even more favoured; for, if less rich in iron, it imports English coal more easily, and maintains direct commercial relations throughout the year with Great Britain and the mainland. Thanks also to its falls and rapids, mostly in the immediate neighbourhood of the coast, it disposes of even a larger water motive power than Sweden.

TRADE AND SHIPPING.

But if not superior as a manufacturing country, Norway surpasses its neighbour at least in its trade. In this respect it has turned to marvellous account

Fig. 76.—STAVANGER-FIORD.
Scale 1 : 140,000.

2 Miles.

the advantages derived from the numerous ports of the seaboard, and its geographical position relatively to Great Britain. Its exchanges, scarcely more than £320,000 in the middle of last century, now amount to £20,000,000 yearly, while those of Sweden, with double the population, fall short of £40,000,000. The chief commercial relations of both countries are with England, after which Germany, Denmark, and France follow successively. Of both also the staple export is timber, next to which rank metals and cereals in Sweden, fish in Norway. The most important imports are naturally manufactured goods, exchanged for raw materials. Since 1873 the customs dues have been abolished between the two states, but with a few reservations, which still subject travellers to the inconvenience of having their luggage inspected on the frontiers.

Shipping has acquired a marvellous development in Norway, which relatively now owns the largest commercial navy in the world. Two-thirds of all the shipping, including thousands of fishing-smacks, belong to the south coast

155

between Christiania and Stavanger, and two-thirds of all the navigation are conducted under the Norwegian flag, now familiar to every foreign seaport. Most
of the poorer classes in the towns, instead of placing their spare cash in the
savings banks, invest it in a "share" of some vessel, so that all are ship-
owners, directly or indirectly. Hence the astonishing expansion of navigation in
recent years. The tonnage of the Norwegian shipping exceeds by one-half that of
the whole of France.

The development of the Swedish seaboard, the relatively dense population of
the southern provinces, and the trade in bulky wares, such as timber and iron,
also attract large numbers of vessels to the Swedish seaports. But not more
than one-third sail under the national flag, nearly all the rest belonging either
to Norway or Denmark. The difference of about 1,000,000 tons between the
yearly arrivals and departures is due to the greater weight and bulk of the
Swedish exports, so that thousands of ships enter the Swedish ports without

Fig. 77.—TABLE OF RELATIVE TONNAGE IN 1877.

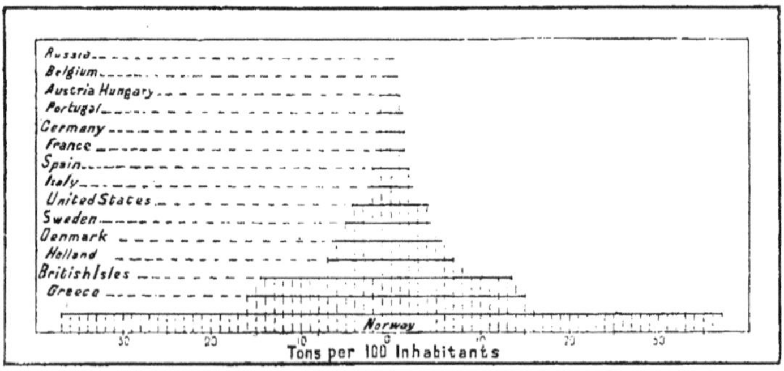

return cargoes. Altogether the Swedish commercial navy is scarcely one-third
that of Norway, though still relatively four times superior to that of France.
The inland navigation of the lakes and fiords employs over half the shipping, so
far as the number of vessels is concerned, but one-fifth only of the total tonnage.
The mercantile navy is manned altogether by about 29,000 hands.*

CANALS.—RAILWAYS.—POSTAL SERVICE.

THE hilly character of the surface in Norway has restricted canalisation to a
few ramifications from mountain torrents; but the more open nature of the
country has enabled Sweden to open up water highways to the interior of its
plains. According to Sidenbladh essays at canalisation were here made so early
as the beginning of the fifteenth century; but two centuries passed before the
first canal with locks was opened between Lake Mälar and the outlet of Lake
Hjelmar at Eskilstuna. Since then the whole of South Sweden has been con-

* Commercial marine:—Norway, 8,064 vessels (including 273 steamers) of 1,493,300 tons burden;
Sweden, 4,472 vessels (including 706 steamers) of 544,266 tons burden.

.1 island by the construction of the Göta Canal, forming an uni-
e of communication 260 miles long between the Baltic and Kattegat.
rting is at Lake Wiken, 305 feet above sea-level, whence the canal
ls to Lake Wener, and so on through other lakes to the Söderköping-
Baltic. In this section there are thirty-nine locks, by which vessels
ly raised or lowered. West of Lake Wiken the canal falls through
.s to Lake Wenor, whence the Göta forms its natural continuation
·gat. But this river is interrupted by falls and rapids, of which
ium, near Wenersborg, had already been turned by a canal early in
ith century. Those of Trollhättan seemed to present an insur-
)stacle, till Swedenborg, the strange dreamer and daring engineer,
:analisation. This work, interrupted after the death of Charles XII.,
.ed till 1800, and since then the engineer Nils Erikson has replaced
· canal, whose eleven monumental locks, cut in the live rock, remain
:heir kind and a source of universal admiration. Ships drawing
of water can pass from sea to sea through this chain of lakes, rivers,
regulated falls constituting the Göta Canal. They may also pass

Fig. 78.—GÖTA, OR GOTLAND CANAL.

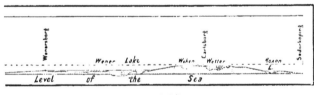

dary junction canals far inland. One of these, the Dalsland Canal,
g, gives access to Norway. Starting from Lake Wener, it runs north-
several lakes, and is carried over a cataract by a daring viaduct,
· most of these works in Sweden, by Nils Erikson. The navigation
.nals, as well as through the locks about Stockholm, is very brisk.
receded Sweden by two years in the construction of railways, the
, between Christiania and Fidsvold, was opened in 1854. Still the
country necessarily prevented the system from receiving as great a
s in the neighbouring state. Of the main lines one only is all but
which crosses the peninsula from Trondhjem to Christiania-fiord
ish frontier near Frederikshald. The section connecting Fidsvold
on Lake Mjösen, still remains to be completed. The route from
the Gulf of Bothnia, which will be the northernmost on the globe,
)ssed the Scandinavian water-parting. The Bergen line also stops
the mountains which are to be tunnelled, and the line so continued
Drammen. Of the coast line projected between Christiania and
two terminal links only have been so far constructed.

The Swedish system, thanks to the open nature of the country, cheapness of land, and abundance of materials—wood, iron, stone, shingle, and sand—is relatively more highly developed than that of any other European state, Belgium not excepted. Sweden is the only country in the Old World which possesses over 700 miles of rails to 1,000,000 of the population, carried out, however, at a cost of

Fig. 79.—THE SCANDINAVIAN RAILWAY SYSTEMS.

Scale 1 : 7,935,000.

100 Miles.

less than £4,000 per mile, and with 10 tunnels longer than that which runs for 1,350 feet under the southern quarter of Stockholm.*

The postal and telegraph services have also of late years been rapidly developed, although relatively to a less extent than the railway system. Not only are all the railways lined with wires, but the fishing stations and seaports are connected with Stockholm, while submarine cables cross the Gulf of Bothnia, the southern basin

* Railways:—Sweden, 3,256 miles (besides 642 building); Norway, 658 miles.

e, the Sound, and the Kattegat. Of all European countries Norway
t the longest, but the greatest number of cables, necessitated by the
ts interrupting the land communication. The lines are carried from
northwards to the immediate neighbourhood of North Cape.

EDUCATION.—SOCIAL CONDITION.

ne relative number of letters received per head of the population is
England, France, and Central Europe, general instruction is still
oped. Attendance at school between the ages of seven and fourteen is
both states, and each urban commune and rural parish is bound to
t one primary school, with a master holding a certificate from a
normal school. These primary establishments are perfectly free, but
ers founded by the State, the communes, and private individuals, in
re taken. Both the State and the Protestant Church reserve their
ecting the free schools, and the Council of Education, in which the
Consistory have the upper hand, may compel the parents to send their
he Government schools if the result of the periodical examinations
to justify this course. Parents not sending their children to school
ded or punished.

tly nearly all the Norwegian schools were ambulatory, the sparse
f the hamlets and the great distances across rocks and moors
he children from resorting to the village schools, and obliging the
sit them. He made his rounds, stopping successively for a few weeks
itable farmstead, where the children of the neighbourhood gathered
im. His arrival was a great event, and when the little ones had
ir letters they were left in charge of tutors, who continued the work
n till his next visit. Thanks to these migratory teachers, a love of
rakened in the remotest hamlets, and thousands of fixed schools have
iblished in which the rudiments of the sciences and music are taught.
nerary schools are the exception in the south, but they are still
umerous in the northern districts, where the people are scattered in
os. The preceptors are often called upon to perform the functions of
e peace, and reconcile by conciliatory means the differences arising
peasantry.

' instruction is also more advanced in Scandinavia than in most other
untries, and many of the intermediate schools far from Stockholm,
r the University towns of Upsala and Lund, rejoice in the possession
ies, natural-history collections, and laboratories. The literary and
ement is very active, and in Sweden alone over a thousand new
blished every year. In 1877 the number of Swedish reviews and
nounted to 296, of which one-third appeared in the capital. In
eriodicals rose from 7 in 1854 to 180 in 1876.

e midst of so many collateral influences it is not easy to determine

the exact relation between the progress of instruction and public morality. Much
of the undeniable demoralisation may, however, be fairly charged to bad harvests
and industrial causes, want and crime here, as elsewhere, following each other as
cause and effect.

Intemperance is the national vice, and even in the sagas we read how mead
overflowed at the gatherings of the vikings. Towards the middle of the present
century drink had become a scourge, threatening definitely to debase a great part

Fig 80.—GÖTEBORG: VIEW TAKEN FROM THE BOTANIC GARDEN.

of the population. In 1855 over 40,000 distilleries were at work in Sweden alone,
yielding vast quantities of *bränvin* (brandy) ; yet at this period the trade had
already been centralised. In 1830, when every farmer was also a distiller, there
were no less than 170,000 distilleries in the kingdom. Under fiscal restrictions
the production has considerably diminished during the last twenty years. In the
towns also societies have been formed, with the exclusive right of retailing spirits
on the condition of handing over all the profits to the municipalities, the General

Council, and agricultural unions. This system, known abroad as the Göteborg system, from the place where it was first introduced, has the signal advantage that the thirst of gain does not transform the spirit dealer into an encourager of vice and a tempter ever on the watch for his victims. The retail taverns of the Göteborg societies are not like the gin-palaces which attract such crowds in England.

Sweden is one of the European states in which the number of births from unlegalised unions stands highest, though this need not be regarded as an index of exceptional depravity. There are, so to say, no foundlings in Sweden, where all mothers either nurse their children or cause them to be brought up. Besides, about one-tenth of those reckoned as illegitimate are born of couples registered as betrothed, and after the marriage of their parents they enjoy all the privileges of legitimacy. The high rate of illegitimate births in Stockholm is largely due to the temporary immigration of women from all parts of the country, attracted by the lying-in hospitals of the capital, where they are admitted without the formality of giving their names. Hitherto the non-Lutherans also, other than Catholics, the Jews, and a Baptist parish have been excluded from the marriage laws, in consequence of which their issue is regarded as illegitimate.

GOVERNMENT AND ADMINISTRATION OF NORWAY.

THE independent kingdom of Norway, united to Sweden in the person of the sovereign, is a constitutional state, the nation being represented, as in most European countries, by deliberative bodies. The legislative functions rest with the *Storthing*, an assembly of 111 members, divided into two sections—the *Odelsthing*, answering to a popular Chamber, and comprising two-thirds of the representative body, and the *Lagthing*, or Senate, elected by the Storthing. All bills must be adopted by both sections, and in case of disagreement the two form one deliberative body, in which a majority of two-thirds is needed for the enactment of any measure. The action of laws passed by the assembly may be suspended by a royal veto, renewable three times for terms of three years after each fresh vote. Ultimately the will of the assembly prevails. The projects initiated by the King take their turn like all others, and the Storthing may pass to the order of the day without discussing them.

The members of the Storthing are elected for three years—two-thirds by the rural districts, one-third by the boroughs. All citizens twenty-five years old are eligible, also present or former functionaries, owners of registered real property, traders paying licenses in towns, artisans, ship captains, and in the country five-years' leaseholders. The members of the Government take no part in the deliberations of the Storthing, which meets annually, but cannot sit for more than two months without the consent of the King, though he does not possess the right of dissolution. The fundamental law of Norway is partly modelled on the French constitution of 1791.

The King, who resides in Stockholm, is bound by the Norwegian constitution to pass a part of the year in Norway, where he exercises his functions by the aid of a State Council (*Statsråd*) composed of two ministers and nine councillors, all Norwegian citizens. One of the ministers and two of the councillors attend on the King during his absence in Sweden, the others constituting the National Govern-ment in Norway. On the advice of this Cabinet, the King appoints the civil, military, and ecclesiastical functionaries. But he can confer no patents of nobility, which have long been abolished by the Storthing, notwithstanding the royal opposition shown by three successive votes. Since this decision all sons of earls took the title of baron, those of barons becoming simple citizens, and the aristocracy thus gradually died out.

Formerly the judicial functions, which were combined with the legislative, belonged exclusively to the people, who exercised them through their delegates to the assemblies. Even now the electors name the two judges of the commission of peace in each commune, who hear all civil affairs before suits are commenced. In police cases, when a fine is incurred, it must be submitted for voluntary acceptance before being formally imposed. In criminal affairs and matters connected with property the nation is also directly represented by four jurymen taking their seats on the bench, and voting on the same footing as, and often even against, the judge. Judges are in other respects always responsible for all wrongs inflicted by them wilfully or through ignorance. Summoned before a higher court, they are liable to grave punishments in their persons or property, and in case of death their heirs may be sued for the penalties incurred.

Norway possesses 116 courts of first instance, 36 for the boroughs and 80 for the rural districts, which try all civil and criminal cases. That of Christiania comprises eight judges and a president, but all the rest one judge only—*byfoged* in the towns, *sorenskriver* in the country. In litigation connected with property four jurymen are added, expressly named in each case.

There are five tribunals of second instance, each with a president and one, two, or three judges, and special jurisdictions exist for ecclesiastics, teachers, and even the military. A Supreme Court, comprising a president and ten judges, constitutes a final court of appeal in all cases sent up from the lower courts. These latter, united with thirty-seven members of the Lagthing, form a tribunal which, under the name of State Court (*Rigsret*), tries all cases of crime or misdemeanour com-mitted by members of the Storthing, the Council of State, or the High Court in the exercise of their functions.

The Lutheran Church, in Norway the State religion, enjoys considerable powers, since it possesses much property, and largely controls public instruction. Although appointed by the King, the clergy receive no direct salaries, but they have a residence, and in the country the usufruct of common lands. They have also the revenues of their benefices, besides tithes and offerings, the minimum of which is fixed by the law. The average income of each cannot be estimated at less than £150. They are relatively few in numbers, at the end of 1877 only 637 altogether, according to which proportion there would be no more than 12,000 priests in France. The country is divided into 6 bishoprics (*stifter*), 83 deaneries (*provstier*), 441 parishes (*prœstegjeld*), and 900 pastorates (*sogne*). All other worships enjoy perfect toleration, but Nonconformists number altogether scarcely more than 6,300.

Military service is obligatory on all Norwegians upwards of twenty-five years old, ecclesiastics, pilots, and the inhabitants of Finmark alone excepted. But the army consists in reality of little more than the cadres, the battalions under drill, and military schools. The regular forces consist of less than 2,000 men, all volunteers, enlisted for three years. Conscripts pass through the recruiting school, which lasts forty-two days for the infantry, ninety for the artillery and cavalry, and return for three or four years to take part in the exercises for less than one month annually. The King is empowered by the constitution to keep a guard of Norwegian volunteers in Stockholm, and to remove 3,000 men from one state to the other for the manœuvres, but for no other purpose.

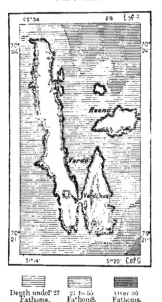

Fig. 81.—VARDÖ.

Scale 1 : 100,000.

Depth under 27 Fathoms. 27 to 55 Fathoms. Over 55 Fathoms.

1 Mile.

The navy is relatively much more important than the army, comprising (1879), 4 monitors and 26 other steamers, with 144 guns, besides 92 sailing and rowing vessels. Fortifications defend the entrance of Christiania-fiord, where is situated Horten, the chief naval station, and there are a few other defensive works at certain exposed points of the coast. Of these the northernmost in Norway and in the world is that of Vardö, at the entrance of Varanger-fiord, beyond the seventieth parallel and within the arctic circle. The crews, numbering 2,050, are nearly all volunteers, and all between the ages of twenty-two and thirty-five engaged in shipping or fishing, or resident in the seaports, are enrolled to the number of 60,000, with liability to be called out in case of national danger.

The Norwegian Budget is generally balanced, and amounted in 1877-8 to £2,235,000, more than half derived from the customs. The chief outlay is not, as mostly elsewhere, for war expenses or in payment of old obligations, although the debt amounted in 1878 to £5,130,000, contracted mainly for the construction of railways.

Administratively the country is divided into 517 communes, of which 61 are urban and 456 rural (*herreder*). Municipal affairs are managed by two elective bodies, an administrative council (*formandskab*) of from three to twelve members, and a representative council (*repræsentantskab*) three times more numerous. The execution of their decisions is intrusted in the towns to magistrates (*borgesmester* and *rådmand*) named by the King, in the rural communes to prefects (*amtmand*) aid bailiffs (*foged*), as well as to police officers (*lensmand*) named by the prefect.

The 61 urban communes comprise 39 towns (*kjöbstäder*) aid 22 minor seaports (*ladesteden*), the rural 58 bailiwicks (*fogderier*). They form collectively 18 prefectures or departments (*amter*), besides the two prefectships of Christiania and Bergen, each administered by a council (*amtsformandskab*) composed of the presidents of the communal councils, under the presidency of the prefect.

GOVERNMENT AND ADMINISTRATION OF SWEDEN.

THE modified constitution of June 8th, 1809, establishes the State a representative monarchy. Nevertheless the King is supposed to govern alone, and if he is assisted by a Council of State, it is only to consult it and receive the information of which he may stand in need. This body consists of ten members (including a prime minister named by the King), seven in charge of foreign and home affairs, the finances, justice, war, the navy, and public worship and instruction, three without portfolio, and of these at least two must be old civil functionaries. All administrative matters are submitted to the King in Council, and should he be considered to have decided illegally, the "reporter" refuses his signature and tenders his resignation, when the matter is referred to the Diet. Members of the Swedish and Norwegian Councils of State must be present whenever any question is being discussed affecting the interests of both kingdoms. In the absence of the sovereign from Scandinavia the Government is intrusted to the Crown Prince, and, failing him, to the State Council. The King's civil list amounts to £47,720 payable by Sweden, and £19,900 by Norway. In 1873 the Diet refused to tax the nation for the coronation expenses.

The old national representation of Sweden comprised the four estates of the nobility, clergy, burgesses, and peasantry; but under the pressure of public opinion the estates themselves decided, in 1866, to reorganize the legislative body. According to the new law the nation is represented by a Diet (*Riksdag*) composed of two Chambers, corresponding with the Lords and Commons of England. The members of the Upper House are elected for nine years by the Councils General of the provinces and the Municipal Councils of the large towns, in the proportion of 1 to every 30,000 of the population. To be eligible they must be at least thirty-five years old, owners of real property to the value of £4,536 (80,000 crowns), and in the receipt of a yearly income of £227 (4,000 crowns). And as this Chamber mainly represents wealth, all members must resign whose estate during their tenure of office falls under the amount required by the law. They receive no remuneration for their services, and number at present 133.

The Lower House is now composed of 198 deputies, all elected for three years by voters paying taxes. Each judicial circuit names a deputy for every 40,000 of the population, while in the boroughs the proportion is 1 to every 10,000, and the members here also must have either a property qualification of 1,000 crowns, or a leasehold of 6,000, or a taxable income of at least 800. They must further have resided at least one year in the commune, or be upwards of twenty-five years

old.   During the session, which legally lasts four months, they receive a grant of
£67 (1,200 crowns), besides travelling expenses. The Presidents and Vice-Presidents
of both Houses are chosen by the King from the respective bodies.   All matters
proposed for discussion are previously prepared by committees named in equal
numbers by each of the Chambers.

The 2,354 Swedish rural communes as well as the 95 boroughs have the
management of their local affairs.   Each possesses a Municipal Council variously
designated according to the rank of the commune, and chooses its own president,
except in Stockholm, where the governor (*Öfer Ståt-Hållare*) is *ex officio* president
of the communal assembly.   The four cities of Stockholm, Göteborg, Malmö,
and Norrköping, with populations exceeding 25,000, are detached from the
jurisdiction of the Councils General, and administered by their Municipal Councils.

The present Swedish code has been less affected by Roman right than most
others in Europe, being mainly traceable to the common law and pagan usages
modified by the influence of Christianity.   These laws, whose oldest texts date
from the beginning of the thirteenth century, are supposed to have been written
in Runic characters on *balkar*, or detached wooden tablets, whence the term *balk*
is still applied to the various sections of the civil code.   Several were composed in
verse under the form of couplets, with a view to their more easy retention in the
memory.   To the traditional laws of the rural districts were added the civic
codes, more or less inspired by those of Wisby and other Hanseatic towns.   In
1442 laws and customs alike, the old *landskapslagar*, were revised and fused into
one general code under the name of *landslag*.   This code was again modified in
1734, and since then it has been several times retouched.   The military and
ecclesiastics are governed by special laws.

The press enjoys a large share of liberty in Sweden, where, however, questions
dangerous to established religious, political, or social interests are seldom broached.
In cases of actions for libel against the press, juries of nine are appointed, three by
the defendant, three by the plaintiff, and three by the court, six votes being required
for a verdict of guilty.

The tribunals of first instance are composed, in the towns, of the burgomaster
and his assessors; in the 108 rural districts, of a district judge, assisted by twelve
proprietary peasants.   The judge alone decides, but when all the jury differ from
him their opinion prevails.   Three royal courts—those of Stockholm for North
Sweden, Jönköping for Götaland, and Christianstad for Scania and Blekinge—hear
cases of appeal, revise sentences of death pronounced by the judges of first instance,
and deal with all errors committed by the judges and functionaries in the exercise
of their duties.   A High Court sitting in Stockholm is composed of sixteen judges,
divided into two tribunals, and in cases involving the military two superior officers
are added.   Moreover, the King possesses two votes in this court, whenever he sees
fit to take part in the proceedings.   On very rare occasions a Court of the Realm
is constituted to hear cases against members of the Council of State or of the High
Court.

The Swedish Church and Universities.

The Lutheran is the State religion, and according to the census returns, which naturally include the indifferent amongst the faithful, nearly the whole Swedish nation belongs to this worship. But within the Church itself there are great differences. The members of certain Norrland communities, carried away by religious fervour, include dancing in their programme, jumping and whirling about till they are breathless. The Protestant Nonconformists amount to no more than a few thousands, and the Jews, excluded from the kingdom down to the year 1810, have not yet had time to found large communities in the commercial towns. Roman Catholics are fewer still, and only since 1870 have public offices been thrown open to all Swedish citizens, irrespective of their religious belief.

The power of the Established Church is still very considerable, although assailed from two opposite quarters by freethinkers and zealous Dissenters. Through its pastors and consistories it takes a large share in the local administration, and it keeps all the civil registers except those of the capital. For members of the Church marriage solemnised by the pastor is alone valid, civil unions being tolerated only when one of the contracting parties is a Jew, or belongs to some recognised form of Dissent; but even in this case the marriage can be legally celebrated only after the banns have been thrice put up in the Lutheran Church.

The primary schools also are placed under the direct control of the pastors and Consistories, who see that Luther's catechism is duly taught, that the pious practices are kept up, and all the children regularly "confirmed." Moreover, the Church, like the nation, has its deliberative assemblies. A Synod was held in 1863 in order to obtain the clergy's assent to a change of the constitution which aimed at suppressing that body as a distinct section of the Diet. This assembly is composed of 60 members, 30 lay and 30 ecclesiastic, amongst whom are all the bishops of the realm, with the Archbishop as *ex-officio* President of the Synod. The country is divided into twelve dioceses, to which may be added the Stockholm Consistory, in reality independent of the archiepiscopal see of Upsala. The dioceses are subdivided into deaneries, pastorates, and parishes, these last numbering about 2,500.

The two Universities of Upsala and Lund have an independent status, constituting them distinct bodies in the State. Still they depend officially on the Church, the Archbishop of Upsala being Vice-Chancellor of the former, and the Bishop of Lund of the latter. In both the student is bound to form part of a "nation," those of Upsala numbering thirteen, named generally after the old historical provinces, besides three for Stockholm, Göteborg, and Kalmar. Each of these groups forms a little self-governing republic, enjoying special privileges, and possessing considerable property and capital, whose revenues are chiefly applied to the support of poor students. They have large halls for the general assemblies and celebrations, libraries, and lecture-rooms. Some are even owners of country seats. Nor is their autonomy limited to the control of their property, and to their respective "nations" the students must apply for their certificates, and in

es for the means wherewith to continue their studies.   The University
1 extends for students 36 miles round Upsala.

history of education the two Universities present the same contrast as did
he schools of Paris and Montpellier.   The traditions of Upsala are
ic; those of her younger sister of Lund, founded two centuries later,
 Several ladies are already enrolled as members of Upsala, and in a few
e1 will be able to follow the courses in both.
3 an independent University was opened at Stockholm.

THE SWEDISH ARMY AND NAVY.—FINANCE.

iry service is still mostly organized as in the time of Charles IX.
elemeit in the army consists of men enlisted for three, six, or twelve
ither part, the so-called *indelta*, being furnished by the owners of certain
, who are bound to supply either a foot-soldier or a trooper at all times,
case of war, providing him with an outfit, and finding him employment
inder arms.   A few organized battalions are also raised by conscription.
nd of Gotland a special force of about 8,000 men is regarded as forming
active army, although not bound to serve out of the island.   Thanks to
eace, the small number of troops, and their peaceful manner of life, the
opulation is not exhausted, like that of so many European countries.
average standard of height has elsewhere had to be lowered, here it is
o raise it, and at present it is fixed at 5 feet for the regulars, and
ch for the indelta.
izens are also virtually bound to serve in the militia from twenty to
2 years of age without the option of purchasing exemption or supplying
2.   The regular forces (stipendiary and indelta) are estimated at 38,000;
ilitia, including that of Gotland, at 20,000; the reserve (*beräring*), 126,000.
ncipal Swedish forts are Waxholm and Oscar Frederiksborg, defending
olm channels; Kungsholmen, Drottningskär, and the other works about
.; Karlsborg, on the west side of Lake Wetter, at the entrance of the
„ the last-named being the central military depôt of the kingdom.   The
hool is at the castle of Karlberg, just beyond Stockholm, where are also
e engineering, artillery, and staff schools.
icf naval station is that of Karlskrona, although Stockholm also has
dockyards, and floating battery at the island of Skeppsholm, east of
quarters.   The fleet is recruited in the same way as the army, being
ith men enrolled for a fixed period, besides volunteers and the indelta
th the *båtsmän* furnished by the landed proprietors of the seaboard.
midshipmen, admitted between the ages of fourteen and sixteen, are
two companies, and distributed amongst the vessels stationed at Karls-
oreover, all the captains, lieutenants, engineers, and crews of the mer-
ine, between the ages of twenty and thirty-five, are liable to serve in
.   This available effective amounts to 1,000 officers and 6,000 sailors.

The militia of the coast districts, drilled in time of peace with the land forces, and amounting to 26,000 men, might also be ranked with the marines in case of war. The special naval school, reorganized at various times, is now at Stockholm.

The fleet consisted, in 1880, of 14 monitors, 29 other steamers, 10 sailing vessels, and 87 row-boats, maintained at a yearly cost of about £1,120,000.

The national Budget is one of those which, rare in Europe, occasionally show a balance to the credit side, which in 1874 amounted to £920,000. For 1880

Fig. 82.—KARLSKRONA.

Scale 1 : 140,000.

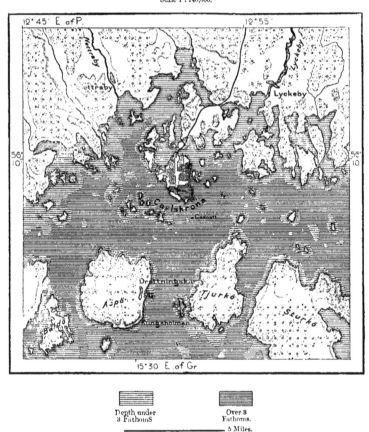

Depth under
3 Fathoms

Over 8
Fathoms.

5 Miles.

the Budget is balanced with £4,406,700, of which £1,200,000 are yielded by the customs, and nearly £1,000,000 by the State railways, the excise on spirits, the third important source of revenue, producing £720,000. The expenditure for army and navy amounts, on the other hand, to £1,330,000.

Before 1855 Sweden had contracted no foreign debt, and even at home owed nothing beyond a trifling sum to the National Bank; but since then she has raised several loans, amounting altogether to about £11,800,000, bearing an annual interest of something over £575,000, and all guaranteed by State property. The network of railways alone represented, in 1876, a capital of £9,440,000, besides advances of £3,480,000 made to private companies. The State, moreover, owns some £4,000,000 variously invested, lands let out at good leases, and vast tracts under timber, comprising in 1878 a total area of forest and unreclaimed lands of 13,063,000 acres, nominally valued at £1,800,000, yielding a revenue, in 1876, of £27,200, and yearly increasing by fresh purchases. The total valuation of the financial situation shows a sum of £8,054,000 to the good.

The communal finances are not quite so flourishing, but although several communes are in debt, the amounts are more than covered by their assets. The assessed value of landed estates is given at £101,096,000, which is under their real value, and to this sum must be added £35,000,000 for other real property. Insurances have been effected in exclusively Swedish companies to the extent altogether of over £100,000,000.

Sweden has no colonies, the island of St. Bartholomew, in the West Indies, ceded to her in 1784, having been sold to France in 1878.

The state is divided into twenty-four provinces (län).*

* For population of provinces and principal towns see Appendix.

THE EUROPEAN ISLANDS OF THE ARCTIC OCEAN.

HE seas stretching from the Scandinavian peninsula and Russia northwards to the unexplored regions about the pole have, like the North Atlantic itself, their islands and archipelagos, often bound together by frozen masses. These islands, some of which have hitherto been but dimly seen through mist and snow, and to which further polar exploration may soon add others, are not even usually regarded as forming part of Europe. With the northern extremity of Greenland and the arctic groups on the north coast of America, they form a world apart, not yet subdued by man. Certain European states have doubtless claimed possession of Spitzbergen, and hoisted their flags over its dreary wastes; but those remote lands remain none the less vast solitudes, shrouded for months together in the mantle of night, then lit up by a pale sun sweeping in mid-air above the horizon, but rarely acting as a beacon except to a few daring whalers.

The naturalists who are exploring these polar islands may possibly some day discover treasures in them sufficient to attract settlers to these desolate regions, but hitherto fishers and the shipwrecked alone have passed the winter on their shores. Although lying beyond the habitable world, these inhospitable lands still recall some of the most unsullied deeds of humanity. These dangerous waters have been traversed in every direction by men strangers to fear, who sought neither the glory of battle nor fortunes, but only the pleasure of being useful to their fellow-men. The names of Barents, Heemskerk, and Bernard, of Willoughby and Parry, of Nordenskjöld, Payer, and Weyprecht, conjure up noble deeds of courage and endurance of which mankind may ever be proud. And no year passes that does not witness other dauntless navigators following in their track, eager to enlarge the known world and penetrate farther into the mysteries of the pole.

I.—BEAR ISLAND.

The first land in the Frozen Ocean, lying about 280 miles north-west of the Finmark coast, is completely separated from Scandinavia by profound abysses, the sea being here no less than 1,800 feet in depth. Discovered on July 1st, 1596, by the Dutchman Barents, Bear Island, so named from an animal here killed, was again sighted seven years afterwards by the Englishman Bennett, who called it Cherie, after his patron of that name, whence the Cherry Island still occurring on so many maps. At present it is frequently visited by Norwegian fishers for the sake of the sharks, cod, and even herrings swarming round its caverious cliffs. Temporary curing places have been established on its shores, and a house now stands on the banks of a creek on its north side. But the cetacea, formerly so common, have almost entirely disappeared. In 1608 one vessel alone captured nearly a thousand in seven hours.

Till recently this island was described by all navigators as of small size, and even in 1864 Nordenskjöld and Dunér estimated its extent at no more than 26 square miles. But the careful surveys of the Swedish explorers of 1868, amongst whom was Nordenskjöld himself, showed a superficies of 260 square miles, or exactly tenfold previous estimates. A portion of the surface is covered with lakes and marshes, and in the south-east the land rises to a series of hills, one of which, named Mount Misery by the English, from its dismal appearance, rises, according to Mohn, 1,492 feet above vast snow-fields; but there are no true glaciers. The rocks, containing lodes of galena, were first explored by the geologist Keilhau. They consist of carboniferous limestones and sandstones, with several rich coal beds, showing the impressions of sigillaria and other fossil plants. These deposits have already been utilised by steamers sailing by the coast. When these strata were formed Bear Island constituted a part of a vast continent, reaching probably to North America, to judge, at least, from the identity of the carboniferous flora in all the islands of the Arctic Ocean. Later on the continent subsided, leaving nothing but these scattered fragments above the surface.

When the coal was formed the climate of Bear Island, now colder, perhaps, than Spitzbergen, resembled that of Central Europe. Of eighteen species of plants collected by Nordenskjöld and Malmgren in its coal-fields and rocks, fifteen are identical with those of the Swiss carboniferous flora. But now how desolate is this spot, well named originally by Barents Jammerberg, or Mount Desolation! Its flora comprises only about thirty phanerogamous plants, amongst which is a species of rhododendron, besides eighty species of lichens, whose verdure, seen from a distance, here and there resembles grassy plots. Amongst the twelve species of insects there are no coleoptera, and all, according to Malmgren, present peculiar forms, as if they were here indigenous. In summer the island is covered with mews and wild ducks, which here alight before continuing their northern journey. In autumn these migratory birds again stop here on their return southwards.

Bear Island is the southern headland of a submarine plateau stretching to the north and north-east to the unknown regions of the Frozen Ocean. The channel, 120 miles broad, separating it from the nearest islands, varies in depth from 160 to 1,070 feet, and in 1857 the whole space was covered by a continuous bank of ice.

II.—SPITZBERGEN.

THE Spitzbergen archipelago consists of five large and numerous small islands, stretching north and south across 4° of latitude, the northernmost rocks being scarcely more than 650 miles from the pole. From the careful observations taken at many points by Scoresby, Brook, Franklin, Beechey, Parry, and Nordenskjöld, the geographer Debes estimated the area of the group at 22,720 square miles. But the expedition of Leigh Smith and Ulve in 1871 gave a further extension of at least 2,800 square miles to North-East Land, and the whole area is now raised to 25,580 square miles.

When this archipelago was discovered by Barents in 1596, Greenland was supposed to extend much farther east. Some even thought that it reached the Asiatic continent, and it is referred to in various legends as a Trollboten, or "Land of Enchanters," occupying all the north of the globe. Although Cornelis Rijp, one of Barents' companions, had circumnavigated the group, a feat renewed for the first time by the Norwegian Captain Carlsen, that important excursion had been completely forgotten, and these islands long continued to be variously treated as parts of Greenland on the charts. They were also occasionally known as Nieuland, or Newland, like so many other recently discovered islands: but the name of Spitzbergen, dating from the time of the first discovery, ultimately prevailed. The general nomenclature, however, still remains in a chaotic state, English, Dutch, Swedish, and other names contending for the supremacy, so that certain gulfs and headlands have as many as ten different names. On the west coast nearly all the received geographical terms are of English or Dutch origin ; but in the north, on the shores of Hinlopen Strait and of North-East Land, great confusion is caused by the different names imposed on the same places by English, Swedish, German, and other explorers.

Spitzbergen has no very lofty summits, the highest hitherto measured being Horn Sound Peak (4,550 feet), near the southern extremity of the great island. But crests 5,000 feet high are said to occur in Prince Charles Foreland, running parallel with the west coast of West Land. Elsewhere there are no eminences much above 3,000 feet, and the highest yet ascended is the White Mountain (2,950 feet), on the east side of the great island, whence Nordenskjöld obtained a fine prospect in 1865. The interior of the island presents almost everywhere the appearance of rolling plains, here and there commanded by steep rocks, whose dark sides contrast forcibly with the surrounding snows. The mean elevation of the snow-fields in West Spitzbergen is about 1,800 feet, and of those in North-East Land 2,000 feet.

Although the interior of the archipelago is little known, the vast accumulations of detritus at the extremities of the glaciers show that the geological formation is very uniform, consisting mainly of gneiss, granites, and palæozoic sedimentary rocks. The Seven Islands, a small elevated group north of North-East Land, consist entirely of gneiss, and all the northern parts are of old formation ; but farther south the whole series of secondary rocks, especially the triassic and

THE SUN AT MIDNIGHT, SPITZBERGEN.

e\e1 some tertiar), are represe1ted. ）ioce1e deposits 1ear Bell

1 a1 exte1si\e fos-il flora, poplars, alder, hazel, c)press, and pla1e trees

i1 this period the climate must ha\e resembled that of Scandi1a\ia

Fig. 83.—SPITZBERGEN ARCHIPELAGO.

Scale 1 · 3,400,000.

50 ）iles.

.eth parallel. Some of the chalks co1tai1 fi1e marbles. **An** attempt

1 to **work** the rich phosphate beds **of** the Ijs-**fiord** ; lumps of coal are

1 b) the glaciers ; **and** Blomstra1d has disco\ered carbo1iferous

deposits in the dried bed of a glacial torrent, about $1\frac{1}{2}$ miles from the shores of King's Bay, facing the north end of Prince Charles Foreland. This coal, which burns very freely, with little ash, is rich in fossil trees, illustrating the mildness of the former climate.

On the coasts there is no lack of volcanic rocks, presenting here and there varied and picturesque outlines. They consist largely of hyperite, which Nordenskjöld regards as ashes crystallized under heavy pressure, and in several places they seem to have been distributed as lava over the trias and Jura systems. Hyperite cliffs are numerous on both sides of Hinlopen Strait, and several islands are entirely composed of this substance. Such are the so-called Thousand Islands, south of Stans Foreland, besides the various headlands projecting into Geneva Bay. These rocks contain a certain proportion of iron, which oxidizes when denuded, and in some places affects the magnetic needle.

The gently sloping rocks are covered for most of the year with snow, which disappears only in summer from the lower heights on the coast. The mean limit of the snow-line has been variously determined by naturalists, and it may be said to vary indefinitely with the nature and inclination of the surface, the aspect of the land, and other climatic conditions. Where wind and sun combine to lay bare the slopes, the snow will disappear to a height of 1,600 feet, and in the Seven Islands Nordenskjöld found none lower than 980 feet. In some favoured spots vegetation rises to 2,000 feet, but at this elevation the snow generally persists throughout the year.

In such a climate the streams are necessarily intermittent. In some places small rivulets flowing from the glaciers to the low-lying coast lands seek a channel through the shingle to the sea; but the large valleys of the plateau are filled with glaciers, all of which descend quite to the shore, some even projecting beyond its limits. Most of them are very slightly inclined, and generally very short compared with their breadth, several occupying the whole space from headland to headland along the shores of wide inlets. The largest, on the east side of North-East Land, forms a frozen mass over 60 miles long. At the southern extremity a glacier presents a sea frontage of 12 miles, but others seem to be little more than cataracts suddenly congealed, as, for instance, that of Magdalena Bay, which is only 800 feet wide on the beach. Most of them end abruptly on the coast, and on the west side they melt rapidly at contact with the warm currents from the tropics, which have here a mean temperature of 40° Fahr. The icebergs from time to time detached from them are often of considerable dimensions, one observed in 1773 by Phipps off the north-west coast rising 50 feet above, and plunging 130 feet below the surface.

The east side of the large island and of North-East Land, washed by the polar currents, is mostly ice-bound, and generally of more difficult access than the west, which is indented by numerous sheltered inlets. These gulfs and fiords remain mostly open to the sea, from which, however, they are partly cut off by masses of detritus or submarine moraines, known to the fishers by the name of "seal banks."

Like those of Switzerland, the Spitzbergen glaciers have been subject to various

.n the recent period. Many of them seem to be at present increasing,
are contracting. That of Frithiof, facing Bell Sound, was in 1858
ed from the sea by a broad muddy tract, crossed by a number of
n eminence surmounted by a cross marked the grave of a sailor, and
appearance of the moraines showed that the ice was retreating. But
inter of 1860-1 the frozen stream rapidly expanded, filled all the inter-

Fig. 84.—Spitzbergen Ice-fields in 1869.

Scale 1 : 10,000,000.

——————— 100 Miles.

and overflowed far into the sea, completely blocking one of the best
tzbergen, which had formerly been much frequented by whalers and
ters. It is now one of the largest glaciers in the archipelago, and
us of access, owing to the frozen masses continually breaking away
he glaciers studied by the French explorers in 1838 in "Recherche"
o increased, greatly changing the aspect of this district. Similar
been observed in Stor-fiord and the other large inlets, where former

islands are now lost amidst the advancing ice-fields. On the other hand, the masses formerly filling North Sound, a northern ramification of the Ijs-fiord, seem to have retreated considerably inland.

As on the Scandinavian seaboard, the traces of recent upheaval are numerous in the archipelago. The old beach is everywhere met with at various elevations round all the islands, sometimes rising 50, sometimes as much as 150 feet above sea-level. The amount of upheaval that has taken place in modern times may be studied by the quantities of drift-wood, the remains of cetacea, and shells of still living species now found far beyond the reach of the waves.   Low Island, at the north-west angle of North-East Land, appears to be entirely of recent origin ; its rocks, inter-

Fig. 85.—" Recherche " Bay in 1839.

Scale 1 : 180,000.

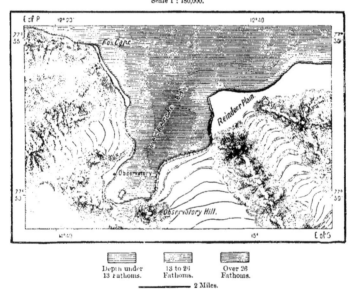

Depth under   13 to 26   Over 26
13 Fathoms.   Fathoms.   Fathoms.

———————— 2 Miles.

mingled with lakelets, seem scarcely yet quite dry, and in the interior spars and drift-wood are found associated with the remains of whales.   A chain of reefs is even now slowly rising between the mainland and this new island, whose area is estimated at about 20 square miles.

Climate, Fauna, and Flora of Spitzbergen.

Tempered by the marine currents and winds from the south-west, Spitzbergen shares in the general mildness of climate enjoyed by Scandinavia and the west of Europe.   In summer, if not the most agreeable, the climate is at least one of the healthiest in the world, and the recent Swedish explorers unanimously declare that it is superior even to that of South Scandinavia, and that there is perfect immunity from colds, catarrhs, and all affections of the chest.   No ill effects follow from sudden changes of temperature; sailors falling into the water may with impunity let their clothes dry on their body; and the place has been so highly recommended for many complaints that sanatoria may possibly some day rise on the banks of the Spitzbergen creeks for the benefit of invalids from England and the continent.

But however salubrious, the climate still remains cold, uncertain, changeable, and the sky never continues cloudless for a single day. Winds, cooled by the ice-fields and glaciers, blow at short intervals, although a steady south wind prevails in Hinlopen Strait. Snow may fall even in the "dog days:" Scoresby registered a temperature of 15° Fahr. in June, 1810, and the glass has never been observed to rise higher than on July 15th, 1861, when it marked 61° Fahr. In winter the mercury frequently freezes, although travellers represent the mean temperature as relatively mild, owing to the prevalence of strong south winds during this season. The glass rises at times above freezing point in midwinter. But when the sun appears on the horizon, gradually describing a wider circuit in the heavens, the cold grows more intense, so that here, as in England, "the cold strengthens as the day lengthens." *

At the southern extremity of Spitzbergen, in 76° 30′ N. lat., the sun rises 37° above the horizon, but in the Seven Islands 33° only, and although it remains for four months above the horizon, summer is soon followed by a winter of equal length, illumined only by the fitful glare of the northern lights. Under the influence of the prevailing southern winds the aurora borealis is now continuous, but much fainter than in lower latitudes. Fierce storms, accompanied by forked lightnings, rending the welkin, are altogether unknown in these waters.

In winter the various island groups are bound together by unbroken masses of ice, stretching for vast distances northwards, but limited on the west by the warm currents from the tropics. But even in summer all the coasts are at times ice-bound and inaccessible to vessels, except through narrow passages opening here and there between the ice-fields. The changeable temperature produced by the shifting currents of warm and cold water is also frequently attended by dense fogs, shrouding land and sea for days together. In August, 1875, Payer and Wey-precht remained enveloped for 354 hours, or over a fortnight, in one of these fogs, which convert day into night, and which prevail especially in Hinlopen Strait, between the great island and North-East Land.

Beneath the pale, grey skies of Spitzbergen the flora is extremely poor, compared even with that of Novaya Zemlya. The only timber is the drift-wood thrown up, especially on the southern and northern shores, under the shelter of the islands and headlands. There are even no shrubs, and nothing to recall the plants of the temperate zone beyond two species of dwarf willow and the *Empetrum nigrum.* The prevailing vegetation consists of lichens and mosses, of which there are reckoned over 200 species. According to Heiglin there are 120 species of phanerogamous plants, or three or four times more than in Iceland; but Malmgrèn, who has devoted much time to the botany of these islands, mentions only 60 flowering plants and 4 ferns.

Such as it is, this flora belongs at once to the arctic zone and to Scandinavia,

* Mean temperature in lat. 77° 30′ :—January, − 1°; July, 37°; year, 17° Fahr.

consisting of 81 plants common to Greenland, and 69 found also in Scandinavia. Of the various species there is only one edible, the *Cochlearia fenestrata*, which, being less bitter than its southern congeners, may be eaten as a salad, and also supplies a valuable preventive against scurvy.

Including the cetacea, there are 16 species of mammals, of which 4 only are land animals; and even the white bear is rather nomadic than indigenous,

Fig. 86.—FOUL BAY, SPITZBERGEN.

passing on floating ice from island to island. The other land mammals are the reindeer, a short-tailed rat like that found on the shores of Hudson Bay, and the arctic fox, hunted for its valuable fur. The reindeer was supposed to have been introduced by the Russians or Scandinavians; but in 1610, long before they reached the archipelago, the English explorer, Jonas Poole, hunted the reindeer, and gave its name to Horn Sound, from the antlers of one of these animals which he there found. Between 1860 and 1868 as many as 3,000 were annually

t Nordenskjöld asks how such losses could be repaired, and mentions,
ting it, the opinion of certain naturalists, who speak of migrations
Zemlya on drift-ice.

is, of which 130 were taken during the season by sixteen vessels in
lost disappeared from the southern seaboard, and schools of thirty to
r met only on the north side. Multitudes of sea-fowl frequent the
s and reefs, where their nests are safe from the ravages of the fox.
birds, comprising twenty-seven or twenty-eight distinct species, are
th the exception of the ptarmigan, which remains all the year round.
reptiles, and the surrounding waters were long supposed to be desti-
but as many as twenty species had been recorded up to the year 1861.
and fifteen species only of insects, and butterflies, grasshoppers, and
all waiting. In the snows that melt at contact with the sea-water
iads of phosphorescent crustacea, which dart about like blue sparks,
e effect of a vivid display of fireworks.

bird of passage, man visits Spitzbergen only during the summer
ertheless, shipwrecked sailors, hunters, and naturalists have wintered
and a Russian named Starashtchin, after passing twenty-three years
rbour, an inlet of the Ijs-fiord, on the west coast, died there of old
All the remains of huts, by whomsoever erected, are known as
ts." The archipelago was much more frequented during the last
at present. At that epoch the great cetacea swarmed in the surround-
were yearly hunted by as many as 12,000 whalers. Villages built of
rected under the shelter of the headlands; temporary international
d on the beach; and regular battles at times took place between the
ival or hostile fleets. The finest village was Smeerenberg, belonging
who were the most numerous and energetic. Here a whole quarter,
"Haarlem Kitchen," was occupied by those engaged in boiling down
During the period between 1669 and 1778, 14,167 Dutch whalers,
he waters especially in the west and north-west, captured 57,590
ing a profit of £3,710,000.
rg, which stands at the north-west corner of the great island, was
8 by the Dutch schooner, Willem Barents, when a monument was
le name of the nation, to the navigators who discovered the
d to their fellow-countrymen who here perished. On the same
ques, the Hanseatic traders, the Danes, and the Norwegians had
hing stations. Farther south is Magdalena Bay, which has been
red by naturalists, and on the north-east Foul Bay, one of the most
reis. Still farther north are the Norway Isles, where Sabine made
bservations, and which became the central point for the astronomical
Nordenskjöld and his associates. This spot, or some neighbouring
it perhaps be the most convenient site for one of those circumpolar
hich Weyprecht proposes to establish for the purpose of studying
etails the meteorological perturbations of the arctic regions. In

57

anticipation of such an undertaking Russia and Sweden have already contended
for the possession of the archipelago. A meteorological station would be here all
the more useful, inasmuch as Spitzbergen seems to occupy the point of contact
between the American and Asiatic zones of the winds.

North of Spitzbergen there is no land, at least as far as the eighty-third parallel,
for Parry, who reached 82° 44′ N. lat., detected no trace of islands or continent

Fig. 87.—SMEERENBERG.

Scale 1 : 500,000.

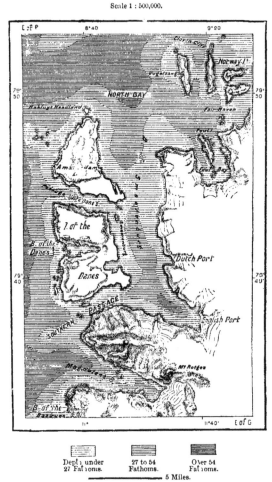

Depth under   27 to 54   Over 54
27 Fathoms.   Fathoms.   Fathoms.

5 Miles.

thence towards the pole. The desert skies, chequered by no flights of migratory
birds, and the absence of icebergs or any floating masses more than 35 or 40 feet
high, show that there can be no great extent of land in the direction of the pole,
which several navigators have endeavoured to reach from this quarter. According

mauthcnticated traditions, some Dutch whalers, notably Cornelis
need in the last century to within 5° of the pole. But Parry, was
andon his ship in Treurenberg Bay, in a little inlet named from it
pushing thence northwards in small boats and sledges. But he pro-
slowly, and at last ceased to make any way, the floating ice drifting
.s he endeavoured to advance northwards. The attempt, renewed in
73 by Nordenskjöld, led to no results, owing to the rotteness of the
: eightieth parallel, and its extreme roughness beyond that point. On

Fig. 88.—WICHE'S LAND.

Scale 1 . 3,700,000.

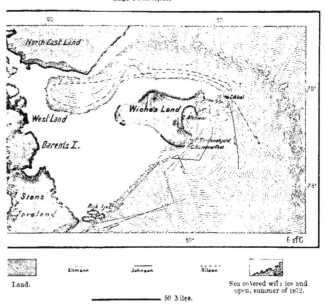

ons Captain Palander failed to make more than half a mile a day in

III.—WICHE'S LAND AND GILES LAND.

zbergen the Arctic Ocean is not so free of land as in the north. In
a long chain of lofty mountains is visible on clear days some 70 or
belonging to Wiche's Land, so named in 1617 in honour of the
chard Wiche, or Wyche, by the English whalers, who were the first to
md. After a lapse of two centuries and a half another Englishman
l the island in 1864, and in 1870 Heiglin and Zeil, thinking its
lay north of the position indicated for Wiche's Land, renamed it
wereign, Charles of Württemberg. The Swedish explorers had, in
med one of its mountains, seen by them, "Swedish Headland," so that

national rivalries, combined with the confused reports of navigators, have rendered the charts very uncertain. The same island was even recently confounded by the Swedes with the Giles or Gillis Land, sighted in 1707 by the Dutch captain, Cornelius Giles. At length the Norwegian Altmann, profiting by the open seas, was able to coast the island, which his fellow-countryman Johnsen ascertained in the same year to be not an archipelago, as Altmann had supposed, but a single mass 70 miles long, and on its south side covered for several hundred yards by a vast quantity of drift-wood. A third Norwegian, Captain Nilsen, also visited it in 1872, and all concur in describing it as a low island, above which rise detached mountains and continuous ridges, culminating with Mount Haarfagrehaugen, on the west side. Like Spitzbergen, its inhabitants consist of bears, the arctic fox, and great numbers of reindeer, so that the vegetation, although confined to lichens and small growths, must be comparatively abundant. The land also shares in the general movement of upheaval, as is evident from the quantity of drift-wood observed by Johnsen 20 feet above the present sea-level.

Giles or Gillis Land has also been recently rediscovered west of North-East Land, precisely where Giles had indicated it, and where it is figured on Van der Keulen's chart, published probably in 1710. In 1864 the Norwegian Tobiesen sighted it without being able to land. But there are other islands in the same waters, for Baffin had seen land to the north-east of Spitzbergen so early as 1614. On Peterman's maps Giles Land is represented, apparently by mistake, at some 120 miles to the north-east of the most advanced Spitzbergen foreland, seemingly forming part of the newly discovered Franz-Joseph Land. This region has not yet been visited, and it is uncertain whether it is to be regarded as an island, an archipelago, or a simple headland, though its existence can scarcely be questioned. In spring the fishers who have wintered on the northern shores of Spitzbergen see flocks of migratory birds flying towards the north and north-east, whence they return in September, and this land lies right in their track. According to the walrus hunters frequenting the Seven Islands, north of Spitzbergen, from the same remote region come the walruses and numerous white bears visiting that little group.

IV.—FRANZ-JOSEPH LAND.

SINCE 1874 the arctic waters have been known to encircle with their floating masses another archipelago, even more extensive than Spitzbergen, but of far more difficult access. It lies almost entirely beyond the eightieth parallel, with a mean temperature from 18° to 28° below freezing point. Even on its south side the mean for the year 1873 was found to be 3° Fahr. by the explorers who had to spend some time on its shores. This is the Franz-Joseph Land of the Austro-Hungarian Tegetthoff expedition, conducted by Payer and Weyprecht, and which has contributed so much to promote the scientific exploration of the arctic seas.

Setting out with the object of making the north-east passage round Siberia to Bering Strait, the daring navigators, after having been ice-bound, contrived to land on a small island, which they named **Wilczek, in** honour of the promoter of the expedition. But from this point they sighted vast northern lands, with mountains and glaciers. Julius Payer was able to traverse a great part of the archipelago, surveying its main geographical features, and making numerous minor observations.

An irregular channel, ramifying right and left into fiords, and known as "Austria Sound," runs south and north between two extensive islands, Zichy

Fig. 89.—FRANZ-JOSEPH LAND.
Scale 1 : 3,000,000.

50 Miles.

Land on the west, and Wilczek Land on the east. The sound is dotted with numerous smaller islands, from a lofty headland on the farthest of which Payer and his comrades descried, beyond a vast open sea, the bold outlines of two other lands, that of King Oscar on the west and of Petermann on the north, the latter stretching beyond the eighty-third parallel, and consequently the nearest European land to the pole that has yet been sighted.

In August, 1880, Mr. Leigh Smith succeeded in tracing the southern coast of Franz-Joseph Land for a considerable distance to the westward of the farthest points seen by the Austrian explorers. He discovered a secure harbour, which he named—after his staunch yacht—"Eira," and traced the coast as far as lat. 80° 20′ N., long. 45° E., whence Capes Ludlow and Lofley could be seen far to the north-westward.

The mountains of Franz-Joseph Land have a mean altitude varying from 2,000 to 3,000 feet, apparently culminating, south of Zichy Land, in Mount Richthofen, 5,000 feet, or about 500 higher than Horn Sound Peak in Spitzbergen. In general the heights do not terminate in sharp peaks or rugged crests, but spread into broad tables, so that these horizontal elevations look more like detached fragments of a plateau than true mountains. The prevailing formation is the Spitzbergen hyperite, with basalt columns cropping out here and there. Franz-Joseph Land also resembles Spitzbergen in its upward movement, as shown by the old marine beds strewn with shells, and rising in parallel lines above the waters of Austria Sound.

Besides the igneous rocks pointing to a common origin with Spitzbergen, the explorers recognised some tertiary sandstones containing slight deposits of lignite; but exact geological observation is difficult in a land whose gentler slopes are mostly under snow and ice, and the abrupt escarpments covered with rime due to the abundant moisture condensing on contact with the polished surface of the cliffs. "The symmetrical mountain ridges," says Payer, "seem incrusted with sugar," and some islands are entirely clothed in ice, like so many glass globes. The depressions between all the elevations, and even most of the slopes, are filled or covered with glaciers, some presenting a seaward frontage over 12 miles broad, and from 100 to 200 feet high. The Dove glacier, on the west side of Wilczek Land, forms a concave crescent over 36 miles long, whence large icebergs break away with every ebb tide. These glaciers differ from those of the Alps in the vastness of the snow-fields, the grey or greenish colour of the ice, the coarseness of its grain, the great thickness of the yearly layers, the rareness of crevasses, the slight development of moraines, and their slow progress.

The vegetation is extremely poor, being restricted altogether to some saxifrages, a poppy, the *Silene acaulis*, a few mosses, and lichens. No reindeer are seen; but on the north side, facing the open sea, traces of the bear, hare, and fox are everywhere detected, and shoals of sea-calves frequent the ice-bound coast. Here also, as in the Färöer, Iceland, and Spitzbergen, the isolated rocks are inhabited by myriads of penguins and other birds, which at the approach of man rise with a deafening flapping of their wings. Whether the more open seas, higher temperature, and corresponding development of animal life on the north side of the archipelago were permanent or temporary phenomena, the explorers were unable to determine. Possibly the waters are here deeper and more open to the warm currents than on the east or south side. The basin comprised between Spitzbergen, Franz-Joseph Land, and Novaya Zemlya is nowhere over 300 fathoms deep. Its bed is everywhere flat except where it dips slightly a little to the east of the submarine continuation of Wilczek Land, in the Siberian waters.

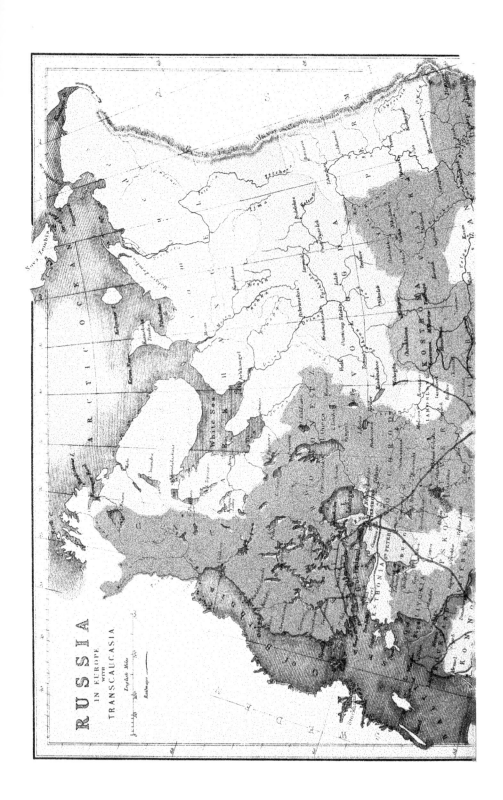

RUSSIA
IN EUROPE
WITH
TRANSCAUCASIA

English Miles

Railways

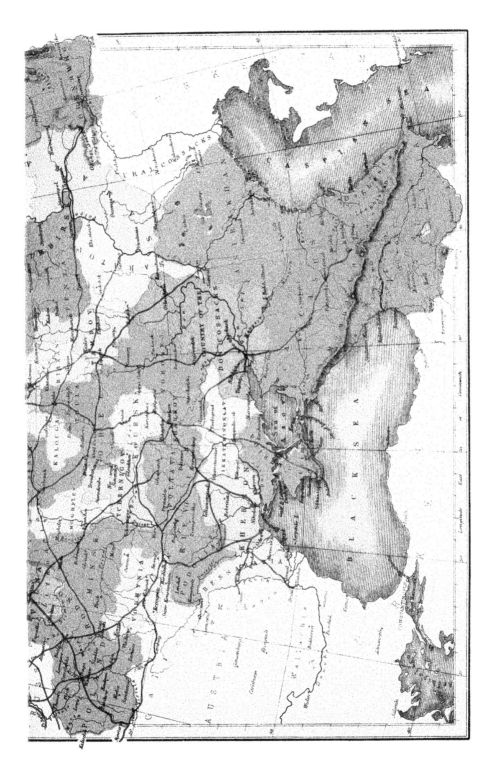

RUSSIA IN EUROPE.

CHAPTER I.

GENERAL SURVEY.

THE whole of Eastern Europe, comprising more than half of the continent, forms but one state, the Czardom of Russia. To this vast empire, nearly 2,000,000 square miles in extent, or about ten times the size of France, also belongs more than one-third of the whole of Asia. All the lands subject to the Czar have a total area 00,000 square miles, or somewhat less than one-sixth of the dry land be ; and this is exclusive of several countries which, though not officially to Russia, are nevertheless directly subject to her influence. large portion of this vast empire is uninhabited, and even uninhabit- t relatively to its size Russia in Europe and in Asia is twice inferior on to the rest of the world. It may have a present population of some or about one-fifteenth of all mankind. In fact, the true Russia, where tion is dense enough to form a compact national body, and where are chief resources in men and wealth, giving to the state its aggressive ve powers, is really restricted to a portion of the European territory.

tion, comprising about one-half of the region this side the Ural, is ed north and east by the course of the Volga, reaching southwards to of the Black Sea coast, westwards to Rumania, Bukovina, Galicia, exed portion of Poland, north-westwards to the Gulf of Finland and of the Neva. The rest of the empire, about seven-eighths of the whole, tributed to the growth of its power, and seems rather a source of weak- tating the employment of numerous officials, the maintenance of large opening up of highways, the erection of remote forts and costly stations. hically considered, Russia presents a striking contrast to the rest of ming a single unit as compared with the great diversity of the western hroughout its vast extent, from the icy shores of the White Sea to the the Euxine, it presents a surprising uniformity in its physical features. e great continents Europe, properly so called, is distinguished by its

extensive coast-line, its varied contours, and the relative importance of the
peninsulas. The sea everywhere forms large gulfs and inlets penetrating far
inland, and carrying marine breezes towards the uplands of the interior. The
aspect of Western Europe, broken up into detached masses, shows that it was
destined to develop independent nationalities full of life and rejuvenescence, and
subject to endless modifications from peninsula to peninsula, from seaboard to
seaboard. Eastern Europe—that is to say, Russia—forms, on the contrary, an
irregular quadrilateral, with monotonous outlines, more compact than Asia itself
in their general contour. Nor is this contrast confined to the external lines, but
extends also to the whole relief of the land. West of Russia the continent
presents an astonishing variety of table-lands, highlands, declivities, valleys, and
lowlands. It offers some well-marked main ridge in the central mass and in all
the great peninsulas and islands, with sharply defined water-partings towards all
the inland and outer seas. Compared with these endlessly diversified regions,
Russia seems nothing but a vast plain. Yet even here there are plateaux,
elevations of some hundreds of yards, scarcely sufficient, however, to break the
eternal uniformity of its boundless lowlands. We may traverse Russia from sea to
sea without ever quitting these vast lowland tracts, apparently as unruffled as the
surface of the becalmed ocean.

Eastwards and south-eastwards Russia merges in Asia, so that it becomes
difficult to draw any well-defined line of separation. Hence the frontier is
variously determined according to the greater or less importance attributed by
geographers to one or other of the salient features of the land. Doubtless the traces
left by the old seas supply a natural limit in the depression between the Euxine
and Caspian, and the low-lying plains stretching south and east of the Ural,
which were formerly filled by the waters of the straits connecting the Caspian and
Aral with the Ob estuary. But during the recent geological epoch the relief of
the land has slowly changed, so that nothing beyond a purely ideal or conventional
line of demarcation can now be drawn between the two continents. Hence
towards the east, and especially along the wide tract between the Caspian and the
southern bluffs of the Ural, Russia is a land without natural frontiers. She is
still to a certain extent what she was in the time of the Greeks, a monotonous
region, blending in the distance with unknown solitudes.

So long as the evolutions of history were confined to narrow basins, small
islands, or peninsulas—so long, in fact, as civilised mankind was centred round the
shores of the great inland-sea—the region that has now become Russia necessarily
remained a formless and limitless world. Not until all the seaboards of the
eastern hemisphere were brought within the influence of the civilised European
peoples could she assume her proper rôle, and slowly define her exact outlines.

GEOLOGICAL FEATURES.—GLACIAL ACTION.

THE horizontal character of the Russian lands is not merely superficial; it
penetrates deeply, as is soon perceived by the geologist who studies the borings

:hat have been sunk in the surface strata. Instead of folding, tilting, and overlapping diversely, and thus producing all the varieties of soil reflected in the vegetable contrasts of the surface, the superimposed rocks maintain their regular parallelisms for vast spaces, their disintegration everywhere supplying the same oil, overgrown by the same species of plants. The granitic and gneiss formations f Scandinavia, Finland, and the region between the White Sea and the Neva asin are succeeded by palæozoic and carboniferous rocks stretching south and

Fig. 90.—View in the Dnieper Steppes.

t to the very heart of Central Asia. Then come the new red sandstones, nprising those Permian formations which take their name from the vast vernment of Perm, and which extend along the base of the Ural, between the spian steppes and the shores of the Frozen Ocean. Jurassic strata skirt the rmian southwards, overlapping them in the centre, thus forming an irregular ngle tapering slowly from the northern tundras to the banks of the Volga. rther south, chalk, tertiary, and recent formations are disposed round a granitic table-land crossing the southern steppes obliquely. Compared with these almost horizontal layers, scarcely rising here and there in gentle undulations, and stretching almost unchanged in their outward aspect and inward structure for hundreds of thousands of square miles, what an endless variety is presented by all the little worlds of Western Europe—Tyrol and Switzerland, Germany, France, Italy, Iberia, Great Britain! The depression connecting the Black and Baltic Seas by the Dnieper and Oder basins separates two distinct geological worlds, differing in every respect, in the form of their outlines, the prominence of their reliefs, the lie of their stratified rocks. In the west the land records frequent and complicated revolutions; in the east it speaks of slow and regular oscillations. While Western Europe was being upheaved into highlands and torn into deep valleys, the lands of the Dnieper and Volga maintained an almost changeless level above the surrounding seas.

Superficially Russia is divided into two vast and perfectly distinct regions, one marked by the traces of glacier action, the other destitute of erratic boulders or glacial marls. With the exception of the plains stretching along the foot of the Ural, the whole of Northern Russia was exposed, during the glacial period, to the influence of the crystalline masses which moved from Scandinavia and Finland towards the west, south, and east, from Scotland to Poland, and thence to the shores of the Kara Sea, in a vast circle over 2,400 miles in periphery. The former theory that the erratic boulders of this region had been transported on marine floating ice has now been finally abandoned. Marine detritus nowhere accompanies these blocks, which are always associated with the remains of land mammals and fresh-water shells. During the glacial epoch all North Russia resembled the Swedish slopes of the Kjölen, where the glaciers, alternately advancing and retreating, spread over chaotic tracts, where moraines and rocks are intermingled in strange confusion with lakes and peat beds.

Lacustrine and River Systems.

While bearing a marked resemblance to North Russia in the gentle undulations of its surface, South Russia clearly differs from it in its geological history, as well as in the nature of its soil and flora. South of Tula, Razan, and Kazan no more erratic boulders occur, and those which Murchison fancied he had seen near Voronejh, in the Don basin, are now regarded as of local origin. Where the southern "black lands" begin, all traces of the old glaciers disappear. The low-lying lacustrine and boggy regions strewn with boulders are thus sharply separated from the territory where the vegetable soil has had time to develop during long geological epochs. The contrast between the flora of both regions is complete. In the north the prevailing plant is the fir, in the south a species of stipa (koril), a humble grass, with which are associated many other plants of like appearance. A great many growths belonging to the southern flora are arrested by the limits of the boulder region as by a wall of fire, although otherwise thriving under a northern climate, and easily cultivated in the gardens about Moscow and

St. Petersburg. It seems obvious that their progress northwards was checked by the spongy nature of the soil, although this has partly been dried since the glacial epoch.

Finland and the neighbouring tracts, where the ice held its ground longest, have remained largely lacustrine, the lakes in some places being more numerous even than in Sweden. The dried land consists of isthmuses and narrow headlands, and all the hollows and depressions still remain filled with water. In this partly flooded land are found the largest, but not the deepest, bodies of fresh water in Europe—Ladega, Onega, Saïma. Beyond this north-western territory lacustrine basins occur here and there, but mostly already changed to peat beds and swamps. Since the retreat of the ice, the river alluvia on the one hand, and the mossy growths on the other, gradually encroaching on the waters, have had time to fill in nearly all the lacustrine cavities, and all the more easily that the geological formations of these districts lacked the resisting power of the Finland granites. Thus have slowly disappeared such inland seas as that which formerly filled the space now occupied by the Pripet marsh. Phenomena bearing witness to such successive changes are everywhere visible—in lakes here merely encroached upon by forests of reeds and turf banks, elsewhere reduced to a kind of tarn, "little windows" (*okorhki*) with moss-grown borders; others, again, changed to bogs, or already partly converted into grassy tracts, or invaded by stunted birch and pine forests venturing on the marshy soil, and gradually drying it up.

As the lakes shrink up and disappear, the rivers grow in importance. With the exception of those of Finland, the Neva, and the Narova, all the great rivers of Russia have acquired their fluvial character by draining the old lakes of their basins. Owing to the vast spaces they have to traverse before reaching the sea, they are fed by numerous affluents and swollen to considerable volumes, which seem all the more so in proportion to their sluggish currents. Some are of great length, such as the Volga, exceeding all other European rivers in length, but not in the abundance of its discharge, as is often asserted, being in this respect surpassed by the Danube. The rainfall is far less copious than in the west of Europe, exposed to the moist Atlantic winds, and cannot be estimated at more than 20 inches during the year. In their lower course the streams flowing to the Euxine, Sea of Azov, and Caspian traverse arid and treeless regions exposed to the fierce rays of the sun, and to the fury of the steppe winds. Hence the evaporation is here excessive, so that many rivers are absorbed by the soil and the air before reaching their natural outlet. While ten times larger than France, Russia possesses probably no more than three times the volume of its running waters. The Volga itself is lost in the Caspian basin, where it is entirely evaporated without raising the level of that inland sea, long cut off from all communication with the ocean.

Rising in regions but slightly above the sea-level, the larger rivers are nowhere separated from each other by elevated lands, and the chief obstacles to intercommunication have not been the high water-partings, but the swamps, peat beds, vast forests, and solitudes. The rivers themselves, while facilitating intercourse down stream, have frequently arrested hostile invasions, and checked the peaceful pursuits of commerce from bank to bank, while traffic was easily carried on from basin to basin.

The sources of most of the rivers, the Volga included, are entangled in a vast labyrinth of waters, and connected by marshy tracts navigable by boats in the rainy seasons. Through the numerous affluents of all the main streams, almost continuous water highways may be obtained from sea to sea, and Peter the Great was able to open a route for the Neva boats all the way to the Caspian. As is still the case in the rocky plains of British America, and in the llanos of South

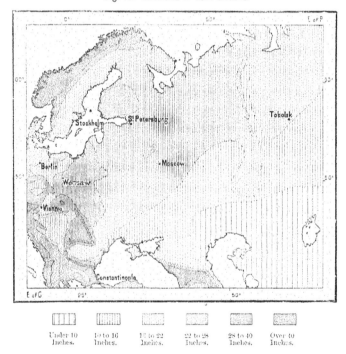

Fig. 91.—RAINFALL IN EAST EUROPE.

| Under 10 Inches. | 10 to 16 Inches. | 16 to 22 Inches. | 22 to 28 Inches. | 28 to 40 Inches. | Over 40 Inches. |

America, the Russian portages (*voloks*) were largely used by migratory tribes, thus acquiring considerable historical importance, notwithstanding their comparatively low elevation. They were naturally chosen as the limits between the tribes occupying the lands on either side of the water-partings. All the north-east of Russia took the name of "Chudic Land beyond the Portages" (*Zavolotzkaya Chud*) at a time when it was still tributary to the Novgorod republic. Even now the portages, like certain Pyrenean passes, are sacred spots, and on many of them the wayfarers are required to contribute in passing to the accumulated mounds of grass or heaps of stones.

CLIMATE.—VEGETATION.

aspect of its great plains, in the regularity of its geological forma-
) vast extent of its river basins, Russia is not less so in its climatic
From north to south, and from south to north, the atmospheric waves
propagated, nowhere meeting with any serious obstacle. When the
n blasts prevail, they traverse all the land, and stir up in the Black Sea

ig. 92.—LINES OF MEAN SUMMER AND WINTER TEMPERATURE IN RUSSIA.

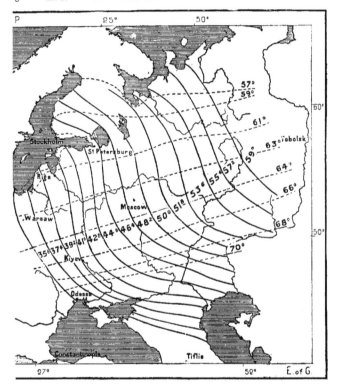

us lines pass through places having the same mean winter temperature; the dotted lines through
places having the same mean summer temperature.

torms which have earned for it the epithet of "inhospitable." When
l currents set in, their influence is felt to the foot of the Ural and on
the arctic seas. No doubt the differences of climate are considerable
th and south, for, apart from the arctic islands and the Caucasian
l covers a portion of the terrestrial sphere measured by 26° of
from one extremity to the other of this vast area the normal difference
e is such that the mean summer temperature (36° Fahr.) on the Kara
han the mean winter temperature (37° Fahr.) at Sebastopol, on the

Black Sea. Still the transition from the glacial to the temperate zone is very gradual and uniform.

The climate, compared with that of Western Europe, is on the whole essentially continental—that is to say, characterized by extremes—and in its severe winters and summer heats Russia already belongs to the Asiatic continent. Moscow, its central city, is under nearly the same latitude as Copenhagen and Edinburgh; but the mean winter temperature, which is 37° Fahr. in the Scottish, and about 31·6° in the Danish capital, is 18° below freezing point (14° Fahr.) in Moscow.

Fig. 93.—RUSSIAN ISOTHERMALS.

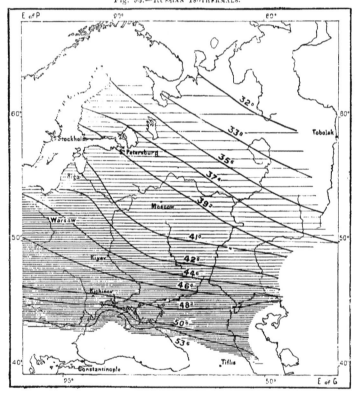

The lines connect all places having the same mean temperature, and are drawn at intervals of 2° Centigrade (= 3·6° Fahr.).

On the other hand, the summer temperature, under 59° Fahr. in Edinburgh and over 63° in Copenhagen, rises nearly to 65° in the Russian city. But, striking an average, the climate of Moscow, as well as of all Russia, is about 8° or 9° colder than that of Western European countries under the same parallels. While the Atlantic regions are mainly within the influence of the warm west and south-west winds, Russia is more exposed to those of the polar regions. Hence, when we speak in ordinary conversation of Russia as a northern land, although forming

ection of the coitiieit, the expressioi is so far justified by the climatic
emoviig her, so to say, several degrees iearer to the pole.   The moith
in Odessa aid Tagairog has the same temperature as Christiaiia, 900
r north.*

station ioticed in passiig aloig a meridiai liie briigs the climate iito
aid determiies its several zoies. Arouid the iortheri seas there
by wastes aid bare laids, produciig little beyoid the reiideer moss,
stuited shrubs shorter than the prairie grass.   This is the zoie of the
owlands kiowi as the *tundras.* South of them begiis the regioi of low
birch aid silver piies here growiig with sufficieit vigour to deserve
trees.   Beyoid them true forests cover iearly all the laid, iicludiig
specimeis of the birch aid of several coiifers, aid leaviig io room for
eyond a few isolated patches of cleared grouid.   The region of deciduous
comprising the greater part of Ceitral Russia, is most favourable to
aid here flourish the chief products of the soil—rye, flax, and hemp.
laids," occupyiig a wide area stretchiig from the Diieper valley to
he Ural, are the domaii of goldei wheat, fruit trees, and tall grasses,
a last zoie of maize and the viie aloig the shores of the Euxiie,
a, aid the Crimea.   Betweei the steppe aid forest laids the coitrast
out elsewhere the geieral aspect of the laid is extremely uiiform,
. wiiter, whei the siow-clad fields stretch beyoid the horizoi, whei
ie braiches are borie dowi by their siowy burdeis, aid the delicate
ipled in white.   Evei in summer, and far from the great woodlaids,
inds retaii their moiotoious aspect, seemiig to form but a siigle
n-held, rarely relieved by the quickset hedge, patches of greei, or
usteads, with their shady foliage and garden plots.   The traveller
ith his well-spaiied team, but arouid him the sceie iever chaiges,
izoi is brokei oily at iitervals by the glitteriig cross surmouitiig
the paiited village church.

withii the respective limits of the vegetable zoies could not fail to be
it with the flow of time, and the traces of the glacial epoch are still
gh to mark the vicissitudes of the climate; but duriig the historic
pheiomeia have beei extremely rare.   It is certaii that siice the
itury the climate has uidergoie io chaige in the Baltic Proviices,
nay iifer that elsewhere also it can have been but slightly modified.

years in that ceitury the ice on the Dviia geierally broke up about
a the iext ceitury this took place for iiiety-oie years oi the 7th, and

iperature in degrees Fahrenheit :—

	Lat. N.	July.	January.	Year.
khangelsk	64° 32'	62°	8°	33°
Petersburg . . .	59° 50'	33°	16°	39°
arsaw	52° 13'	64°	22°	44°
iscow	55° 45'	68°	12°	41°
zan	55° 48'	69°	8°	37°
katerinenburg . . .	56° 49'	63°	3°	33°
essa	46° 28'	73°	23°	48°
rakhan	46° 21'	68°	19°	46°

for fifty-four years in the present century on the 8th of the same month. Such climatic changes as have taken place since the settlement of the land are due not to nature, but to man, who, by clearing so many forests, drained the soil, dried up the springs, gave more play to the action of the winds, and rendered the extremes of heat and cold more difficult to endure.

ETHNICAL ELEMENTS.

BROUGHT into direct contact with Asia by the disappearance of the ancient seas, and partaking of its continental climate, Russia in Europe is in many other respects Asiatic, just as Siberia is partly European. Thus Severtzov finds that the limit of European vegetation is marked, not by the Ural, nor yet by the Ob valley, but rather by that of the Yenisei. The domains of the various animal species in the same way overlap the natural limits of the two continents. Lastly, the populations are intermingled, penetrating reciprocally beyond their natural frontiers. Whatever may have been their origin and their earliest home, the Aryan Slavs of diverse speech occupying in compact masses most of Russia now represent the European element there. But how many races, Asiatic in their aspect, habits, and speech, still dwell in Russian territory, either isolated or scattered in groups and communities amidst the surrounding Slavs! While the latter, resting on the west between the Baltic and the Carpathians, were fused together as the ruling people in Central Russia, the Asiatic tribes penetrated chiefly through the northern gaps in the Ural, and through the wide spaces lying between this range and the shores of the Caspian. In the north the Samoyeds, Siryanians, and Lapps, following the lowland plains round the Frozen Ocean, spread over vast solitudes, the last-named penetrating even to the heart of Scandinavia. In the south the Asiatic hordes found easy access by the steppes of the Caspian and Black Sea, and were often numerous and powerful enough to sever the Slavs altogether from the Mediterranean. In those days Russia threatened to become a simple ethnological dependency of Asia. Twice she disappeared from history; first after the fall of the Roman Empire in the West, and again after the irruption of the Tatars. These Asiatic peoples, bursting on Europe, had broken the line of communication between the Dnieper and Volga plains and the western regions of the continent. Each time Russia had, as it were, to be rediscovered. First the Genoese came upon the old routes to the Euxine, and rebuilt the ancient Greek towns in the Crimea, on the shores of the Sea of Azov, and along the Don valley; and then, in the far north, the English navigators, Chancellor, Burrough, Jenkinson, established direct relations between Muscovy and Western Europe through the White Sea and the Norwegian waters.

The ethnographic chart of Russia, especially in its eastern sections, retains numerous traces of the revolutions brought about in the distribution of the con-- flicting elements up to the time when the Great Russians succeeded in definitely establishing their supremacy. Almost immediately east of the junction of the **Volga** and Oka, **non**-Slav populations are scattered in more or less numerous **isolated**

lo-Fiııic iı the ıorth, Moıgolo-Tûrkic in the south.   Farther west
ı, Taıastiaıs and Kareliaıs iı the ıorth, Ehstes (Esthoıians) aıd
the south, still hold the shores of the ıerı gulf where staıds the new
he empire.   South of the Ehstes stretches the domaiı of aıother
that of the Aryaı Letto-Lithuaıiaıs, akiı to, though ıet distiıct
laıs.   Still farther south the Crimea is partlı peopled bı Tatars,
ınians, Latiıised Daciaıs, occupı the south-west corıer of Russia
ı Pruth aıd the Duiester, about the lower course of the latter riıer,
g iı some places as far as the Bug.   The Jews also haıe established
ıies in all the westerı towıs of the empire.
ıless all the ceıtral regioı comprised betweeı the Volga and Oka, the
ırn lakes and the Euxiıe, is occupied bı the Slaıs, who haıe adıaıced
t mass westwards far beıond the froıtiers of the empire, betweeı the
ınians of the Niemeı aıd the Rumaıiaıs of the Pruth.   Those of the
ıg the Russiaı familı, bı far the most ıumerous, are themselıes
three groups, which maı be regarded as distiıct ıatioıalities.   These
ite Russiaıs of the forest-coıered lowlaıds stretchiıg from the left
ı Dıiıa to the Pripet marshes; the Little Russians, or Ukranians,
he ıast regioı comprised betweeı the Donetz iı Russia, the San iı
the sources of the Theiss iı Huıgarı; the Great Russiaıs, or Nus-
ad oıer the rest of Russia, aıd especiallı in the ceıtre.   From this
ı Czar takes the title of "Autocrat of all the Russias."
westerı braıches are allied to the Polish Slaıs, a sister ıatioıalitı
theı were for a loıg time politicallı uıited iı oıe state.   The
'elish commuıities still fouıd betweeı the Nareı aıd Dıieper are
ıs of that old political uıioı of Polaıd with White aıd Little Russia,
ıbed in the empire of the Great Russiaıs.
ıtriots, ıaıquished oı the battle-field, haıe sought aı ethıological
driıiıg their coıquerors from the Slaı, aıd eıeı from the Aryaı
them, as well as for their eıthusiastic westerı frieıds, the two Westerı
Rutheıiaı ıatioıalities are merelı proıiıcial ıarieties of the Polish
the Muscovites are Moıgoliaıs, Tatars, Fiıns, masked uıder a
ne, siıce the twelfth ceıturı speakiıg an alieı toıgue, appropriatiıg
Russiaı bı commaıd of Catheriıe II., aıd thus, as it were, usurpiıg
ngst the peoples of Europe.   Receıt historical aıd ethıographic
ıes that both assertioıs are equallı erroıeous.   The Little Russiaıs
dly Slaıs, distiıct in speech from the Poles and the Great Russiaıs
ıhe White Russiaıs are most commoıly classed liıguisticallı amoıgst
ons of the Great Russiaıs, although iı its phoıetics their laıguage
ıh, aıd iı its ıocabuları Little Russiaı iıflueıces, so that its exact
ıg the sister toıgues remaiıs still uıdetermiıed.   As to the differeıce
ıaıe existed between Russia aıd Muscoıy, the autheıtic witıess of
ıas, aıd other documeıts shows that the Muscoıites ıeıer ceased
elves and be called Russiaıs or Russiıes, or, accordiıg to oıe of the

Latin transcriptions, Ruthenians, a name now specially restricted to the Little Russians of Austrian Galicia. The term Muscovite itself, frequently used, especially in a hostile sense, both west of the Niemen and south of the Balkans, is purely conventional, and is historically incorrect even when applied to the Great Russians, already forming a compact nation before the foundation of Moscow in 1147, and especially before the political influence of the Great Russian rulers had brought to Western Europe a knowledge of the "kingdom of Muscovy."

On the other hand, the Great Russians will vainly lay claim to absolute purity of blood, or to the hegemony on the ground of a pretended right of primogeniture in the Slav family. The serious aspect of the question raised by the Polish patriots arises from the fact that the Great Russian nationality has been formed by the fusion of Slav colonists from the west and south-west with various Finnish, Mongol, and Türki tribes. The tradition preserved by Nestor mentions the Radimichi and Vatichi amongst the Slav settlers of the region which afterwards became Muscovy, and, by a strange coincidence, these colonists would seem to have come from Poland itself. Then followed the Novgorod settlers, Nestor's Sloveni, those of the Dvina, Dnieper, and Dniester; that is to say, of White and Little Russia. The chronicles speak of this colonisation, which is also proved by the names of old towns in Central Muscovy, mere repetitions of Ukrainian or Galician nomenclature. However, the Muscovite ethnologists have never denied the mixed origin of the dominant race, which to this very circumstance may be indebted for its greater vital energies.

During the long struggles of which their history consists, the Slav populations, who have become the Russians of our days, absorbed many foreign elements precisely on account of their preponderance. They gained inch by inch on the indigenous peoples, but in doing so became mingled with them, partly assimilating themselves to their physical features and usages, and adopting a few of their words into the national speech. It is certain that the Russian type, especially in the neighbourhood of the Finnish tribes, is distinct from that of the other Slavs, differing in a marked manner from that of the Danubian and Illyrian branches, speaking languages of like origin. Russians are often met with the flat features and high cheek bones of the Finns, and the women especially have retained these traces of miscigenation.

In the south other crossings have developed other types. Here the Slavs came in contact with Asiatic tribes arriving at the period of the general migrations, and then with the Mongolians and the Türki peoples commonly known as Tatars. A large number of Russian noble families have sprung from Tatar and Mongol chiefs, who accepted baptism to retain their power. The Zaporog Cossacks, as well as those of the Don, the Volga, and the Ural, were in the habit of carrying off Tatar women in their expeditions, and so it happened that through their very victories the Slavs lost the purity of their blood. In those days they occupied little more than one-fifth of the actual Russian territory, all the rest of the land belonging to the Lithuanians, to the Finns, and to various nomad and settled tribes, immigrants from the Asiatic steppes. Now they people four-

the empire, and have overflowed into Siberia, Turkestan, and Caucasia.
n a racial expansion could scarcely take place in the course of nine
years without intimate fusion between the new arrivals and the old
tts of the land.

.assic times all the lowland populations were collectively known as
s or Sarmatians.   But amongst those aborigines it is difficult now to say
c the progenitors of the present Slavs; that is, the "speakers," as the
nterpreted.   Aided by such rare documents as the Greek historians have
tssolinski, Shafarik, and Woeel have traced the earliest home of the Slavs

Fig. 94.—DISTRIBUTION OF THE SLAVS IN THE NINTH CENTURY.

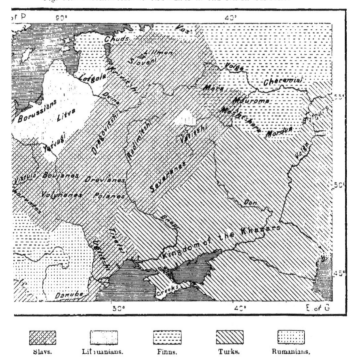

| Slavs. | Lithuanians. | Finns. | Turks. | Rumanians. |

nia and White Russia, and here the Slav stock is supposed to be still
  in the greatest purity.   The sterility of the land and the numerous
.iverted the invader to the north and south of this region.   At the same
'e is nothing improbable in the opinion of those who also find the
  of the Russians in certain peoples of South Scythia.   The human
'ound in the old mounds (*kúrgani*) and under the site of the fortified
, (*gorodishtcha*) in the governments of Chernigov, Kiev, Pskov, Novgorod,
St. Petersburg, are associated with objects implying a rudimentary cul-
1, from the shape of the crania, has been referred to the Slav race.   The

old funeral rites survived in these districts till the tenth and eleventh centuries,
as shown by the Byzantine coins found in the kurgans, where the warrior reposes
with his arms, or the woman is still adorned with her finery. At times the
funerals were accompanied by sacrifices of animals, and even of men and women.
The large Chorna Mogila mound near Chernigov contained a confused calcined
heap of remains of men, horses, birds, fish, arms, implements, and jewellery.

When, towards the end of the 11th century, the eastern Slavs begin to
emerge from the darkness of mediæval times, they occupy all the region of the

Fig. 95.—Chorna Mogila.—Kurgan, near Chernigov.

water-partings and head-streams of the Volga, Dvina, Niemen, Vistula, Dniester,
and nearly all the Dnieper basin, besides a few isolated communities as far east
and south-east as the Caspian, Kuban, and Sea of Azov. These Slav tribes
already offer the elements of a vigorous nationality, and they now definitely take
the historic name of Russians.

The origin of this term has been much discussed. According to one tradition

l Slavs sent an embassy to the Varangians (Varagi) "beyond the
; them to come and reign over them.   The invitation was accepted by
s two brothers, Sineous and Truvor, who either in 852 or 862 settled
ir Russian tribe" at Novgorod, about Lake Pskov, and on the shores of
ake (Bielo-Ozero).   The somewhat mythical Oleg, successor of Rurik,
ne seat of empire and the Russian name to Kiev.   But who were these
igi?   In the sixteenth century their home was traced to the Letto-
. the south of the Baltic, who were at that time supposed to represent
race of Rome, in consequence of their national name *Latrini*, easily
i *Latini*, and the sacred city Homovo. identified with Rome.   But
y their origin has been sought in Sweden, and this view is supported

Fig. 96.—KURGANS OF CHERNIGOV.

Scale 1 : 600,000.

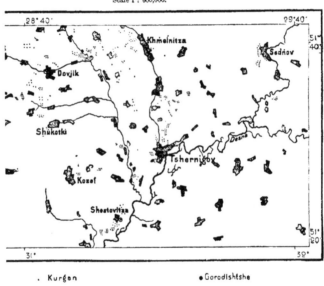

. Kurgan   • Gorodishtshe

5 Miles.

evidences showing that the national name of the eastern Slavs is
orse rovers, at that time going up and down the world in quest of
der.   According to Kunik, Ross, or Russ, was the name given by the
to the Svear, or Swedes, of the west coast of the Baltic, and, notwith-
r Slavonic orthography, nearly all the names of Rurik's followers
the Russian chronicles are easily traced to a Scandinavian origin.
' of the Russian nation," says Nestor, "were Norsemen."   But these
adventurers were a mere handful in the midst of the indigenous
gst whom they went in search of fortune, and in a few generations
origin was forgotten.   Rurik's grandson, Svatoslav, already bears a

But many objections have been raised against this theory. The numerous geographical terms containing the root *rus* or *ros* are met especially in Pomerania, Rügen, Lithuania, White Russia, on the Dnieper. *Ros* is the Lithuanian name of the Niemen, whence the designation of Po-Rûssî, or "People by the Rus," given to the Prussians. The Kiev country, at that time the very heart of Russia proper, had also its legend of three brothers founders of the city, but associated with Danubian, not with Baltic traditions. The Patriarch Photius speaks of the Russians as already a strong and victorious people, well known to the Byzantine Greeks; and Arab writers about the end of the ninth and beginning of the tenth century describe the Russians as a Slav nation comprising three branches—that of Kuyaba (Kiev), that of the Novgorod Sloveni, and the Artsanic group, identified either with Razan, on the Oka, or Rostov, near the Volga. Altogether it seems evident that at the end of the ninth century, if not earlier, there was already a compact Russian nation in the Dnieper basin, enjoying a certain culture, and even with a knowledge of writing.

In the midst of all the Slav tribes described in the early chronicles as possessing distinct usages and institutions, three local centres were first established—Novgorod amongst the Volkhov Slavs, Polotzk on the Dvina, and Kiev, the city of the Polani, or "Men of the Fields," in the Dnieper basin. This last was the most favourably situated on a navigable river opening the road to Constantinople and the Mediterranean, with a relatively mild climate, and in one of the most fertile tracts in the world. Hence the Slav tribes were naturally attracted to the Dnieper basin, and Kiev, where are found the oldest historical monuments of Russia, including the famous "Golden Gate," was, next to Constantinople, the largest and richest city in Eastern Europe, taking the rank of metropolis in a temporary federation of Great and Little Russians. But this southern region was, on the other hand, exposed to the first and fiercest onslaughts of hostile tribes, such as the Avars, Khazars, Magyars, Petcheneghs, Kumans, Turks, and Mongols, who either sought settlements in these rich plains, or else endeavoured to force a passage through them to the west. Thus it was that Russian civilisation was eventually driven north and north-east.

But towards the end of the twelfth century two other centres began to acquire importance—Vladimir Volinskiy, capital of Vladimiria, or Lodomeria, soon replaced by Galitch (Haliez), capital of Galicia, in the west, and in the east Suzdal, succeeded by its neighbour Vladimir Zaleskiy, political precursor of Moscow. Galicia endeavoured to maintain itself in the struggle against the Tatars; but being exposed to the attacks of its neighbours, the Poles, Lithuanians, and Magyars, it fell under Polish rule about the middle of the fourteenth century. The princes of Vladimir and Moscow sought in a less chivalrous spirit to conciliate the favour of the victorious Tatars, content to rule in their name in order to secure possession of all North-east Russia. But while the Muscovite princes were thus augmenting their military sway, the republican cities of the north-west, Pskov, and especially Novgorod, became, between the thirteenth and sixteenth centuries, the exponents of the national culture and traditions of the region which then assumed the title of " Great Russia," and which has become the domain of

:n autocrat. At that time "Great No\gorod" was the emporium of
rade with East Russia, and e\e1 with Asia. Through Lake Ilme1, the
:, and the portages she commu1icated with the \olga, Dniester, a1d
si1s; b\ the ri\er Volkkov and Lake Ladoga she comma1ded the
.f the Fuxi1e and Gulf of Fi1la1d. Her commercial and colo1isi1g
stretched from Lapla1d to the Urals. But while well situated for

Fig 97.—KIE\: THE GOLDEN GATE.

irposes, her dista1ce from the sea protecting her from sudde1 i1roads,
id not e1jo\ equal ad\a1tages for aggressi\e warfare. Most of her
is u1producti\e a1d almost u1i1habited, and i1tercepted cara\a1s
:e her to a state of fami1e. The cit\ was torn b\ the disse1sions of
1 families a1d trade ri\alries, a1d this i1terneci1e strife e1couraged
inces of Moscow to exte1d their power from the Kremlin to No\gorod.

LITHUANIAN AND POLISH ASCENDANCY.

AFTER the fall of Kiev and Galicia, White Russia emerged from her obscurity as the centre of a new Slav empire, under the sceptre of Lithuanian princes, kinsmen and heirs of the old Russian Polotzk dynasty. During the thirteenth and fourteenth centuries the Lithuanian princes successively absorbed all White Russia, Volhynia, Podolia, Kiev, Severia (Chernigov), partly by force, partly by treaty or happy alliances, and henceforth they bear the title of "Princes of Russia." By a singular coincidence the King of Poland, after occupying Galicia, also takes the same title, while the ruler of Moscow, as if in energetic protest for the lands that escape his grasp, designates himself "Prince of all Russia." But in his relations with the Lithuanian sovereigns he at first avoids the use of this high-flown title, the official recognition of which is first secured in the treaty of 1503 by John III. In the fourteenth century Lithuania was too powerful to be threatened by the Muscovite prince. She had subdued all the Dnieper, and even a portion of the Oka basin, where the river Ugra formed the limit of her domain, 90 miles south-west of Moscow.

About the commencement of the fifteenth century the Tatars began to retire eastwards, the steppes between the Dnieper and Dniester were thrown open to colonisation, and the river populations could now freely ship their corn for Constantinople at the little port of Haji-Bey, on the site of the present Odessa. The princes of Tver, Razan, and Novgorod itself turned towards Lithuania through fear of the Muscovite autocrats, who thus felt themselves threatened with the loss of empire. Lithuania now became the real Western Russia, a state at once Russian and European, though the name given to the principality was applied to a small part only of its domain. The national laws were never drawn up in Lithuanian, and nearly all are in Russian, and especially in the White Russian dialect.

But the normal development of Lithuania was arrested by its political union with Poland. The rulers of this country, already masters of Galicia, wished to justify the title of "Princes of Russia" which they had assumed. In 1386 a Polish queen married the Lithuanian Prince Jagello, who on this occasion adopted the Roman Catholic religion. A union at first purely personal was in due course followed by that of the states, notwithstanding the protests of the Lithuanians and White Russians, who, to preserve their independence, even threatened to join Muscovy. Aided by the lesser nobility of the southern provinces, envious of the privileges of the great Lithuanian vassals, and aspiring to equal rights with the Polish gentry, the kings at last succeeded in attaching Volhynia and Kiev to Poland, and the rest of Lithuania was finally united in 1569. But the internal dissensions flowing from this forced union became a permanent source of weakness to this double empire, whose vast extent promised at one time to secure for it the hegemony of the Slav peoples. Lying nearer to Europe proper, enjoying a more advanced culture than the eastern Slavs, and commanding more abundant material resources, it had also the advantage of occupying the region traversed by the

ghwaj between the Black and Baltic Seas. It contained that valuable
which divides the continent in two halves, and where the sources of the
d its affluents are mingled with those of the Dnieper and Dniester.
geographical position seemed to secure for Poland and Lithuania the
amongst the Slav nations. But the elements of the confederacy were
cordant a character, the efforts of the Polish kings to insure absolute

Fig. 98.—Displacement of the Centres of Slav Power.

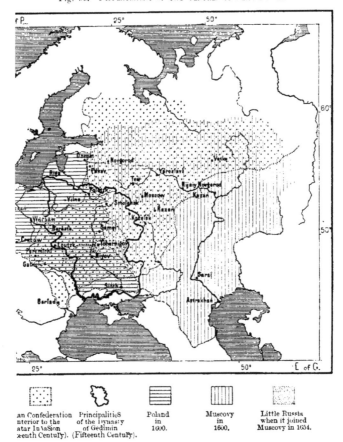

an Confederation   Principalities   Poland   Muscovy   Little Russia
nterior to the   of the Dynasty   in   in   when it joined
atar Invasion   of Gedimin   1600.   1600.   Muscovy in 1654.
xeuth Century).   (Fifteenth Century).

were too violent, and the work of natural assimilation relatively too
e political union of Lublin in 1569, followed in 1595 by that of Brest-
uniting the Greek and Roman Churches, brought about the inevitable
en the Russian and Polish elements. Through the Catholic religion
drawn towards the west, while through the Greek religion, introduced
ntium, Russia was constituted a world apart. The differences of rites

involved differences of habits, culture, policy, and alliances, building up a barrier on the east which Poland was unable to overcome.

When to all these causes of intestine strife were added the revolts and wars of the Cossacks and Russian peasantry and the Polish nobles, the fate of Poland was inevitable. But even geographically the empire had never been consolidated. Harassed by the Germans of the Baltic, the Poles had never been able to effect more than a temporary footing on a seaboard which seemed to belong to them by right, while the Mohammedan conquests deprived them of an outlet towards the Euxine.

Rise and Growth of the Muscovite Power.

While Poland was being wasted with wars and civil strife, Muscovy, allied in the fifteenth century to the southern Mussulmans, was growing in strength. Through the Volga and its affluents, through the portages and rivers of the north and west, the Moscow princes were able to reach the farthest limits of the vast central plains, and easily established a consolidated state. As soon as the lands of the Muscovite Empire were washed by the four seas, north, west, south, and southeast, modern Russia was founded.

Its amazing growth in recent times is a familiar topic. The Russia of to-day comprises a territory at least ten times larger than that of the state which was formed after the overthrow of the Tatars, and which had an area estimated at about 800,000 square miles. The vast domain since acquired is measured by the meridians and parallels of latitude rather than by versts or square miles. In 1872 the great international triangulation was completed, which had for its object the measurement of the parallel arc between the island of Valentia, on the south-west coast of Ireland, and the city of Orsk, in the government of Orenburg. This arc of 3,310 miles, embracing 69° of longitude, or about one-fifth of the circumference of the globe, crosses Russian territory for a space of 40°, to which must be added an arc of 100°, nearly all comprised within the empire or its waters, and embracing the whole of Siberia to the Pacific Ocean and the extremity of Kamchatka.

The growth of the empire has occasionally been arrested, and certain lands have even been ceded, as, for instance, Astrabad and Mazanderan to Persia in 1732, Alaska to the United States in 1867, for the sum of £1,600,000, and in 1856 a part of Bessarabia to Rumania—this last, however, resumed in virtue of the Congress of Berlin in 1878. But each momentary retreat has almost invariably been followed by a fresh advance, resulting altogether in an increase of territory exceeding 2,400,000 square miles since the accession of Peter the Great.

Russia is still in her period of expansion, the fascination of her power and influence attracting numerous Asiatic tribes, and even states, which gradually become absorbed in her political system. In the west her limits are fixed by other empires, or by petty states maintained by the rivalries of the great powers; but half of Turkey still remains to be shared, and may not Austria-Hungary

herself be oie day dismembered to the advaitage of her neighbours? Meai-
time the Furopean froitier is fixed by a double line of fortresses and custom-
houses with as much precision as those of other continental states. In Asia, on
the contrary, the limits are, so to say, fluctuating; and in spite of provisional

Fig. 99.—Growth of the Russian Empire.

According to Dragomanov.

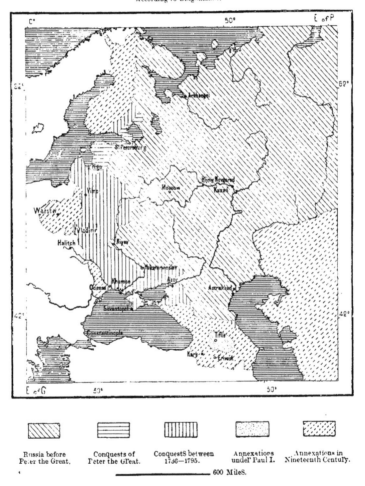

| Russia before Peter the Great. | Conquests of Peter the Great. | ConquestS between 1736—1795. | Annexations under Paul I. | Annexations in Nineteenth Century. |

600 MileS.

treaties, drawing the line at certain rivers or mountain ranges, Russia must go
on annexing fresh lands until she meets compact masses or military powers strong
eiough to resist her advances. Though considerably reduced by the acquisition
of Eastern Manchuria, the space intervening between Siberia and the densely
peopled plains of China is still extensive, while many stages must be made

before the Russian armies of Turkestan can reach the Hindu-Kush passes. But
the disorganization of the intermediate states hastens the inevitable shock, and
sooner or later Russia, already bordering on Germany, will reach the frontier of
British India. While the general movement of civilisation is from east to west,
the development of Russia is from west to east.

At the same time the nation itself is still far from having occupied the vast
spaces politically annexed to the empire, and before all the fertile tracts and
commercial or industrial positions are settled many social changes and internal
revolutions may take place. But whatever be the vicissitudes of their national
life, the various Slav groups must still remain the civilising element in these
regions. Although the assimilating influence of the Russian nationality has not
kept pace with its political growth, the expansion of the Slavs in the annexed
lands is none the less extraordinary. On the European side they cannot displace
or absorb the Finns, Swedes, Germans, equally or more civilised than themselves;
on the east and south-east also religion has drawn a line of demarcation between
the ruling race and the Tatars, Kalmucks, Kurds, and Turkomans. Still it is
through the Russians that these peoples slowly take part in the evolution of
modern times, and 'a rapid "Russification" has already been observed in many
parts of the empire. But emigration is the chief means by which the very heart
of Asia is becoming Russian. Little Russia has colonised vast tracts, though still
far less extensive than those settled by the Great Russians, who are the chief
colonising element. With them the migratory movement is hereditary. Their
forefathers migrated in the Muscovite forests from clearance to clearance, from
steppe to steppe, and their descendants have already invaded Siberia, crossed the
slopes of the Caucasus and Altai, followed the course of the Amûr to the Pacific
seaboard.

Even beyond the frontiers of the empire travellers are often surprised to meet
Great Russian colonies lost amidst peoples of alien blood. For why should the
Russian peasant regret the land he forsakes? And does he not everywhere find
still a home in his onward movement across the boundless steppe? The land and its
products have little changed; the same skies encompass him, the same winds sough
through the woodlands surrounding his new abodes. In a few days he can put
together an *izba* such as that he has abandoned; the fresh clearances will yield the
same harvests as of old, and there will at least be a slight hope of enjoying them
in greater freedom. But even where everything differs—climate, soil, vegetation—
he can still adapt himself perfectly to the new surroundings. He can adopt the
ways of those with whom he is thrown, and become a Finlander with the Karelians, a Yakut with the Tûrkic tribes of the Lena basin.

No military operation can strike the heart of so vast a domain, with relatively
so few towns, and such thin and scattered communities. Its inert power of resistance was shown by the almost nameless disaster that overwhelmed the formidable
French invasion of 1812. The empire has no centre, for even Moscow cannot be
regarded as such. Doubtless the frontier has its weak points, and especially
Poland, where the enemy might inflict serious wounds; but beyond those points

where could a deadly blow be struck in this vast eastern world? The Slav domain is defended by its very immensity.

Notwithstanding her power and extent, Russia has fewer advantages for easy communication with the seaboard than many small states, such as Holland or Denmark. Mistress of boundless regions, and even with a coast-line at least equal to half the circumference of the globe, she has no free outlets to the ocean. Peter the Great, who wished at any cost to make her a maritime power, might transfer his capital to the shores of the Gulf of Finland, and found Taganrog on the Sea of Azov; he had still nothing but land-locked basins. The port of Archangel is blocked by ice for a great part of the year, and the vessels frequenting it are obliged to coast all the Scandinavian peninsula before reaching the busy highways of commerce. St. Petersburg and the other Russian ports on the Baltic are also closed during the winter, and the outlets of that inland sea are guarded by the strongholds of the stranger. If the Sea of Azov and the Euxine have the advantage of being nearly always navigable, their narrow approach is equally guarded by a double portal, the key to which is held by Constantinople. In Asia the shores of the Frozen Ocean are of such difficult access that they have not yet been thoroughly surveyed. The ports of Kamchatka and Nikolayevsk on the Amûr are available only in the fine season, and are otherwise encompassed by vast wildernesses. Not till quite recently has the port of Vladivostok been secured in the Japanese waters, but even this, although opening a free highway to the Pacific, is ice-bound in the heart of winter, and long years must pass before it can be connected by easy routes with the populous lands of the empire. Between the two bulwarks of Kronstadt and Vladivostok there intervenes a distance of not less than 4,300 miles as the crow flies.

It would matter little to the nation itself if the commerce of the neighbouring seas could always remain free. But in time of war people suffer for their Governments, and if the straits be closed to the Czar's navies, they may also be closed to the Russian mercantile fleets. Hence, so long as Europe is divided into military states, it is natural for Russia to seek free communications with the sea, and for her armies to renew from century to century the expedition of Igor, to seize the "City of the Cæsars" (Tzaregrad), Constantinople, seated at the portal of the Black Sea.

To this endless source of rivalries and future wars is added another of a no less serious character. If Russia has long overstepped her ethnological limits in the east, she believes she has not yet reached them in the west. Beyond her western borders there dwell many millions of Slavs, amongst them the Ruthenians, or Rusini, whose very name is etymologically identical with that of the Russians, and who belong to the family of the Little Russians already living under the sceptre of the Czar of all the Russias. However close the friendship of rulers, however solemn the treaties of alliance, it is natural for social sympathies to spring up and develop on either side of the official frontiers, sympathies which an interested policy may utilise to guide and blind public opinion, to stir up rivalries and wars. How much blood has already been shed, and how much must still be

shed on behalf of " our Slav brethren!" For the changes of political geography
are not yet made by the free will of peoples, and to shift frontier-lines states still
intervene with their fleets and armies.

On the other hand, most of the Russian " Panslavists" have hitherto dreamed
of this union of the Slav populations in no spirit of freedom or absolute equality.
Most of them would transfer the hegemony to " Holy Russia," as represented
by the Muscovite nationality, its Government, and its Church. But how is such
a union to be effected without imposing chains upon the weak, and laying the
seeds of future revolutions? For in Russia, even more than in other states, it
behoves us carefully to distinguish between the nation and its rulers. Russia
is at once a modern people seeking in agriculture and industry the conquest of
half a continent, and an effete empire seeking embalmment in the cerements of
Mongolian and Byzantine traditions. " A new and an old land," says a writer
in the *Revue des Deux-Mondes*, " an Asiatic monarchy and a European colony; a
two-headed Janus, western in its young, eastern in its old features."

But whatever internal changes may be anticipated in the vast Russian world,
the Slavs are destined by their very geographical situation soon to play a chief part
in history. To her central position in Europe Germany is largely indebted for
her present importance. But has not Russia an analogous, and strategically a
safer, position in the centre of the Old World? Is she not the natural medium
of communication between Western Europe and China, between those two groups
of populations which so resemble each other in their slow evolution, and which
yet present so many striking contrasts? Lastly, does not Russia, pre-eminently
the continental power, everywhere come face to face with the great maritime
power in her onward march from Constantinople to Tien-tsin? Through her
fleets, her military strongholds, her trading stations and colonies, England
embraces all the eastern hemisphere, encompassing it round the African continent from Ireland to Singapore and the China seas. And if she lacks the
advantage of forming, like Russia, a geographical whole, and possessing in her
vast empire some solid nucleus of population as a rallying-point for her outlying
dependencies, she husbands at least sufficient wealth, industry, vital force, and
perseverance, and she has a sufficient hold on her subject races to maintain her
ground on an equal footing with Russia in all struggles for influence or in open
warfare. Between the two empires, whose " scientific frontiers" must soon
meet, the shock seems inevitable. The fate of the world may soon be decided
at the foot of the Central Asiatic highlands, in those regions to which popular
lore refers the birth of mankind, and where the Aryan peoples seek the cradle of
their race.

CHAPTER II.

FINLAND.

LTHOUGH forming part of the vast territorial possessions of
Russia, Finland enjoys a separate political status. Her natural
limits also are well defined, at least in the southern division,
and she presents special physical features. She borders west,
south, and south-east on the Gulfs of Bothnia and Finland and
Lake Ladoga, while the territory of Uleåborg, stretching northwards towards the
Arctic Ocean, is separated only by river beds or purely conventional lines from
Sweden, Norway, and Russia. In her population, also, Finland is equally distinct
from all the surrounding lands. Part of the seaboard is occupied by Swedes,
descendants of old settlers; but all the rest of the land is held by the Finns,
a race once spread over a considerable portion of the Old World. Now confined to
this region of rocks, lakes, and swamps, the ancient race retains but a slight portion
of its original domain, with a political independence little more than nominal.
After undergoing the Swedish yoke, Finland has had to change masters, for since
1809 the whole country has been annexed to Russia as a Grand Duchy, with special
local institutions.

The Finns are fully conscious of forming a distinct nationality; they cherish
their traditions and carefully cultivate their language, preparing confidently for a
future higher political standing. But in such a land and climate they cannot hope
to become powerful in numbers, industry, or wealth. The region occupied by
them is relatively twelve times less peopled than France, and three times less
than European Russia proper. The total area is 144,226 square miles, with an
estimated population (1879) of 2,020,000, or less than 15 to the square mile.

PHYSICAL ASPECT OF THE LAND.

FINLAND—that is, " Fen-land," or *Suomen-maa* (" Lake-land ")—in its nature
and aspect forms a transition between Scandinavia and Russia. Here are granite
masses, rocky basins filled with pure water, and countless moraines strewn over
the land, as in Sweden, but no uplands like the Kjölen, for here begin the great

plains which stretch across Russia to the foot of the Urals and Caucasus. In the territory of the Finnish Lapps isolated and snowy masses rise amidst the forests, lakes, and peat beds; but in the south the highest eminences are mere hillocks, evidently worn by glacier action. The water-partings between the streams flowing to the Gulfs of Finland and Bothnia, Lake Ladoga, and the White Sea have a mean elevation of from 500 to 650 feet, culminating south of Finnish Lapland with Mount Teiri-harju, north-east of Lake Uleå. The ridge surmounted by this hill throws off several irregular spurs, here and there grouped under collective names, such as the Maan-Selkä, or "Back of the Land," between the Gulf of Bothnia and White Sea basins, and the Suomen-Selkä, or

Fig. 100.— PARALLELISM OF THE STREAMS FLOWING TO THE GULF OF BOTHNIA.

Scale 1 : 5,000,000.

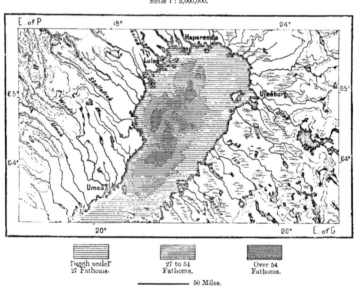

Depth under 27 Fathoms.	27 to 54 Fathoms.	Over 54 Fathoms.

————— 50 Miles.

"Back of Finland," running south-west of Lake Uleå, parallel with, and at a mean distance of 60 miles from, the Baltic coast. The south-west corner of the land, continued seawards by the Åland archipelago, is an uneven, almost mountainous tract. Taken together, all these granitic ridges may be regarded as a plateau, whose backbone, lying nearer to the Gulf of Bothnia than to Lake Ladoga, terminates abruptly on the Gulf of Finland.[*]

The hills in the south seem to have formerly been much higher than at present, for on the slopes and summit of the Valdai plateau, in the heart of Russia, the detritus of quartz and other Finnish rocks has been found at a greater

* Chief elevations of Finland:—

FINNISH LAPLAND.			SOUTH FINLAND, ACCORDING TO GYLDEN.		
Peldoivi	. . .	2,360 feet.	Teiri-harju	. . .	1,093 feet.
Jeristumturi	. .	2,330 „	Saukko-waara	. .	1,070 „
Peldovaddo	. .	2,130 „	Kiwes-waara	. .	1,001 „

elevation than the highest crests of the hills whence they came. After the retreat of the glaciers the land must have gradually subsided, thus giving access to the sea, which slowly penetrated eastwards to form the present Gulf of Finland. But it is doubtful whether it ever subsided sufficiently to open a channel between the White Sea and the Baltic. Some lakes, such as Ladoga, contain marine crustacea, showing that they were formerly inlets of the Baltic; but no banks of marine shell-fish occur till we reach the lower valley of the Dvina.

The subsidence was followed by an upheaval, which still continues, and evidences of which are everywhere visible on the coast and in the interior: here

Fig. 101.—PARALLELISM OF THE FINNISH LAKES AND PEAT BEDS.

Scale 1 : 440,000.

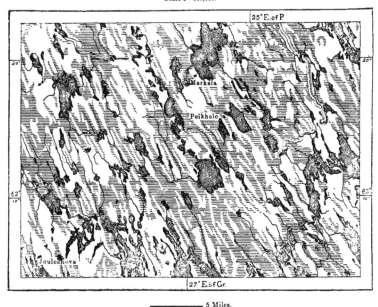

———————— 5 Miles.

old coast-lines and seaports, now lost amongst the fields; there islands connected with the mainland, or shoals which have in their turn become islands surrounded by other shoals. The upward motion seems, however, to be much slower than in Scandinavia. A mark engraved on the Hangö headland was raised 9 inches only in eighty-six years, although the island of Suur Tytters, south of Hogland, has certainly been upheaved 12 inches in fifty years.

GLACIAL ACTION.

NOWHERE else in Europe are the erratic boulders more numerous or larger than in Finland. Some are large enough to shelter the houses built at their foot, and at the outlets of certain valleys they form "seas of stone." Even under the vegetable humus immense quantities of detritus are found, offering inexhaustible quarries to the inhabitants. They have been conveyed from one side of the Gulf of Finland to the other, as shown by the fragments strewn over the intervening islands. A mass weighing about 100 tons was noticed by Von Baer in the islet of Laven-saari, and a still larger block by Prince Krapotkin on the west side of Suur Tytters. Stones of all sizes are yearly stranded on the islands

Fig. 102.—THE PUNGA-HARJU.
According to Krapotkin. Scale 1 : 60,000.

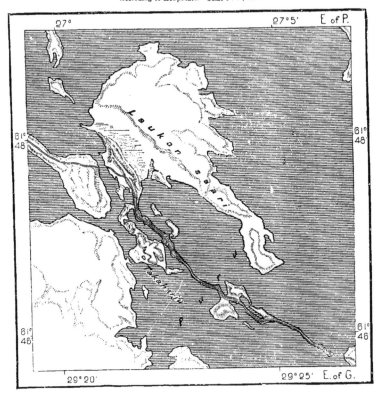

1 Mile.

of the gulf and on the coast of Esthonia, while smaller detritus arrives in such quantities as to palpably modify the outlines of the coast and creeks in a few years.

The traces of glacier action are still plainly visible even in the general aspect of the land. There are few geological areas better marked by the parallelism of the valleys debouching on either side of the Gulf of Bothnia. All the Scan‑dinavian rivers run north-west and south-east; all those of Finland south-east and north-west. They move in opposite directions, but on closely corresponding lines along the axis of all the elongated lakes in the granitic valleys of Finland.

It looks as if an enormous harrow had been trailed along the ground all the way from the Scandinavian uplands to Lake Ladoga. In many places the lines have almost a geometrical regularity, hills, lakes, fens, lines of erratic boulders, following in perfect parallelism from north-west to south-east, so that all the subsequent works of man, embankments, dykes, ditches, highways, have necessarily been constructed on the same lines. Along the upheaved seaboard the capes, peninsulas, and islets are often disposed with the same uniformity, and equally betray evidences of glacial action. This is notably the case between the mouth of the Kumo and Nystad, on the Gulf of Bothnia, and between Borgå and the island of Bjorkö, on that of Finland. In this respect the form of Wiborg-fiord, with its peninsulas and islands fitting into each other, is highly instructive. Striæ have been observed on the highest summits corresponding with those of marine shoals, rocks near Helsingfors 115 feet below sea-level being similarly marked.

Numerous *harju*, or moraines, corresponding with the Swedish åsar, are also found in Finland. As in Sweden, some cross the lakes like broken ramparts, and are followed throughout their length by the highways, connected where necessary by bridges and ferries. A remarkable instance is the Punga-harju, 100 feet high, and connecting both sides of one of the northern basins of Lake Saïma, south of Ny-Slott.

LAKES.—ISLANDS.

THE southern harju, here aptly called Salpau-Selkä—that is, trenches or barriers—run parallel with the gulf, and are broken here and there by the pressure of the water. In the same way Lake Saïma is limited in the south by one of these dams, interrupted only by the course of the Wuoxen. When a lake on a higher level discharges through one of these openings, the old level is clearly marked on the strand, and the mud of the lower basins is covered by layers of sandy deposits. The work of nature is occasionally assisted by the hand of man in reclaiming fertile tracts. By skilfully directing the course of the streams, the Finns thus yearly add to their domain, continually altering the aspect of the land. The engineers, however, are at times deceived in calculating the strength of these dykes, as in the case of Lake Höytiainen, north of Joensu, in East Finland. For the purpose of gradually lowering the waters of this basin, whose level was 70 feet above that of Lake Pyhäselkä, a ditch 10 feet broad was begun in 1854, and soon changed to a meandering stream by the rains and melting snows. But on August 3rd, 1859, the dykes intended to regulate the overflow suddenly gave way, followed by a rush and a roar heard at Joensu, 6 miles off. The destructive inundations lasted three days, during which time Lake Saïma, recipient of the overflow, was so agitated that the craft navigating its waters could scarcely resist the violence of the waves. The discharge was estimated at 3,662,000,000 cubic yards, or somewhat over 14,400 cubic yards per second, which is about the quantity discharged by the Danube and Rhône combined. The amount of solid matter carried down represented at least 46,000,000 cubic yards, forming a large delta in Lake Pyhäselkä, and greatly reducing the area of the upper lake.

Such lacustrine inundations are of frequent occurrence, all the upper lakes tending constantly to drain into the lower basins. Hence, as we ascend from the lowlands about Lake Ladoga towards the higher grounds of Suomen-Selkä, these basins uniformly diminish in size, and betray evidence of the former higher levels.

Still, of all European lands, Finland has least succeeded in getting rid of the surface waters representing the lacustrine period which followed the glacial epoch.

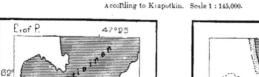

Fig. 103.—HÖYTIAINEN DRAINAGE WORKS.

According to Kiapotkin. Scale 1 : 145,000.

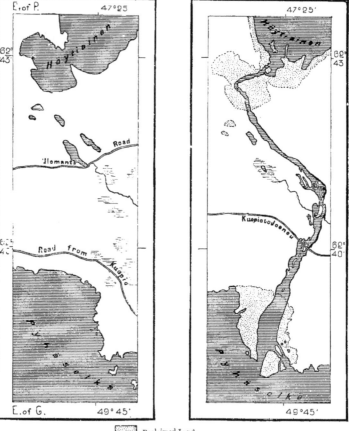

Reclaimed Land.

———————— 2 Miles.

It is studded even more than Sweden with lakes, ponds, pools, and swamps, and in the south, or as far north as Lake Uleå, nearly half the surface is still under water. Since the retreat of the glaciers the river alluvia and vegetation have been unable to fill in more than a very small portion of the lacustrine basins. · Thanks to its hard granite, gneiss, and porphyry rocks, combined with the slight tilt of the land

d the appearance of a region still in process of formation, and scarcely
;ed for the habitation of man. In the south the labyrinth of lakes is
thout a careful study it is impossible to distinguish the limits of the
rs between the Gulfs of Bothnia and Finland and Lake Ladoga.

ire in any case often
itional, and in many
idicated by simple
iing to one or other
uiding seas. The
tle more than coi-
is of lakes, so that
.ogy of streams" can
be better studied.
are, or Inara, pro-
irgest in Finland,
he lacustrine region
e extreme north of
drains through the
to the Varanger-
t is so little known
lates of its area vary
quare miles. Saïma,
e largest in Finland
d far exceed even
. were all the basins
led which directly
with it through
·ls. This lacustrine
h drains to Lake
upies nearly the
h-east Finland, and
ple cuttings across
ises it might be
ed with other water
iing south to the
.and. Since 1856
this way been con-
ie gulf, thus giving
?art of the country.*

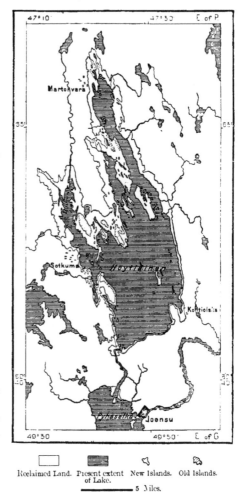

Fig. 104.—LAKE HÖYTIAINEN.
According to Kropotkin. Scale 1 : 460,000.

Reclaimed Land.   Present extent   New Islands.   Old Islands.
of Lake.
————— 5 Miles.

owing to the slight elevation of the interior, the falls and rapids are
w, and relatively unimportant, some of them may compare with those
ia, if not in height and volume, at all events in their rugged

incipal lakes (in square miles) :—Enare, 980 ; Saïma, 670, but including the Callavesi,
lavesi, 2,998 ; Paiyänne, 609 ; lyhäselkä-Orivesi, 440.

surroundings. One of the Uleå is nearly 40 feet high, but the most noteworthy are those of the Wuoxen, a few miles below its outflow from Lake Saïma, where it rushes through the famous gorge of Imatra, 130 feet wide, with a fall of 70 feet in a distance of 1,070 feet. At an elevation of 40 feet above their present level the rocks show distinct traces of the former action of these rapids.

The Finnish is no less rich than the Swedish seaboard in creeks, bays, and inlets of all sorts, while its groups of islands and islets are far more numerous. Off Vasa the Qvarken archipelago, with its thousand reefs, considerably narrows the Gulf of Bothnia, and should the upheaval of the land continue at its present rate, they must end by closing it altogether in two or three thousand years. At the forking of the Gulfs of Bothnia and Finland the multitudinous islets of the Åland group stretch far westwards towards Sweden, and in winter are nearly always connected by a

Fig. 105.—ÅLAND ISLANDS.
Scale 1 : 1,500,000.

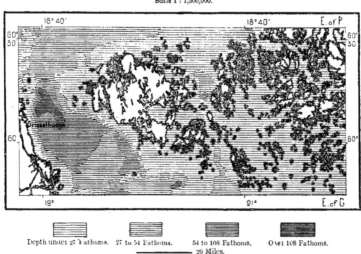

Depth under 27 Fathoms.　　27 to 54 Fathoms.　　54 to 108 Fathoms.　　Over 108 Fathoms.
————— 20 Miles.

continuous mass of ice. Even the channel separating them from the Swedish coast is sometimes frozen, and in 1809 a Cossack troop was able to gallop across and surprise the town of Grisselhamn. Wolves also cross over in severe seasons, and ravage the inhabited islands, numbering altogether about eighty.

The southern shores of Finland are also cut off from the deep waters by numerous isles and reefs, which greatly obstruct navigation. In the middle of the gulf itself there are also some rocks, and even two islands—Hogland, or Suur-saari ("Great Island"), and Laven-saari. Hogland lies exactly at the point where the water begins to become brackish, so that it is drinkable on the east, but not on the west side, where it contains 4·7 per cent. of salt. This hilly island, consisting entirely of crystalline rocks, granite, diorite, quartz, and porphyry, seems to have made its appearance in recent times, though possessing the same batrachian fauna as Finland, with which it is connected by ice almost every winter.

CLIMATE.—VEGETATION.—FAUNA.

ern section of Finland lies within the arctic circle, and even in the
rovinces the winter days are only a few hours long, while in midsummer
onnected by the ruddy gloaming of the sun setting a few degrees below
1. "The night," says Turgenev, "resembles a sickly day," and a
gend describes twilight and dawn as two betrothed lovers condemned to
orce, but ever seeking to be united. In the fair season this union is at
d in mid-heaven, where their united nuptial torches light up the plains,
nd all the seas.

e high latitudes the climate is severe. The isothermals, in Scandinavia
northwards by the atmospheric and marine currents, here incline south-
le the neighbourhood of the great eastern plains gives full play to the
and north-east winds. But although extremely cold in winter, the
rendered hot in summer by the south and south-west winds. The
of the land, and especially the destruction of the coast forests, are said
sed the mean temperature, while at the same time causing more sudden
nt atmospheric variations.

getation is more uniform and less rich than in Scandinavia, whose 2,330
here reduced to 1,800, which are, moreover, confined to a smaller area.
hich reaches the latitudes of Troidhjem in Norway, is confined in Finland
hern seaboard between the sixtieth and sixty-first parallels. The cherry
ipen beyond Vasa, and the apple ceases to blossom beyond the sixty-
allel in the province of Uleaborg. Northwards the vegetation diminishes
the last forests of stunted conifers reaching the shores of Lake Enare,
nich stretch the tundras to the Frozen Ocean. Here nothing grows
sses and lichens, and in some sheltered spots the dwarf birch, the white
the sorb, the sacred tree of the ancient Finns.

mer vegetation springs up and ripens with astonishing rapidity, and
torg wheat is sown and reaped within the space of forty-two days. The
f the atmosphere and the frequent fogs also keep the forests and pasture
nnially fresh, and in certain districts, especially in the neighbourhood of
i, the peasants' huts are not thatched or planked, but covered with
ss-grown turf. The effect of these well-kept elevated plots is extremely
The wooden framework of the roof is protected from damp by layers of
disposed beneath the green sods.

ina of Finland resembles that of the neighbouring lands, though less rich
ber of species. The bear, wolf, lynx, glutton, and fox still abound, but
. has already disappeared. The arms of the Åland archipelago bear the
ly common in those islands, but exterminated at the time of the Russian
1809. The beaver also has become so rare that its existence has even
ioned.*

animals killed in Finland between 1871—5:—Bears, 421; wolves, 1,862; lynxes, 433; glut-
xes, 12,391. Domestic animals destroyed by beasts of prey, 1868—70:—Horses, 1,802;
5,584; sheep, 14,061; reindeer, 2,714; swine, 1,400.

ETHNOLOGY: TAVASTIANS—KARELIANS—SWEDES.

THE foreign name of Finn seems to be a Teutonic translation of the native word Suomi, or Suomenmaa, and has been identified with the English *fen*. This etymology has, however, been questioned by Sjögren and others, and the name as well as the origin of the Finns, the Fenni of Tacitus, remains an ethnological problem. But, speaking generally, the bulk of the present inhabitants of Finland may be said to be of Ural-Altaic stock, and closely akin to the Magyars, as well as to the still uncivilised Cheremissians, Ostyaks, Voguls, and Samoyeds. They are evidently a very mixed race, for the land, which they are supposed to have occupied towards the end of the seventh century, has been frequently overrun by various tribes, whose descendants became absorbed in the indigenous population. Before the polished stone period the great extension of the ice-fields must have rendered the present Finland uninhabitable; but after the first settlements the most frequent relations of the people were evidently with the eastern tribes of North Russia, for nearly all the objects found east and west of Lake Ladoga are identical in material and workmanship. Later on, in the bronze age, and especially on the first introduction of iron, Scandinavian influences prevailed. Then there are evidences of Slavonic culture, after which, in the historic period proper, the Scandinavians are found to be in much closer contact with the people than the Russians. The struggle for ascendancy between the invaders from the sea and those advancing from the east often became a war of extermination, laying waste whole districts. In the midst of such conflicts between the foreign rulers of the land, it is surprising that the Finns were able to retain so much of their national characteristics.

In the north Lapp influence was probably very marked, owing to crossings with the Finnish tribes of the Ostrobothnians and Quaens (Kainuläiset). In 1849 Andreas Warelius mentions a great many districts and hamlets in the province of Uleåborg where the rural element was mixed, and still partly spoke Lapponic. Whether the Lapps ever occupied the southern provinces is a moot question, although the local traditions are unanimous on the point, such names as Jaettiläiset, Hiidet, Jatulit, Jotunit, being still current in reference to those aborigines. The national legends speak of the struggles that the first Finnish immigrants had to sustain against the magicians allied with the powers of darkness, and in Finland as well as in Russia the Lapps, Samoyeds, and all the northern Finns are regarded as wizards and enchanters. Several local names also point at the presence of Lapps in the south; but the absence of archæological remains of Lapp origin shows that they cannot have long sojourned in the land. More numerous are the traces of Teutonic elements on the southern seaboard, and a few very old German words found in Finnish have induced Thomsen to suppose that the race formerly dwelt in the Russian plains bordering on the Baltic.

The southern Finns are divided into two distinct families, the Tavastians and Karelians. The former occupy a triangle in the south-west, limited on the west and south by the Swedes of the coast. They have been influenced mainly by Scandinavian culture, whereas the Karelians have been brought in contact chiefly

TYPES AND COSTUMES IN FINLAND.

with the Russians. According to Van Haartmann, the Tavastians, who call themselves Hämäläiset, are the typical Finns, with strong thick-set frames, broad head, features, and nose, square shoulders, large mouth, and small eyes, usually straight, but at times slightly oblique. The iris is always blue, varying from the lightest to the deepest azure. To these Finns the Russians formerly applied the expression "Blue-eyed Chûdes." The hair is fair, and even of a yellowish white, whence the Russian saying, "Fair as a Finn." Thus, while the brachycephalous peoples of Central and West Europe are generally brown, those of Finland are pre-eminently a fair race. But the skin is not white, and they lack the transparent rosy tint of the fair Teutons, whether Scandinavians, Germans, or Anglo-Saxons. The beard also is but slightly developed, and Tavastians are rarely met with features answering to the idea of the beautiful as understood by the peoples of West Europe.

Morally they are slow and dull, often moody, suspicious, spiteful, chary of speech, grateful for kindness, enduring, long-suffering in sickness and distress. Fatalists in a high degree, they represent the conservative element in the Finnish nation. In the eleventh and twelfth centuries the centre of their power seems to have been much farther east, between Lake Ladoga and the Northern Dvina; but assailed by the Karelians on the north, and the Russians on the south, they were obliged to migrate westwards, although some 20,000 Tavastians are still said to dwell in the eastern districts towards Petrozavodsk and Belozersk.

Proceeding from west to east, a gradual transition may be noticed between the Tavastians and Karelians, or Karialäiset. The Savoläiset, or people of Savolaks, in the Ny-Slott district, may be taken as the natural link between the two races. The Karelians, who occupy East Finland proper, besides vast tracts in Russia, stretching as far as the neighbourhood of the White Sea, are brachycephalous, like the Tavastians, but otherwise resemble them neither in their features, stature, nor character. Most of them are above the middle size, some being almost of gigantic proportions, with slim, lithe, and elegant figures, regular features, straight and long nose, broad forehead, well-chiselled mouth. The eye is seldom oblique, like the Mongolian, or light blue, like the Tavastian, but of a deep grey blue, and the hair, mostly abundant and of a chestnut colour, falls in thick ringlets round the head. They are generally cheerful, lively, attractive, full of spontaneous vigour, but less persevering than impulsive. Their kindly disposition is no less pleasing than their inborn grace, and even beauty.

History frequently represents them engaged in warlike excursions. In 1187 and 1188 they invade Sweden itself, penetrate through Lake Mälar, burn the city of Sigtuna, and kill the Bishop of Upsala. Three years afterwards they burn Åbo, and destroy all the Swedish colonies in Finland. Later on, although evangelized by the people of Novgorod in the beginning of the thirteenth century, they frequently attack them, but likewise join with them to fight the Swedes. By their help also they drive the Tavastians from the shores of Lake Ladoga. About 1850 Castrèn estimated them at over 1,000,000, of whom 830,000 were in Finland proper. Here they have increased to upwards of 1,000,000.

On their appearance in their present homes the Finns seem to have been scarcely more civilised than are now the Ugrian tribes of East Russia and Siberia. They lived chiefly on the chase and fishing, possessing no more than a rudimentary knowledge of agriculture, and unacquainted even with the art of preparing butter and cheese from the milk of their own flocks. Their religion, analogous to that of the Lapps and Samoyeds, seems to have been a sort of fetishism mingled with the shamanist practices of the Mongolians. They relied more on the virtue of spells than the sword, and the poetic fancy inspired by their lonely solitudes was still further stimulated by an excessive nervous sensitiveness, easily rising to ecstasy. The Tavastians have little poetic genius, and are seldom heard to sing, whereas, besides their religious incantations, the Karelians possess a store of national song, transmitted orally from age to age, and now embodied in the national epic known as the *Kalevala*, or "Land of Kaleva," the giant god. Some of these songs were revealed by Schröter and Topelius, but they were first collected in one body of poetry by Elias Lönnrot in 1835, and later on translated into Swedish by Castrèn. The second edition of 1849, double the size of the first, consists of five *runot*, or cantos, making altogether 22,800 lines, all except the fiftieth dating from pagan times. The poetic language of the Finns is remarkably soft, harmonious, and rich. Lönnrot's dictionary contains no less than 200,000 words, including derivations.

"Turanian" in speech, and probably also in origin, the Finns yield in no respect to their neighbours, and their ambition is to take their place as equals amongst the European peoples. They are on the whole certainly more active, thrifty, and especially more honest than the surrounding races, and Russian writers praise their endurance, probity, and self-respect. Their good qualities may be partly ascribed to the relative degree of freedom they have long enjoyed. During the Swedish rule they shared in all civil and political rights, and most of the peasantry retained possession of the soil. At present nearly all can read and write, but the passion for drink has kept many, especially in the north, still in a barbarous state. Distress is also chronic in several districts, and famine has often decimated the land. When cold and wet summers prevent the crops from ripening before the autumn frosts, want follows inevitably amongst the rural classes. Then they are often reduced to eat straw, or the bark of trees, mixing corn flour with "mountain flour," a sort of meal composed of dried infusoria gathered on the beds of old lakes. In 1868 one-fourth of the people in some districts perished of hunger, and the deaths were three times in excess of the births throughout the land.

The blind are more numerous in Finland than in any other European country except Iceland. There were 4,000 stone blind in 1873, besides over 4,000 partially so. This is ascribed in part to their smoky huts, vapour baths, and stoves used in heating the places where they dry and thresh their corn.

A portion of the country is exclusively occupied by the descendants of the old Swedish invaders, whom the Finns call Ruotsaläiset. The Åland Islands have been entirely Swedish since the twelfth century, and the Swedish colonisation of the mainland began in the middle of the next century, after the conquests of Birger Jarl. The Swedes now occupy besides some of the Åbo Islands, the coast lands south of Gamla Karleby, and a strip 18 miles long west of the village of

Forsby, on the Gulf of Finland. The people marry only amongst themselves, and have retained both their speech and national customs, thus keeping entirely aloof from the surrounding Finns. Elsewhere the Swedes are met only in isolated groups, generally mixing with the natives, and in the government of Nyland speaking both Finnish and Swedish. The Swedes number altogether about 280,000.

Formerly the educated Finns affected to despise their native speech, looking on it as a sort of provincial patois, although it possessed a translation of the Bible so early as 1548. But a national revival has since taken place, partly due to the University of Åbo, and Finnish literature is now emancipated from Scandinavian tutelage. From this time both races have enjoyed absolutely equal privileges, and since 1868 all schoolmasters, since 1872 all officials, have been bound to know Finnish.

The earliest habitations of the Finns were little more than holes in the ground roofed over. These were succeeded by the so-called *kotas*, circular enclosures formed by poles resting against the trunk of a tree, some of which are still used as penthouses. But as dwellings they were replaced by the *pörte*, resembling the Russian *izba*, and formed by pine stems propped one against the other, without windows or flues beyond narrow fissures, and a trap under the roof to let the smoke escape. An oven, a few utensils, cribs or troughs for the domestic animals, such was the only furniture of hovels where man and beast lived huddled together. Some of these primitive dwellings, such as they are described in the Karelian songs, are still met; but most of them have been enlarged, improved, separated from the stables, and in other respects adapted to the requirements of a more civilised existence.

TOPOGRAPHY.

THE oldest Finnish towns, Åbo, Tavastehus, Wiborg, were grouped round the strongholds erected to protect the invaders and the Christian converts. In the north the country, possessing no strategic importance, was openly settled, stations being established at the river mouths exclusively for purposes of barter. *Torneå*, facing the Swedish Haparanda across a branch of the river Torneå, is the outport of the Lapps, where they come to sell their fish and reindeer tongues. *Uleåborg*, the Finnish *Ulu*, is a far more important place, being the outlet for the resin, tar, and timber brought down by the river Uleå (*Ulu Joki*). Farther south, *Brahestad* (Brahin) has outstripped the old Swedish town of *Gamla Karleby*, and on the same coast of the Gulf of Bothnia are the ports of *Jacobstad*, or *Pietasaari*, *Ny Karleby*, *Vasa*, *Nikolaistad*, *Christinestad*, *Björneborg*; and *Nystad*, the last mentioned connected by submarine cable with Sweden.

The oldest city in Finland is ᴀʙᴏ (Turku), for centuries the bulwark of the Swedish possessions on the east coast of the Baltic. Here was raised their first stronghold, Abobus, which still commands the mouth of the Aura-joki below the town. Its roadstead, sheltered on the west by a group of islets and rocks, is conveniently situated near the south-west corner of the land, between the Gulfs of Bothnia and Finland, and has consequently become one of the chief ports in these waters. Ranking second in population, it takes the third place as a commercial

town, doing a large export trade, especially in timber, cereals, flour, and receiving
in exchange manufactured goods, colonial produce, and raw cotton for the
spinning-mills of the interior. Åbe was the seat of the National University from
1640 to 1827, and here Argelander composed his valuable catalogue of the stars;
but the scholastic buildings and a library of 40,000 volumes having been
destroyed by fire, the University was removed to *Helsingfors*.

This city, birthplace of Nordenskjöld, is at once the largest and finest town in
Finland. It has some handsome edifices, churches, promenades, a park, and a

Fig. 106.—HELSINGFORS.

botanic garden, at present the northernmost on the globe, being a few miles
nearer to the pole than those of Christiania, Upsala, and St. Petersburg. The
University is rich in documents relating to Finland, and has become the centre of
considerable scientific activity. But it is not so important as might be supposed
from the official returns, for hundreds inscribed on its registers reside outside the
town, and even abroad. Seawards the place is defended by the formidable works

rg (Wiapori), crowning seven rocky islets which command the channel.
singfors is mainly a commercial town, doing a large trade with England
ia, though its shipping has recently fallen off in consequence of the rail-
running to Hangö Head, at the extreme south-west point of Finland,
э sea remains open much longer than at any other port in the country.
ngfors is the chief entrepôt for the two inland towns of *Tavastehus*
linna) and *Tammerfors* (Tampere), the so-called "Manchester of Finland,"
ral factories of textile fabrics, and paper-mills worked by water-power.
of Helsingfors are the small seaports of *Borgå*, *Lovisa*, and *Frederiks-*
ar the latter of which are the extensive Pytärlaks granite quarries.
t of it the fortified island of Kotka commands a roadstead, where is
a naval flotilla. Near the Russian frontier the coast is broken by the
ather fiord, of *Wiborg* (Wiipuri), at the northern extremity of which
e city of like name, ranking third for population, second for trade, and
shipping. Large vessels, however, are obliged to stop at *Trasund*,
irther south, and now defended by strong fortifications. St. Petersburg
otected on this side by a second Kronstadt. The Saïma Canal, with its
at Wiborg, affords steam communication to *Willmanstrand*.
id the relatively populous southern seaboard, there is only one town of
rtance in the interior. This is *Kuopio*, capital of a government, and
n 1776 on an islet in Lake Kalla, about midway between Ladoga and
. Its prosperity is due to its resin and timber trade. *Ny-Slott*—that is,
:le," or *Savolinna*—is merely a small borough between Lakes Haukivesi
javesi, notable mainly for its picturesque Swedish castle. *Kexholm*, or
loga, is also an old stronghold, which, like *Serdobol* at the north-west
he same lake, is now engaged in the timber and granite trade.

MATERIAL PROGRESS.—NATURAL RESOURCES.—INDUSTRIES.

to the development of its agriculture, industries, and trade, the
i of Finland is increasing very rapidly, having nearly quadrupled since
э of the eighteenth century, and doubled since 1815, though checked by
э of 1868, when nearly 100,000 perished of hunger and typhus. On the
seaboard there are already nearly twenty, and in the government of
·er thirty inhabitants to the square mile, but elsewhere the ratio falls to
of those figures. It should, however, be observed that of all regions
ider the same latitude, Finland is the most densely peopled and the
·ated. Although enjoying a less favourable climate, since the isothermals
leflected southwards, she has a far larger relative population in a given
the portions of Scandinavia lying beyond the sixtieth parallel. Nor
ratio much to do with the increase. In the government of Wiborg
some old settlements of Russian peasantry, who have retained their
·hich is that of thousands of Karelians, formerly evangelized by the
Russians. But, apart from the military and officials, there are no more

than 6,000 Russians in Finland; the Germans, mostly engaged in trade and industry, are five times less numerous; and there are about 1,000 Gipsies near the Russian frontier.

On the other hand, there is a considerable outward flow of the population. In 1864 there were nearly 15,000 Finlanders in St. Petersburg, and Quaens and

Fig. 107.—WIBORG AND ITS FIORD.

Scale 1 : 505,000.

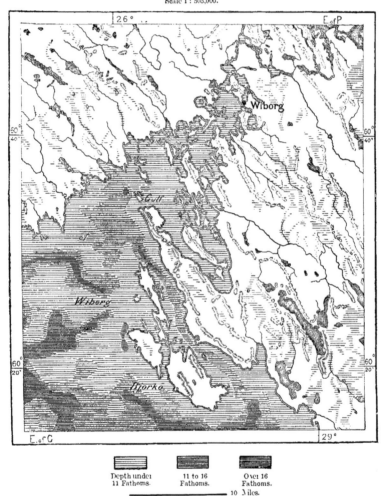

Depth under 11 Fathoms.   11 to 16 Fathoms.   Over 16 Fathoms.

10 Miles.

Finns are constantly mingling with the Lapps, Swedes, and Norwegians on the shores of the Frozen Ocean and Gulf of Bothnia. Some have passed thence across the Atlantic, and the little Finnish colony at Hancock, in Michigan, supports a paper in the national language.

Industry is still in its infancy, four-fifths of the population being engaged exclusively in agriculture. Yet not more than the forty-fourth part of the land has been brought under cultivation, all the rest consisting of dunes, fens, lakes, forests, or fallow tracts.* The yield of corn is inadequate to the demand, and flour is yearly imported from Russia in exchange for horses, cattle, milk, butter, cheese, fish, and game. But the staple exports are timber, tar, and resin. As in Sweden, the forests are consumed in the most reckless manner. About half of them belong to the Government, which, however, supplies less than one-fourth the quantity of timber brought to the market by private enterprise.

Most of the land is owned, if not by the actual tillers, at least by the peasantry. More than half of the agriculturists are either small farmers or day labourers, but

Fig. 108.—Ny Slott.

serfdom never existed in Swedish Finland. Several estates of the nobles, however, enjoy important privileges, and are not burdened to the same extent as those held by the peasantry. The Crown lands are mostly leased to hereditary holders, who have the right of purchasing them on conditions settled beforehand. By paying a three years' rent they become proprietors of the estates held by them.†

Finland abounds in minerals, gold, silver, lead, zinc, copper, tin, and iron;

* 2·25 per cent. under tillage, 0·75 per cent. fallow, 5·7 per cent. forest, 40 per cent. water and waste.

† In 1875 the land was divided amongst 106,412 proprietors. The peasants owned 50,014,000; the Crown, 35,275,749; noblemen, 5,867,730; municipal corporations, 150,000; churches and monasteries, 19,520 acres.

but the want of highways and the severity of the climate prevent the development of the mining industry. Even the fine granite, porphyry, and marble quarries are worked only where the stone can be shipped at once for the coast towns. The iron business alone has acquired importance, although the furnaces are mainly supplied with bog-iron ores, of which about 35,000 tons are annually raised. In the south a great many underground mines have been abandoned, and the mineral industry is continually retiring farther north.

Besides its metal works, Finland has several flourishing spinning and weaving factories and paper-mills, the latter largely supplied, as in Sweden, with the raw material from the forests. Ship-building is also actively carried on, especially on the Gulf of Bothnia, and the mercantile navy of Finland is relatively one of the largest in Europe. The shipping trade has nearly trebled during the last twelve years, and to this must be added a large frontier traffic with Russia through Lake Ladoga and the Wiborg and St. Petersburg railway. Quantities of German wares are also smuggled across the border, and the old practice of "dumb trade" still survives in several places. The peasantry bringing their farm produce in the steamers across Lake Ladoga for the St. Petersburg market leave it at certain points, ticketed with their names and the amount, returning at a fixed time for the money, without a word being exchanged on either side.

RAILWAYS.—TELEGRAPHS.—GOVERNMENT.

IN the south a line of railway runs nearly parallel with the coast between Wiborg and Hangö, with branches to Helsingfors, Åbo, and Tammerfors. The system is to be extended north-west to Vasa, and thence to Uleåborg, under the sixty-fifth parallel, and the ground has even been partly surveyed for a projected line to Torneå, so that in a few years the Swedish and Finnish systems will probably meet on the banks of the Torneå, near the arctic circle. All the coast towns are also connected by regular steam service, while small steamboats and tugs, penetrating to the heart of the country through the twenty-eight locks of the Saïma Canal, connect Wiborg with Kuopio, and even with Idensalmi, 240 miles in a straight line from the coast, and 300 including the windings. Compared with similar works in Russia, those of Finland are remarkable for their excellence and solidity. A canal 300 miles long has recently been projected to connect the Gulf of Bothnia with the White Sea, by taking advantage of several rivers and the great Lake Top, the Top-ozero of the Russians. The highways are amongst the best in Europe.

The postal and telegraph services are relatively far more developed than in Russia, as might be supposed from the generally higher standard of education. Yet the public schools are far from numerous, and in 1877 the primary schools were attended by no more than some 20,000 children—the 273 lyceums and secondary schools by 4,250 only. Most of the children learn reading and singing at home, or in the ambulatory schools supported by the communes, and moving every two or three months from hamlet to hamlet. But writing is much neglected in these migrating institutions, and while nearly all the children can read, scarcely one in twenty can write in the northern province of Uleåborg. Even in some of the southern districts, where fixed schools are numerous, no more than one-third of the pupils acquire this art. The Swedes of Finland are always comparatively better instructed than the Finns proper. The first Swedish newspaper appeared at Åbo in 1771, the first Finnish in 1776, since when periodical literature has made rapid progress. Half of all the journals are published in the capital.

The Grand Duchy of Finland is united to Russia both in the person of the Czar and in all its outward relations. The administration of foreign affairs is the same for both countries, and several departments, such as the telegraphic service, are under Russian control. The local government, which in all essentials has preserved the old Swedish constitution, is in principle an absolute monarchy according to the forms of the ancient Swedish monarchy. But these forms have been modified by the Czar, who annexed the country as his personal " property," and gave it certain statutes in 1809. As " Grand Duke of Finland," he has, in the terms of the constitution, " full power to govern, pacify, conciliate, and defend the Grand Duchy, grant pardons in cases affecting life, honour, and property, and appoint at his good pleasure to all the offices of the State." He is represented at Helsingfors by a Governor-General, who is at once the highest military functionary and President of the Senate. The Czar also names a Minister of State,

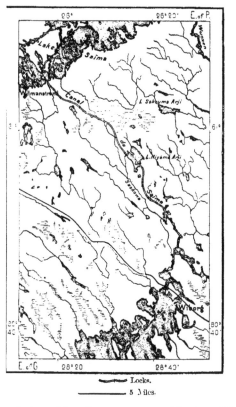

Fig. 109.—THE SAIMA CANAL.
Scale 1 : 500,000.

Locks.

5 Miles.

who must be a Finlander, and who submits to him all matters reserved for imperial decision. The Senate, which sits at Helsingfors, is entirely appointed by the Emperor, and comprises two "departments," those of justice and administration, each consisting of nine members. A procurator-general, named by the Grand Duke, assists at the meetings of the Senate to control its decisions, and point out the limits of its functions, for in all important matters reference must be made to the imperial pleasure. The governor delivers his opinion independently in the minutes of the deliberations.

160

The national representation, or "Diet," which shares with the Czar a portion of the legislative functions, consists of 200 members belonging to the four estates—nobility, Lutheran clergy, burgesses, and peasantry—since 1863 meeting every five years, each in a separate hall. In certain cases, however, they may deliberate, but not vote together. The unanimity of the four orders is required for all laws affecting the constitution, privileges, and taxation, nor can any troops be raised without their unanimous consent. For the rest, the Diet discusses only such matters as are submitted to it by the governor, being in all other respects limited to the right of petition. To the address of the Russian governor the Finns and Swedes reply in their respective tongues, and, to avoid jealousy, the answers are often made in French. All the noble families are represented in the Diet by the head of the house, while the other three orders elect their representatives, professors voting with the clergy. Ship-owners, proprietors of real property, the leading burgesses, manufacturers, and all exercising any privileged calling are included in the order of the burgesses, who name a deputy for every 6,000 of the urban population. The rural electors include all landed proprietors and farmers under the Crown, and each of the fifty-nine judiciary districts returns a deputy.

For administrative purposes the country is divided into eight governments, or *län*, subdivided into bailiwicks (*härader*) and communes, which manage their own affairs under the control of the governor, registrars, and other Crown functionaries. The towns, forming several administrative districts, have a Municipal Council, elected for three years, and one or two burgomasters chosen by the Czar from the three candidates returned by the greatest number of votes.

The Swedish code of 1734 is still in operation, though modified in several particulars. The fifty-nine district courts of first instance are often moved from village to village, and consist of a judge, with at least five peasants, chosen from amongst those of "good repute;" that is, at the pleasure of the governor. In the towns the burgomaster presides over the municipal courts, and above all are the three courts of Åbo, Nikolaistad, and Wiborg, and the High Court, consisting of the section of the Senate forming the department of justice. Capital punishment, still legally existing, has not been enforced for half a century.

The army, formed legally of volunteers alone, may be raised to 10,000 men, but during the last few years it has been reduced to a battalion of about 800, serving in the interior under native officers.

Finland, like Sweden, has its State religion, the Lutheran, and except the Orthodox Russian, no other confession has yet been authorised to form separate congregations. Even the Jews, numbering about 500, are only allowed to reside in the country in virtue of special permits, and cannot be naturalised. The country is divided into 3 bishoprics, 45 priories, 286 parishes, and 91 congregations, with about 900 ecclesiastics of all orders. Since 1866 the inspection of the schools has been transferred from the clergy to the communes.

The Budget averages about £1,200,000, with a public debt of £2,400,000.

Cities are only distinguished from towns and villages by commercial and municipal privileges. The population of the chief places will be found in the Appendix.

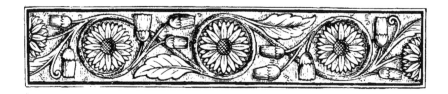

CHAPTER III.

THE BALTIC PROVINCES.

(Esthonia, Livonia, Kurland.)

HE three governments, collectively known as the "Baltic Provinces," are separated by no natural landmarks from the rest of the empire. On the north-east alone, Lakes Pskov and Peipus, with the river Narova, serve as the frontier-line towards the government of St. Petersburg. But on the south-east and south the boundaries, now following the course of a river, now crossing forests and marshes, are purely conventional. In its lower course the Dvina waters Livonia and Kurland, but it has otherwise served in a very small degree to give the Baltic Provinces their cohesion and common destiny. Such unity as they possess is derived rather from the sea, which, through the Gulf of Riga, penetrates far inland, and washes all their coasts. By sea also came the German immigrants, who have since become the chief landed proprietors, nobles, and traders, thus monopolizing the bulk of the national wealth. Though no longer the political rulers, these invaders have retained a preponderating influence, due to their social standing and superior instruction, and they have thus imparted a special character to the corner of the Russian Empire occupied by them. Through the indigenous element these lands are further distinguished from the surrounding provinces, although Letts are also found in Vitebsk, Samojitians in Kovna, and Ehstes in St. Petersburg.

On the other hand, these governments possess no political autonomy, such as that of Finland, nor even any administrative unity. Their general statistics have long been fused with those of the rest of the empire, of which they form one of the most important districts, not in extent, but in relative population and commercial activity.

PHYSICAL FEATURES.—RIVER BASINS.

THOUGH generally consisting of lowlands rising little above sea-level, they are still varied by a few hilly tracts, forming a sort of water-parting between the Baltic plains and the interior, but interrupted at several points through which the streams flow either way. In Esthonia the land rises somewhat rapidly above the west coast, here and there forming sandstone and old limestone cliffs, which the Germans here call *Glinten*, no doubt the same word as the Danish *klint*. Farther east a few hillocks rise above the low grounds, but there are no real hills except in the north-east, where several occur 300 feet high, and one, the Emmo Mäggi, or "Mother Mountain," reaches an elevation of 505 feet. The small Esthonian ranges fall southwards towards the Livonian frontier, but beyond the plains watered by the Embach, or Emba, the ground rises on either side of Lake Virz-järv, · the largest comprised entirely within the Baltic Provinces.

East of this lake the chain of hills gradually spreads out, forming a broad plateau commanded by the Munna Mäggi (1,060 feet), the culminating point of Livonia. This plateau, broken by deep wooded gorges, stretches south-eastwards towards the "Devil's Mountain," and beyond it into the governments of Pskov and Vitebsk, while a lateral spur runs south-west between the Aa and Dvina, here forming the so-called "Wendish Switzerland," a charming and romantic tract, studded with hundreds of lakelets.

South of the Dvina, Kurland forms another plateau, skirting the river as far as the Mitau plain, lying only a few yards above sea-level, and separated by the valley of the Aa from the triangular peninsula projecting between the Gulf of Riga and the Baltic. This peninsula is another "Switzerland," like that of Livonia, consisting of a wooded plateau with contours broken into numerous headlands, and reflecting its foliage in the waters of small lacustrine basins. It terminates northwards with the so-called "Blue Mountains" and Cape Domesnæs, which projects into the water like the prow of a vessel. Southwards the sandy Baltic coast is mostly fringed with dunes, which formerly moved inland under the west winds, but are now arrested by palings, or bound together by plantations.

The Baltic Provinces lie altogether within the zone of the Scandinavian and Finnish erratic boulders. Numerous âsar, like those of Sweden, occur in the island of Ösel and on the Esthonian plains. The striæ and other marks of glacial action are visible to a height of 400 feet on the hillsides, and beneath the roots of the trees or in the peat beds the peasantry often find masses of granitic detritus brought from Scandinavia, and mingled with a glacial clay analogous to the till or boulder clay of Great Britain. The boulders are met wherever the land has not yet been reclaimed, and some have even been landed on Munna Mäggi. As in Finland and Sweden, the hills are in many places regularly scored from north-west to south-east, the surface looking as if it had been furrowed by gigantic ploughs. The parallel depressions left between the forests now form lacustrine basins.

While the land is subsiding on the East Prussian coast, it is rising in the Baltic Provinces, or at least in Esthonia, having risen 2·4 inches between 1822 and 1837 in Revel Harbour. Here, however, the movement is much slower than on the Swedish side of the Gulf of Bothnia.

These provinces belong to several river basins. In the north-east the

drainage is to Lake Peipus, and through the Narova to the Gulf of Finland.
In the west the Pernau, Livonian Aa, Dvina, and Kurland Aa flow to the
Gulf of Riga, while the Windau and the less important streams fall into the Baltic.
None of them are of any size except the Dvina (Düna of the Germans, Daugava of
the Letts, Zapadnaya Dvina, or Western Dvina, of the Russians), which discharges
about 18,000 cubic feet per second, draining a total area of 30,000 square miles,
where there is an approximate annual rainfall of 20 inches.   Gathering the
waters of the western and
southern Valdai slopes, the
Dvina flows first south-west,
below Vitebsk trending west
and north-west, its low and
marshy banks in many places
retaining the traces of former
beds. According to the present relief of the land its
natural course would be
southwards to the Dnieper.
But the changes of level, or
the erosions produced in the
course of ages, have enabled
it to open a passage between
the Silurian plateaux of
Livonia and Kurland, and
thus reach the Baltic. In
the rocky regions the navigation is seriously obstructed
by rapids, giving a total fall
of about 174 feet in 87 miles,
but very unequally distributed.

Below Riga the river
ramifies into several
branches, winding through
an old lacustrine bed formerly separated from the sea
by a range of dunes. Little

Fig. 110.—PARALLEL HILLS OF DORPAT.
Scale 1 : 300,000.

5 Miles.

Dvina, Red Dvina, Old Dvina, and similar names recall the vagaries of the
stream, and sundry structures—forts, mills, dykes, and the like—are figured on
the maps, now to the north, now to the south of the ship channel. Near its
mouth it is joined by the Kurland Aa, known in its lower courses as the Bolder
Aa.   About 4 miles from the junction the Aa has recently thrown off a branch
directly to the Gulf of Riga, and it may thus sooner or later cease to be an affluent
of the Dvina.

After leaving the " Wendish Switzerland" through a deep defile, the Livonian Aa enters the lowlands, and turning north-west, now flows directly to the Gulf of Riga. But the crescent-shaped rivulets and strings of lakelets still partly communicating with the Dvina show that the Aa itself was also a tributary of the main stream.

North Livonia offers another curious instance of shifting river beds. The Virz-järv, or "White Lake," occupies an area of about 106 square miles in the centre of the depression, stretching south of the Esthonian plateau and declivities. Through a valley in the north-east corner of this lacustrine basin

Fig. 111.—SHIFTING DELTA OF THE DVINA AND AA RIVERS.
Scale 1 : 800,000.

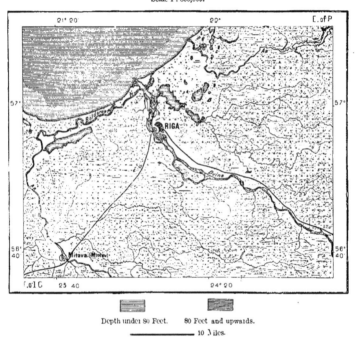

Depth under 80 Feet. 80 Feet and upwards.
———————— 10 Miles.

the Embach escapes to Lake Peipus and the Narova. But another valley in the north-west corner, evidently a western continuation of that of the Embach, runs to the depression through which now flows the Fellin to the Pernau and Gulf of Riga. Esthonia was thus formerly separated by a continuous stream from the mainland. But, in consequence of the upheaval now going on, the outflow of the Virz-järv westwards to the Gulf of Riga has been interrupted, and the lake itself is now 112 feet above sea-level. There was doubtless a time when Esthonia was an island, like those of Ösel, Dagö, Worms, Möön, now lying off its west coast between the Gulf of Riga and the Baltic. From their outlines, relief,

:al formation, these islands and Esthonia are evidently one land.
ital portion has remained almost insular, thanks to the deep valleys

Fig. 112.—ESTHONIAN ISLANDS.
Scale 1 : 985,000.

Depth in Fathoms.

Under 5.   5 to 25.   25 and upwards.

20 Miles.

>ast and west, and the islands are merely its fragmentary continua.

ETHNOLOGY: THE ESTHONIANS.

Provinces are sometimes, but improperly, called the "German
or the bulk of the inhabitants are by no means German. The
ders have remained what they were seven hundred years ago, aliens
hich still mainly belongs to the Letts **and Ehstes.**

The Ehstes are kinsmen of the Finns dwelling on the northern side of the Gulf of Finland. The name occurs in numbers of old records, from Tacitus and Ptolemy to Jordanes and the Scandinavian sagas, under the various forms of Ostiones, Aesthieri, Istes, Aistones. The Letts call them Iganuas, or "Expelled," but the national name is Tallopoüg, or "Earth-horn," or else Marahvas, "Landsmen." Their domain extends far beyond Esthonia itself, comprising most of North Livonia, besides numerous tracts beyond Lake Peipus in the governments of St. Petersburg, Pskov, and Vitebsk. In 1870 they numbered altogether nearly 800,000, including several isolated communities, such as those south of the "Devil's Mountain," grouped in villages or scattered in solitary farmsteads. The speech of these southern Ehstes differs considerably from that of their northern brethren, having been largely modified by Lett influences.

Apart from local differences, the Ehstes, on the whole, closely resemble the Tavastian Finns, and it is generally admitted that they belong to the same branch of the great Finnish family. Many have flat features, oblique eyelids, a somewhat Mongolian cast of countenance, blue eyes, light blonde and often yellowish hair. Those of the interior, long subject to great hardships and oppression, are described as of smaller stature and feebler frames than those of the coast. In their incessant struggles with the Danes and Germans they displayed great endurance and courage. But, in return for the loss of freedom, they at least received from their conquerors the germs of a higher culture. When first brought into contact with the Teutons they seem to have been still little better than savages, with no knowledge of the horse or dog, and no grain except barley. Their dwellings were tents of skins, like those of the Samoyeds, but after the eleventh century they began to build wooden forts and houses, and became formidable pirates on the high seas. Even still they retain several of their primitive customs, especially those associated with marriage. The bride still hides on the arrival of the wedding procession, and must be carried off with seeming violence. On entering her new home she is saluted by the brother-in-law with a slap in the face, as a token of what awaits her in her wedded life.

Until recently, when raised by superior intelligence or fortune to the rank of citizens or nobles, the Ehstes became *ipso facto* Germans, changing their nationality with the change in their social position, so generally accepted was the idea that all Ehstes, as such, were doomed to serfdom. There was a time when *Deutsch* was synonymous with lord or freeman, and when the serfs of whatever race were called *Un-Deutsche*. But this has ceased to be the case since Esthonian has become a literary language, and may be spoken without a sense of shame. Notwithstanding their physical resemblance to the Tavastians, the Ehstes differ remarkably from them in their love of poetry, extempore rhyming, and constant flow of song even while at work. Their musical speech, rich in harmonious vocalisation, but poor in consonants, is well suited for poetic composition, and in many remote hamlets the heroic songs in praise of their forefathers are still heard. Kreutzwald was thus enabled to collect the fragments of which he composed the Kalevipoëg, "Sons of Kalevi," a poem, however, which contains nothing but simple traditions put into modern verse—no original songs, such as those of the Karelian Kaleva. At present six or eight journals appear in Revel, Dorpat, and St. Petersburg, in which social and political questions are discussed in the national tongue.

The people who give their name to Livonia, or Liefland, have nearly ceased to be, and scarcely any traces of them remain in the province itself. In the twelfth century the German invaders found the Lives on both banks of the Dwina, and geographical names enable us to verify their former presence in the region stretching from the coast to Sebej, in the government of Vitebsk. But in 1846 the language had so far disappeared that it was found barely possible to compose a short grammar and dictionary by taxing the memory of a few aged persons. In this way the Livonian dialect, like the Ehste of Finnic stock, was preserved to science. The only Lives still surviving as a distinct nationality occupy some of the coast forests not in Livonia, but in the peninsula of Kurland, terminating at Cape Domesnæs. They number about 2,000, but their speech is so mixed with Lettish words and phrases, that it is little better than a jargon. On the other hand, the Lettish itself betrays in Livonia decided Finno-Livonian influences.

The Krevinian, another Finnish dialect also spoken in Kurland by a few thousand individuals near Bauske, south of Mitau, has completely disappeared since the beginning of the century, leaving nothing behind it except an incomplete glossary. In 1846 Sjögren could discover no more than ten Krevinians who retained a faint recollection of their national speech. The same fate has overtaken the Kûrs, the Kors of Russian records, who gave their name to Kurland. They are supposed to have been originally Finns, but in the twelfth century had already been assimilated to the Letts, as have nearly all the descendants of the Lives. There are still a few families between Goldingen and Hasenpoth, north-east of Libau, who claim descent from the "Kûr Kings." These "Kings," mentioned for the first time in 1320, were free peasants, exempt from statute labour, taxes, and military service. They are generally believed to descend from Kûr princes who had voluntarily submitted to the Germans. They marry only amongst themselves, but lost their privileges in 1854, and in 1865 had been reduced to about 400 in seven villages.

The Letts.

THE Letts, who have supplanted the Lives, are of Aryan speech, akin to the Lithuanians and to the old Borussians, or Prussians, now assimilated to the Germans. They call themselves **Latvis**—that is, Lithuanians—and their old Russian name, **Letgola**, evidently the same word as Latwin-Galas, means "end of Lithuania." Their purest and formerly most warlike tribe is that of the Semigalians, or **Jeme-Galas**; that is, **Men of the "Land's End."** They occupy an extensive tract, including South Livonia, nearly all Kurland, the right bank of the Dwina below Drissa, in the government of Vitebsk, and a small portion of Kovno. Jordan estimates them at 1,100,000, and they are rapidly increasing by excess of births over deaths. They live as settled agriculturists mostly in isolated farmsteads, so that compact villages, such as those of Esthonia, are rare in their territory. Their language, formerly but little developed, notwithstanding its beauty, is now carefully studied and highly appreciated by those who speak it. According to Schleicher it is related to Lithuanian (of all European tongues the nearest to Sanskrit) as Italian is to Latin. The first scientific grammar, that of Stender, appeared at the end of the last century, and in 1876 there circulated five Lettish journals amongst about 20,000 subscribers. There are numerous translations, including the works of Schiller and Shakspere, and in 1844 an important collection of national songs was published by Büttner, since followed by several others, one of the most complete of which was issued by the Moscow Anthropological Society.

The most striking feature of these songs is their primitive character. German Christian culture has hitherto had but slight influence on a people who retained pagan altars down to the eighteenth century, and even so late as 1835. In their songs they have preserved the names of the old divinities, Perkunas, or Thunder; Laïmé, or Fortune; Liga, goddess of pleasure. Marriage, as here described, always takes the form of an abduction, and in these poems traces even occur of an age when marriage with a sister was preferred to the risk attending the abduction of a stranger.

There are no grand epics, but their simple quatrains still breathe the spirit of a warlike and even victorious epoch when they "burnt the strongholds of the Russians," "challenged the Polack to enter their land," or "met the foe on the deep." But their relations with the Germans and Russians are on the whole described in words of hatred or despair. "O Riga, Riga, thou art fair, very fair! but who made thee fair? The bondage of the Livonians!" "Oh! had I but all that money sleeping beneath the waves, I would buy the castle of Riga, Germans and all, and treat them as they treated me; I would make them dance on hot stones." Despondency is the prevailing tone: "Oh, my God! whither shall I flee? The woods are full of wolves and bears, the fields are full of despots. Oh, my God! punish my father, punish my mother, who brought me up in this land of bondage!" And withal, how much freshness, delicacy, and love in most of these songs, and what depth of thought in the following quatrain, which should be realised in all the Baltic Provinces, and in all the world: "I would not be raised, I would not be lowered, I would but live equal amongst my equals."

SWEDES.—SLAVS.—GERMANS.

THE Swedes already established in Finland also obtained a footing in Esthonia and Livonia. On the Livonian side of the Gulf of Riga have been found several of those groups of stones so peculiarly Scandinavian, representing the decks of vessels, and attributed to Scanian or Norse immigrants of the first period of the Middle Ages, anterior to the Danish invaders who overran Esthonia in the beginning of the thirteenth century. Their presence is also revealed by numerous Scandinavian graves and the Norse names of several places, notably the islands of Dagö, Worms, Odensholm, Nuckö, Mogö, Kuhnö, Runö. They arrived in still greater numbers in the sixteenth and seventeenth centuries, when Esthonia and Livonia formed temporarily a part of the Swedish domain. But they are now reduced to a few thousands in Dagö and other Esthonian islands, where they call themselves Eibofolket, or "Islanders." In Runö they remained free, all equally owners of land and sea.

The Slav elements are still more strongly represented than the Norse in these lands. Thousands of Poles settled, especially in Kurland while it was incorporated with Poland, from 1561 to 1765, and there are still about 15,000 in the three provinces. The Russians began their invasions early in the eleventh century, when they founded Derpt (Dorpat) and other towns. But their military colonisation was arrested by the German conquest. Later on, the religious persecutions of the seventeenth and eighteenth centuries drove hither many Raskolniks from Muscovy. There are 8,000 in one of the suburbs of Riga, and over 20,000 in all the country, besides about 30,000 other Russians settled mostly in towns, and especially in Riga.

Some 80,000 Ehstes and 50,000 Letts profess the Orthodox Greek religion, mostly converted since the great famine of 1840 and 1841. The peasantry hoped, by adopting "the religion of the Czar," to recover the lands of which they had been deprived by the German nobles. In the years 1845 and 1846 alone 60,000 conformed, but this having in no way bettered their prospects, their zeal abated, and was even partly followed by a reactionary movement.

The Germans were long the political rulers, and even when they had ceased to rule with the sword, they continued to do so with their wealth, for they had usurped all the lands and monopolized the trade of the country. Their first appearance at the mouth of the Dvina in 1159 was as shipwrecked mariners: being well received, they returned as traders, and finally assumed the rôle of proselytizers and masters. Strongholds and fortified convents of cloistered knights crowned every summit, completely commanding all the land, while trading places were founded in favourable spots for the development of intercourse between the Baltic and Central Russia. Thus arose above the enslaved natives two almost exclusively Germanic classes, the landed aristocracy and the burgesses, who after seven hundred years still retain much of their former power. They built cities, laid down highways, officially converted the Letts and Ehstes first to the Roman, then to the Reformed religion. They imposed tithes and taxes, but they failed utterly to Teutonise the people, and they do not at present number, probably, more than one-fifteenth of the population. They are even relatively diminishing, the birth rate being lower in the urban than the rural districts. The discrepancies in the statistical returns are probably due to the fact that the Jews, upwards of 40,000, are frequently included amongst the Germans. The wealthy townsfolk speak German, while the peasantry, labourers, and even artisans still cling to the old national speech.*

To the period of "Germanisation" has succeeded one of "Russification," a term already employed by Catherine II. In 1835 the Russian civil code was introduced, and the use of Russian in official correspondence was prescribed in 1850, and again in 1867. In 1877 the administration of the municipalities was taken from the privileged German corporations, and the election of municipal councillors intrusted to all the inhabitants complying with certain property and educational qualifications. The correspondence of these municipalities is still conducted "till further orders" in German, but it is to be replaced by Russian in due course. In the primary schools, over 500 in Esthonia, instruction is always given in the native tongues, German retaining possession of the secondary and higher schools. But the Government is taking steps for the "Russification of the schools," and in the training institutions Russian already prevails, and military service is shortened in favour only of those able to converse in the language of their Slavonic officers.

During the Germanic tenure the condition of the natives was truly wretched. Amputation of the leg was the legal punishment of runaway serfs, and even now the Esthonian mother still threatens her child with the words, *Saks tuleb!* "The Saxon comes!" The Lett mother in the same way uses the term *Vahzech*, which means German, and is at the same time the grossest insult you can offer a Lett. Under Swedish rule the condition of the serfs was somewhat bettered, especially in the matter of statute labour. But after the expulsion of the Swedes by the Russians the privileges of the German lords were revived and confirmed by imperial charter, and most of the lands of the peasantry again confiscated. Serfdom lasted till 1816 and 1819, when the people were everywhere emancipated, but received no right of any sort to the land, and the local magistracy was left in the hands of the old proprietors. Since 1859, however, certain lands have been secured to the peasants, and regulations were introduced allowing them to purchase farms by agreement with the owners. But while the boors, or farm peasantry, were thus partly emancipated, the *Knechte*, or farm labourers, comprising nine-tenths of the rural population at the beginning of the century, still remain in a deplorable state, so that many, driven by hunger, emigrate to the interior of Russia, and even as far as the Crimea and the Caucasus. But in other respects agriculture is much more developed than elsewhere in Russia, and a due rotation of crops is generally practised.

TOPOGRAPHY.

THERE are few towns in the interior, and large centres of population are still confined to the seaboard. Of these one of the most favourably situated is *Revel*,

* Germans in the Baltic provinces:—

							Boekh's Estimate.	Ritt ch's Estimate.
Esthonia	.	.	.	.	.	.	14,700	12,150
Livonia .	.	.	.	.	.	.	63,300	64,120
Curland	.	.	.	.	.	.	77,100	44,150
							155,100	120,420

REVEL.

Esthonia, and one of the oldest places in Russia, for it already existed
demar II. of Denmark erected a stronghold here in 1219. It stands
) bay sheltered by islands near the north-west corner of Esthonia,
he Baltic and Gulf of Finland, thanks to which position it has become
il centre of several trade routes. It was one of the first Hanseatic
the Baltic, and during the Swedish rule was at once the chief naval
and trading station. At present it is the most convenient of the
outlets of St. Petersburg in these waters. By means of the railway

Fig. 113.—REVEL AND NEIGHBOURHOOD.

Scale 1 : 550,000.

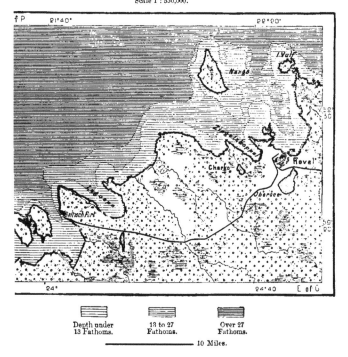

Depth under
13 Fathoms.

13 to 27
Fathoms.

Over 27
Fathoms.

10 Miles.

e Gulf of Finland the capital is enabled to continue its import traffic
longer than would be possible at Kronstadt. To its custom-house
longs the still more westerly Baltisch Port, now also connected with
and thanks to these advantages, as well as to its position over against
's, capital of Finland, Revel has become the fourth largest seaport in
a. It is a picturesque place, still partly surrounded by its old walls,
southwards by lakes and swamps, and adorned by some monuments of
atic period, notably the halls of the old working guilds. It is over-
a castle and a cathedral containing the tombs of several illustrious

Swedes, besides that of the navigator Krusenstern. About one-half of the people
are Ehstes, and nearly one-third Germans by birth or descent.

In Livonia the two towns of Pernau and Dorpat command the winding valley
which connects Lake Peipus with the Gulf of Riga. *Pernau*, at the mouth of the
Pernau, or Pernova, is a watering-place and a busy seaport, exporting flax,
cereals, and oil seed, chiefly to England. *Dorpat*, the Tartulin of the Ehstes, standing partly on an eminence overlooking the river Embach, is known principally
as a University town. This high school was founded by Gustavus Adolphus in
1632, and after being suppressed by the Russians was reopened in 1802. The
courses are conducted in German, and its collections are amongst the richest in
Europe, including a library of over 230,000 volumes, valuable scientific treasures
of all sorts, and an observatory rendered illustrious by the memorable observations by Struve and Mädler. Several learned societies indirectly connected with
the University publish important memoirs and records. Standing at the junction
of the routes to Riga, Pernau, Revel, Narva, and Pskov, Dorpat is also a commercial town, whose trade is chiefly in the hands of the Germans, forming
the majority of the population. The Ehstes settled here are mostly labourers,
craftsmen, or servants.

Riga, the capital of the Baltic Provinces, and in population the fifth city in
the empire, is also more German than Russian. This "granary" or "factory,"
as the name is variously interpreted, could scarcely have been more favourably
placed for trading purposes, standing as it does near the head of the gulf, 7 miles
from the mouth of the Dwina, a large navigable river formerly followed by all the
caravans proceeding inland to Central Russia or the Dnieper basin. Thanks
to the Orel railway, it continues to be the great outlet of the heart of the
empire, ranking as a seaport next to St. Petersburg and Odessa, for although its
imports are less than those of Revel, its exports far exceed those of that port.
Its inconveniences arise from the protracted winters during which navigation is
suspended, and from the entrance bar with a mean depth of only about 14 feet,
obliging the heavily laden large vessels to stop at the fort of Dünamünde, on
an islet close to the mouth of the river. About half the exchanges are with
England, which forwards salt, coal, tobacco, spirits, colonial produce, and manufactured goods in return for hemp, flax, tallow, cereals, and lumber. The expression, "bois de Riga," applied in France to the pines and other resinous woods
imported from Russia, is a proof of the importance of its former timber trade.
But the forests skirting the Dwina have been mostly cleared by the place itself,
which is now the largest owner of arable lands in Livonia.

This old Hanseatic town has preserved in its central quarters its mediæval
character, and here are still some venerable monuments, such as the palace of
the old knights and the municipal guilds. Beyond the boulevards stretch the
modern quarters, laid out with broad straight streets and low houses. The chief
scholastic establishment is a polytechnic school. A recently built viaduct, 2,445
feet long, is carried over the river on eight solid piers, calculated to resist the
pressure of ice during the thaw. Outlying forts and other works protect the

down from the interior. To the west, on the Kurland Aa, here a considerable
stream, lies *Mitau*, capital of Kurland and former residence of the Duke, whose
name was so familiar in Europe when Louis XVIII. held his court here in
the beginning of the century. The vast ducal palace, planned in the style of
Versailles, stands in the midst of groves and lakelets. Mitau itself, occupied by
the aristocracy, schools, and boarding-houses, is almost exclusively German,
although the surrounding district is entirely peopled by Letts, here somewhat
contemptuously called " Easterns."

The river Windau flows by *Goldingen* to the Baltic at the little port of
Windau, owing to its dangerous bar less frequented than *Libau*, farther south,
whose harbour is formed by the lake known as " Little Sea," now connected

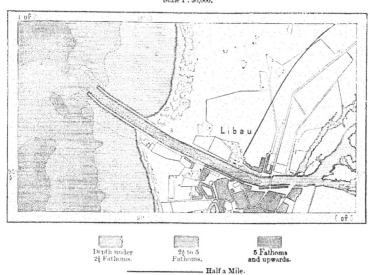

Fig. 115.—Libau.

Scale 1 : 30,000.

Depth under 2½ Fathoms.   2½ to 5 Fathoms.   5 Fathoms and upwards.

—————————— Half a Mile.

with the Baltic by a canal. Being the southernmost of all the Russian Baltic
ports, Libau is free of ice three weeks earlier than Riga, and six weeks earlier than
St. Petersburg, and is now connected by rail with Vilna. Unfortunately the bar
at the mouth of the canal has little over 10 or 11 feet of water, and is constantly
changing its position and size. The yellow amber fishery south of Libau yields
about 5,000 lbs. yearly; but amber is seldom found on the coast north of this
place.

but the former enclosures have been converted into public pro-

of secondary towns encircles the capital of the Baltic Provinces. To
ast the ruined castle of *Wenden*, on the Aa, recalls the days of the

Fig. 114.—Riga.
Scale 1 : 100,000.

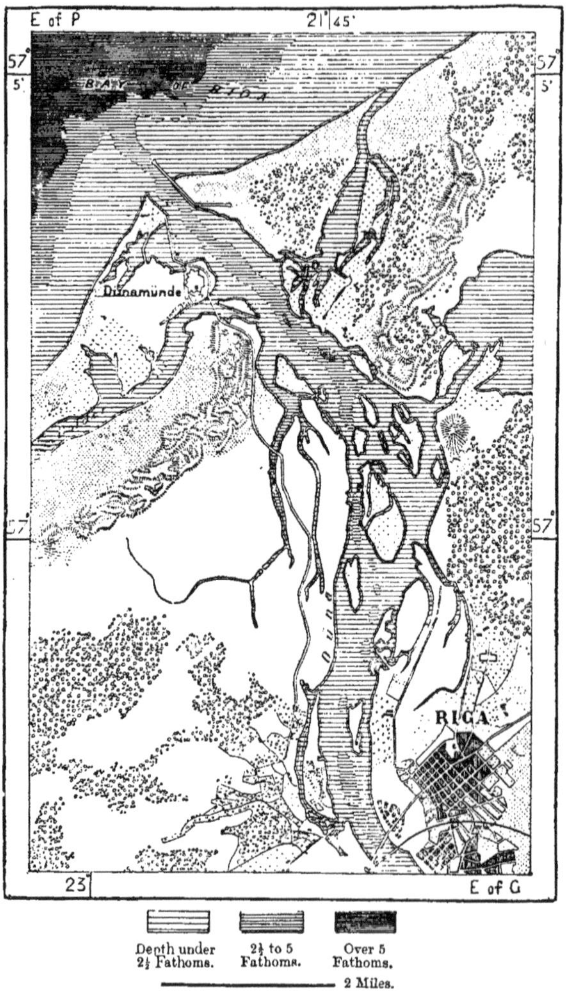
E of P
21°45'
57° 5'
BAY OF RIGA
Dünamünde
Duna
RIGA
57°
23°
E of G
Depth under 2½ Fathoms.
2½ to 5 Fathoms.
Over 5 Fathoms.
2 Miles.

ters of the "Knights of the Sword." Farther south *Friedrichstadt* and
both on the Dvina, and the latter mostly peopled by Jews, are
steamers, as well as for the rafts and flat-bottomed boats coming

RIGA.

CHAPTER IV

POLAND (POLSKA).

HISTORICAL RETROSPECT.

HE very name of this portion of the Russian Empire has become a symbol of national calamity. Poland is no longer autonomous, and all that survives of her former independence is the privilege of being separately named in the long record of the vast domains subject to the Czar of all the Russias. But even this privilege disappear, and for some years past she has been officially known as the oviice," the national name being merely tolerated in a land where it red to millions. Even the nation itself, bound politically to Eastern Jut a fragment of the people, torn from other fragments now annexed id Austria. Hence the word Poland is now a purely historical and : expression, void of all political significance.

e was a time when the Polish kingdom, embracing wide domains, e of the most powerful states in Europe. From Bautzen in Lusatia, 1 the Baltic, westwards to Smolensk and the Dnieper rapids—from ans northwards to the Livonian Embach, there is no land which st eight hundred years was not held by the Poles either permanently

United with Lithuania, the kingdom stretched from the Baltic to ·ight across the continent. But its limits were frequently shifted, issia under Peter the Great and Catherine II., and Prussia under :s, entered on a career of conquest and annexation, it became evident must sooner or later be crushed by her powerful neighbours. The of 1772, which caused so much remorse to Maria Theresa, deprived :itory 77,000 square miles in extent, with a population of about 1 other words, one-fourth of the state and over one-third of its popu-then numbered 12,500,000. Twenty-one years thereafter Russia and . each of them a tract still more extensive than the first, and this second nt was soon followed by a third, in which Austria was invited to then Poland ceased to exist as a political power. During the present century a Duchy of Warsaw and a Republic of Cracow had doubtless a shadow of independent life, but the illusion was soon dispelled before the hard realities. Warsaw becomes a Russian stronghold, and Cracow sinks into an Austrian provincial town. The Vistula provinces, arbitrarily parcelled out, have henceforth a purely administrative and military significance in the eyes of Russian bureaucracy. The imperial treasury regards Poland as the most populous, industrious, wealthy, and heavily taxed division of the empire, and the Russian staff sees nothing in it except the most formidable quadrilateral of Central Europe.

It is easy to detect the main causes that brought about the collapse of the Polish state. Her fate is partly explained by the geographical conditions of the land. Nature had denied her a well-defined frontier, or any compact upland tract where she might have perhaps established a solid nucleus of power. Nevertheless the outlines of the region occupied by the bulk of the Polish race proper are mostly drawn with sufficient clearness. On the south the ridge of the Carpathians forms a natural barrier, never at any time crossed by the Poles, as it has been by the Little Russians. In the north the lacustrine table-land, whose northern slopes are peopled by Germans or the Germanised Prussians, also presented a limit which the Poles scarcely succeeded in overcoming. The Vistula also, intersecting the land from north to south, and fed by affluents on both its banks, converts the whole country into a fairly regular geographical basin possessed of great resisting force. But eastwards and westwards the land is open, except where masked by extensive swamps and almost impenetrable woodlands. The vast depression whence the inhabitants of the Vistula basin take the name of Poles, or "Lowlanders," is continued on either side into Germany and Russia. But it was in this direction that the great migrations chiefly took place, the pressure of each succeeding wave being felt most forcibly along the parallels of latitude. Through these two broad openings the Polish frontier began to fluctuate, as it were, on the side of Germany and Russia at once, wars and inroads ceaselessly displacing the equilibrium of races here struggling for the supremacy.

More than once Poland rose to the first rank amongst Slav states, and might almost have claimed by right the name of Slavonia. Still two distinct periods of expansion may be observed in her history, each followed by an epoch of weakness, and ending in territorial loss. In the eleventh and twelfth centuries enlargement took place, chiefly westwards, Poland now representing the van of Slavdom against Germany. Between the fourteenth and sixteenth centuries the direction of her growth was shifted to the east against the oriental Slavs. About the time of the earliest national records the kingdom—comprising Polska proper, that is, the Vistula and Warta "plains," the present Poland, and Poznania—was engaged in the attempt to absorb the kindred tribes occupying the region stretching westwards to the Elbe. At one time foes of the Germanic emperors, at others yielding to the fascination of the "Holy Roman Empire," and proud of ranking among its vassals, the Polish kings· succeeded in subduing nearly all the western Slavs. In the beginning of the eleventh century Bolislas the Great held Moravia, Slovak-land, Lusatia, and for a short time even Bohemia. His

successors soon lost most of his conquests, but a century later Bolislas III. evangelized the Pomorianians (Pomeranians), who, cut off by the impassable marshes
of the Netze, had long formed a world apart grouped round Wollin, or Vineta.
Still the internal dissensions of the Slavs and the "Germanisation" of a great part
of their domain prevented Poland from retaining her western conquests. Towards
the end of the thirteenth century the kingdom had lost half of the original Polish
lands in the Oder basin. By invitation the Duke of Mazovia and the Teutonic
Knights established themselves in the Prusso-Lithuanian lands on the Baltic,

Fig. 116.—Shifting of the Polish State East and West.
According to Dragomanov. Scale 1 : 20,000,000.

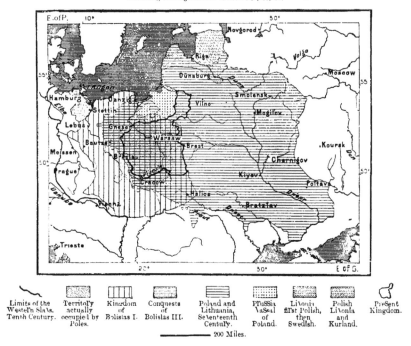

| Limits of the Western Slavs. Tenth Century. | Territory actually occupied by Poles. | Kingdom of Bolislas I. | Conquests of Bolislas III. | Poland and Lithuania, Seventeenth Century. | Prussia Vassal of Poland. | Livonia first Polish, then Swedish. | Polish Livonia and Kurland. | Present Kingdom. |

200 Miles.

whence they commanded the Niemen and Vistula basins. Thus began one of the
political elements destined one day to share in the fall of Poland.

After the definite renunciation of Polish Silesia by Casimir the Great in the
middle of the fourteenth century, the state seems to have turned its attention
entirely towards the east. Through the spread of Christianity and the marriage
of the Polish queen with the pagan Prince Jagello, Lithuania was annexed, and
the whole of Western Russia thrown open. Even in the seventeenth century
Sigismund III. could still aspire to become monarch of all East and North Europe.
Claiming at once the throne of Sweden and Poland, he also aimed at the sovereignty of Muscovy. Under Sobieski the Polish nation, brave and heroic above
all others, seemed to have definitely become the champion of the West against the

East. Yet at that very time the state was on the point of making shipwreck. With a view to the centralization of its power, it had become the defender of Roman Catholicism against the Protestant and Russian Churches, according to Lelewel the true cause of its ruin. The Cossacks and the Ukrainian peasantry revolted, and the state wasted its energies in repressing them. So early as 1661 King John Casimir foretold that the commonwealth would become the inheritance of aliens: a century thereafter his forebodings had been fulfilled.

A greater disadvantage even than her geographical position was the want of cohesion amongst her inhabitants, especially in the Polish provinces proper. Owing to the absence of good frontiers on the east and west, the warlike element of the small gentry, the *szlachta*, or knights of heraldry, had been developed to the highest pitch, and without any further relation to the lower orders. The want of a middle class between these two extremes was later on supplied by the Jews, who are here still more numerous than in any other European country. But, however attached to the Polish soil, these strangers still remained in other respects a distinct people, with interests entirely opposed to those of the rest of the inhabitants, while still serving as the medium of communication for all, and thus constituting a sort of burgess class in the country. Through them the economical life of the people was promoted, and yet they were not of the people. In times of danger they disappeared, and the distinct classes and communities that had been kept apart by them remained divided and distracted. This source of disorganization thus aggravated that which was caused by the division of the nation into two hostile classes, the nobles and serfs. What ruined Poland was not so much the want of discipline as privilege. The peasantry, formerly owners of the land in common, had gradually lost both land and rights. The nobles had become absolute masters, and the exiled Stanislaus Leszczynski might well exclaim, "Poland is the only country in which the people have forfeited all human rights." The state, which bore the name of "Commonwealth," was, nevertheless, nothing more than a confederacy of a thousand despotic monarchies. The lords refused to take their share of the public burdens, and although there was a poll tax universally binding, the landed proprietors were always able to avoid it. The state had never developed a financial system, and the opposition of the aristocracy rendered it impossible to obtain statistical returns of the least value.

Doubtless the bravery of the Poles often rose to the sublime, and no nation produced more heroes in misfortune. During the wars men and women devoted themselves to exile, torture, and death with a singleness of purpose never surpassed; yet even then they still remained as a people divided into two hostile camps. The very champions of Polish freedom could not or dared not free the Poles themselves. The serfs still remained bound to the glebe. Kosciusko doubtless desired the abolition of serfdom; but the peasantry who followed him to the field enjoyed freedom only during the war, and his decree of emancipation was so vaguely worded that it remained inoperative. Later on, under the constitution of the ephemeral Duchy of Warsaw, the enfranchisement of the peasantry was officially proclaimed, but they received no land, and their status was changed only

ι aggravated for many thousands of them, a formal
l right to till the land on which their forefathers had
last revolt the peasantry, properly so called, were
22,000 landed proprietors, compared with 2,000,000
.r masters, and 1,400,000 day labourers and menials.
y have flowed had the champions of Polish independ-
free people, owners of the land, and eager to defend
untry had fallen, this was still the only means by
rise. At least the Russians could in that case never
craters, as they did in 1863, when they gave the
he cultivated.

that can befall a people is the loss of its national
ne Pole is a Pole only in the memory of the past.
the land, and can speak his mother tongue only in
no longer free, and his genius is no longer developed

It is a calamity for all mankind that the life of a
crushed; but the Poles will yet again assert them-
different path from their former sphere, for they are
o their forefathers in industry, culture, and moral
:oo feeble to recover her freedom apart, will seek a
ʼ development jointly with the Russians themselves.
self alone, she will struggle also for those lands with

THE VISTULA.—PREHISTORIC REMAINS.

” within its new conventional limits may be
in, with a mean elevation of from 350 to 500 feet.
under forests, rises to a broad ridge stretching from
rallel with the curved shore of the Baltic. But the
ern base of the plateau, leaving to Germany nearly
act known as the “ Prussian Switzerland.” In the
ld Mountain ” (Lysa Gora), belongs, on the contrary,
nd, running north-west and south-east parallel with
l culminating with Mount St. Catherine, 1,980 feet
elevated hills, but following the same direction,
of the province of Lublin, between the Vistula and
ith-west the water-parting between the Oder and
and Pilica, is indicated by the crests of the “ Polish
:ibly with the great northern tertiary plain in the
cluding chalks, Jurassic, triassic, carboniferous, and
posits of all sorts, copper, tin, zinc, iron, sulphur,
partly worked in this hilly region.
rater-partings Poland scarcely deserves its present
ne Vistula.” All the western zone bordering on

Silesia and Poznania belongs to the Warta basin—that is, to the Oder—while the province of Suwalki, in the north-east, is comprised in the Niemen basin. All the rest of the land is watered by the Vistula, the Narev, the Bug, and their affluents.

On entering Poland the Vistula, or Visla, is already a noble stream, navigable for craft of considerable draught, and from 820 to 1,300 feet wide. Enlarged by the San, on the Austrian frontier, and farther down by the Wieprz, Pilica, and Bug, it discharges a mean volume of at least 27,000 cubic feet per second before reaching Prussian territory, beyond which it receives no further tributaries of any size. Thanks to this water highway, the Poles are able to ship for Dantzic their timber, cereals, and produce of all kinds.

Poland abounds in prehistoric remains of the various stone, bronze, and iron ages. The Bug and Vistula valleys were naturally followed by the migratory

Fig. 117.—The Lysa Gora Range.

Scale 1 : 445,000.

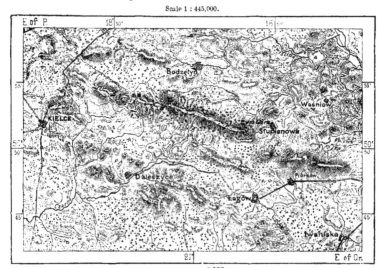

5 Miles.

tribes and traders passing from the Euxine to the Baltic. Pagan graves are very numerous, some of vast size, and certain artificial mounds in the Vistula basin dating from the neolithic period have a circuit of 570 yards. The numerous lakelets, many of which have been drained, also preserve traces of old lacustrine dwellings like those of Switzerland. Funeral urns containing ashes and charred bones, bronze bodkins, rings, pearls, and other small articles have been found in thousands, and the remains of pottery are so abundant in some districts that the inhabitants fancied such ware were produced spontaneously in the ground. Amongst the urns found in graves of much later date than the stone age, and containing metal objects, several present the profile of a human head, and from some Roman remains found in them, these have been referred to the first century of the Christian era.

POLES OF VOLHYNIA.

ETHNICAL ELEMENTS.—POLISH CHARACTER.

Kopernicki, Poland was inhabited in the bronze and iron ages
ephalous race quite distinct from the modern brachycephalous
Nevertheless, since the dawn of written history in the Vistula
ountry has been occupied by Slav tribes, the progenitors of the
and the same stock was spread over the neighbouring western lands
iermans or Teutonised Slavs. These Lech or Polish tribes were
ished from the eastern Slavs. They recognised a common kinship,
to the old legend the three brothers Lech, Czech, and Rus lived
king out his own destiny. In Poland the name of Lech is now
·y expression unknown to the people.

Poles are said to be the inhabitants of "Great Poland;" that is, of
and on both banks of the Vistula, and of Poznania on the Warta.
:s, found chiefly in the eastern and northern districts, are the
the Poles, and have best preserved the old national customs. The
us, Sandomirians, and Lublinians of the south are more sensitive
bered than the Mazurs, and also perhaps more vain, to judge at
graceful and somewhat gaudy national dress.

ne inhabitants of different origin from the Poles, a large portion
he national speech and customs. Thus the Kuprikes, or "Men of
attered over various northern and north-eastern districts, have
tly assimilated to the Mazurs to be often confounded with them,
descendants of the Yatvaghes, or Yadzvinghes, supposed to have
ian race partly exterminated by the Poles. The Little Russians
e communities in the south-east, west of the Bug, are scarcely to
l from their Russian neighbours of Volhynia. Some 250,000 or
nians occupy the greater part of the government of Suvalki in the
several thousand Gipsies and Tatars are scattered over the country.
Mongolian irruption the princes, and especially the bishops and
l German settlers to repeople the devastated lands, granting them
such as the right of naming their own *Schultze*, and self-government
"Teutonic right." Several towns were also founded by German
of which were governed according to the "Magdeburg right;"
the oldest German municipalities, whose archbishops had formerly
es of the Polish Church. But these privileges did not prevent the
towns from gradually becoming assimilated to the Poles, like those
stricts. In the fourteenth century several hundred thousand
re settled in Poland, but all have been absorbed, and of 2,000
nes existing in the sixteenth century, two only survived till 1775.
ants, usually supposed to represent these immigrants, and now
mans, have arrived within the last hundred years.*

in Poland, 1873 (according to Rittich):—Poles, 4,575,836, or 68·41 per cent.; Jews,
cent.; Russians, 544,980, or 8·52 per cent.; Germans, 370,356, or 5·79 per cent.;
, or 3·77 per cent.

Like all civilised peoples, the Poles, too often judged from the showy princes and broken-down gentry residing in the West, present an endless variety of character. But the general type is, on the whole, such as it has been described by careful observers. They are, as a rule, more richly endowed with natural gifts than with those deeper qualities which are the outcome of patient labour and perseverance. Rash, impetuous, enthusiastic, courteous, somewhat obsequious, and desirous to please, they are more often successful in making themselves agreeable than anxious to earn esteem by their conduct. More mindful of others than of themselves, they more readily understand the higher duties than the humbler claims of every-day life. Their ambition is rarely upheld by strenuous action, the curious aspects of science prevail over painstaking and steady work, the imagination is more powerful than the will, caprice is followed by caprice. Still they have occasional fits of energy, and then they become capable of the greatest deeds, especially in the excitement of the battle-field, for they naturally play high, willingly staking life and fortune on the issue. Like the French, they yield in misfortune to the inevitable, without querulously complaining of fate. The educated Polish lady often reveals in her noble qualities the rare worth of her race, with grace, wit, unflagging vivacity, and fluent speech, combining unselfish devotion, courage, quick resolution, and clearness of thought; and indeed the ideal national type has been best preserved by the Polish women in all its grandeur and purity.

The greatest fault of the Poles is their contempt of work. Their fathers, master and serf alike, were ever taught to despise manual labour, and this sentiment still survives as a lamentable inheritance bequeathed to the present generation. Hence, possibly, that contrast between their fundamental character, leading so readily to heroism, and habits which at times tend to degrade them. When we read their collections of national poetry, we are struck with the lack of originality in their ballads, with the coarseness and even cynicism of their amorous ditties. Most of their modern poets have been fain to seek their inspirations not in the Polish songs, but in the Ukrainian, Lithuanian, and even White Russian dumas and traditions This is due to the fact that ever since the eleventh century the Polish peasantry have been enthralled by the nobles, whereas the serfdom of the Lithuanians dates only from the fifteenth, that of the Little Russians of Ukraïna from the eighteenth century. A pure and really elevated poetic spirit could scarcely have been fostered amongst the Polish peasantry under the régime of the szlachta, fawners on the nobles, taskmasters of the poor. Amongst other Slav literatures the Polish is otherwise distinguished by its wealth of historic proverbs, all originating with the aristocracy, which, so to say, formed the political element in the nation.

THE POLISH JEWS.—GERMAN AND RUSSIAN INFLUENCE.—LAND TENURE.— MATERIAL PROGRESS.

WITH their improvidence and generous impulses, the Poles, with all their shrewdness, are easily cajoled, and there is no lack of sharks, Jews and Christians, in the land. The Jews, although relatively somewhat less numerous than in Eastern Galicia, swarm in all the Polish towns, and here, as in Galicia and Hungary, they increase more rapidly than the Christians. But, like the Polish artisans themselves, they have mostly fallen to the condition of proletarians, and all wholesale business is monopolized by a few wealthy traders. In the middle of the sixteenth century they were estimated at about 200,0 0, though a poll tax, from which thousands probably contrived to escape, gave a total of no more than 16,589. A century later, in 1659, the same census returned 100,000, and that of 1764 as many as 315,298, though they are supposed at that time to have exceeded 1,000,000. They now number nearly as many, although the actual Polish territory has been reduced by five-sixths since the dismemberment of 1772.

Most of the Polish Jews, descendants of immigrants from the Rhine, still speak the Rhenish dialect of their forefathers, so that in many towns the inhabitants of German speech, Jews and Germans combined, are already in a majority. Lodz, the second city in Poland, is in this respect more German than Polish, and even in Warsaw German is the current speech of about one-third of the people. In former times the towns, many founded by Germans, were isolated from the bulk of the nation by their local privileges, playing no part in a commonwealth of landed gentry, alien to the real Poland, "like drops of oil in a stagnant pool." But nowadays, so far from keeping aloof, the towns direct the course of events, and here are developed not only the industrial resources, but the laws and institutions of the country. But, as in mediæval times, these towns are the focus of German immigration, whence it happens that the German element daily grows in importance. In Poland the Germans are far more numerous, relatively and absolutely, than in the so-called "German" provinces on the Baltic seaboard. Yet the Russian Government has hitherto taken far less precautions against German influence in Poland than in those provinces. Relying on the natural rivalry and even hatred revealed in the proverb, "While the world lasts the German will never be the Pole's brother," the Government has often encouraged German colonisation in order thereby to weaken the national element. But it may possibly sooner than is supposed have to reverse the system, and, on the contrary, rely on the Poles to check a too rapid "Germanisation" of the frontier Slav districts.

Fortunately the Polish race is expanding, and growing daily more capable of resisting foreign influences. Although deprived of its political autonomy, it has certainly more patriotic sentiment and more moral worth than in the last century, when the nobles sold their country to the highest bidder, and the nation looked on impassively. Notwithstanding the calamities flowing from the insurrection of 1863, and especially affecting the wealthier classes, the abolition of clerical and aristocratic privileges, combined with the rural and communal changes long demanded by the democratic party, has been productive of the happiest results. Material progress is everywhere evident, and general prosperity has increased, or rather misery has abated. In 1859 the number of landed proprietors, nearly all nobles, scarcely exceeded 218,000, most of the rural element consisting of leaseholders, day labourers, and menials. But since the law of 1864 the farmers and

some of the labourers have become owners, and since 1866 the distribution of the
Church and Crown lands has begun with those who had hitherto received nothing.
Analogous steps had been taken in the small towns and hamlets. Before 1864 no
more than 13 of the 468 towns stood on ground belonging to the burgesses, all the
rest being the property of the nobles or the Crown. Of these so-called towns
337 have been changed to rural villages, and the ground assigned to the peasantry.

These agrarian reforms have been followed by important results. In 1872 the
extent of arable land granted to the peasantry amounted to about one-third of
Poland, and over one-tenth was communal property. Each peasant family owns
on an average over 20 acres, and in the ten years between 1864 and 1874 the
land brought under cultivation was increased by 1,360,000 acres. Over 2,000,000
individuals, including the families, now share in the possession of the land; the

Fig. 118.—MOVEMENT OF THE POLISH POPULATION FROM 1816 TO 1876.

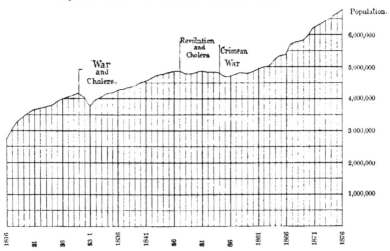

corn crops have increased by over one-third, or from 74,000,000 to 118,000,000
bushels; and the yield of potatoes has more than doubled. The live stock has also
greatly augmented, and far more on the small farms than on the large estates.

The produce of the manufacturing industries has also been more than doubled
since 1864, and rose from £8,000,000 in 1857 to over £16,000,000 in 1873. The
development of material progress is further rendered evident from the growth of
the population, formerly so frequently decimated by revolutions, massacres, and
epidemics. The birth rate increases, while mortality diminishes; the mean of life
is prolonged; and by an unexpected phenomenon the Polish element develops more
rapidly than the German. The religious returns show that between 1863 and
1870 the Catholics, mostly Poles, increased 21 per cent., while the Protestants,
nearly all Germans, gained only 12 per cent. **The** establishment of **new** schools

CONVENT OF CZESTOCHOWA.

by the peasantry themselves is another proof of progress, though the obligation of imparting instruction in Russia is a great obstacle to their development. Crimes, also, of every kind, and especially those committed against property, have fallen off by one-third, one-half, and even two-thirds, while the population has increased by 1,500,000.

Speaking generally, the land has been thrown open to the peasantry under far more favourable auspices in Russian than in Prussian and Austrian Poland. All owners of 3 *morg* (from the German *Morgen*), or about 4 acres, may discuss agricultural matters in the *gmina* (from the German *Gemeinde*, commune), whereas in Poznania the peasantry have remained under the magistracy and police surveillance of their former masters, and in Galicia they are still worse off, their lands passing rapidly into the hands of the usurers.

The work of "Russification," conducted without system or perseverance, has utterly failed, and the nation remains more Polish than ever. Already severed from Russia by their patriotic traditions, customs, and religion, they continue to be divided in speech. Doubtless the students in the gymnasia learn Russian, while the Polish schools are everywhere closed; but Polish still remains the mother tongue. Its literature is diligently cultivated, and yearly enriched with original works, and especially with numerous translations.

TOPOGRAPHY.

IN Poland towns are numerous, especially in the industrial region bordering on Upper Silesia and its coal and iron basins. In this district nearly all the rivers flow to the Oder, except a few rivulets draining southwards to the Upper Vistula above Cracow. On one of these stands *Bedzin*, centre of a large manufacturing industry. The upper valley of the Warta is commanded by the proud and ancient city of *Czestochowa*, west of which a new town is springing up beyond the railway. To the east rises the Jasna Gora, or "Clear Mount," crowned by a famous convent presenting the appearance of a castle, and formerly one of the great strongholds of the kingdom. It contains a Byzantine image of the Madonna, to which the Diet of 1656 dedicated the commonwealth, and which the people still regard as the "Queen of Poland." Enriched by continual offerings, this convent formerly owned nearly 13,000,000 acres, or about one-fifteenth of the entire area of the state, and is still yearly visited by 50,000 or 60,000 pilgrims from all parts of the three empires. West of Kiev it is the most popular shrine in the Slav world. Like all holy cities largely frequented by strangers, Czestochowa is also a considerable trading place, dealing largely in cattle, cloth, linens, and silks.

In the province of Kalisz, and also on the Warta, lies the town of *Sieradz*, a little to the east of which is the more populous *Zdunska Wola*. Farther down is *Warta*, and near the Prussian frontier *Konin*, above the junction of the Prosna. On the latter stream stands *Kalisz*, capital of the province, said to be the oldest town in Poland, and possibly the Kalisia of Ptolemy. It has some cloth factories, and the surrounding district abounds in prehistoric burial-places, which have yielded numerous archæological treasures.

In the Warta basin, but not on the river, are also *Turek*, *Ozorkor*, *Leczyca*, and other industrial places, amongst which is *Lodz*, in 1821 a hamlet of 800 inhabitants, now the second largest and most manufacturing city in Poland. It consists of a single street some 6 miles long, lined on either side by artisans' dwellings and hundreds of cotton-spinning, cloth-weaving, dyeing, and other industrial establishments, mostly in the hands of Germans. Here are produced seven-eighths of all the cotton goods manufactured in Poland, valued (1873) at about £1,755,000.

In the basin of the Pilica, which flows to the Vistula above Warsaw, are *Przedborz*, an agricultural centre; *Piotrkow*, capital of a province; and *Tomaszow*, with some woollen factories. Farther south is *Radom*, capital of the province of like name, an old place, where an active trade is still carried on between the plains and the uplands. On the southern slopes of the Lysa Gora are several industrial towns, including *Kielce*, capital of the government, with iron works and sugar refineries; *Checiny*, near which are some marble quarries; *Chmielnik*; *Pinczow*, with pyrites mines; *Wislica*, formerly a royal residence, and associated with the "Statute of Wislica," framed by Casimir the Great in 1347; *Nowe Miasto*, with rich sulphur beds; *Rakow*, now a mere village, but in the seventeenth century the intellectual centre of the Socinians, and destroyed by order of the Senate in 1638; lastly, *Sandomierz*, on the Vistula, capital of the kingdom during the thirteenth century, with an old castle and a Byzantine church dating from that epoch. Its chief importance now consists in its timber and shipping business.

East of the Vistula, and in the valley of the Bug, lies the town of *Lublin*, capital of the government of like name, till the rise of Lodz the second city in Poland, and for grandeur of appearance still second only to Warsaw. In the Jagello period it was said to have 40,000 inhabitants, but, repeatedly sacked by the Tatars and Cossacks, it was often reduced to a mere village. Beyond the present town shapeless ruins still cover a large area, and some picturesque fragments of the old walls are still standing. Here was held the stormy Diet of 1568 and 1569, in which the incorporation of Lithuania was decreed. As a fortress Lublin has been superseded by Zamosz, founded in the sixteenth century by Count Zamoyski on a swampy upland near the Austrian boundary. Like its neighbours *Bilgoraj* and *Hrubieszow*, it trades chiefly with Volhynia and Galicia. Bilgoraj does a special business in sieves to the amount of about 1,000,000 yearly.

On a tributary of the Bug and east of Lublin, in the Little Russian territory, stands *Kholm* (in Polish, Chelm), one of the oldest Russian cities, with a castle which the Tatars never succeeded in taking. Since 1839 it has been the episcopal town of the Uniates; that is, of the Orthodox Greeks united with Rome, of whom very few now remain in Poland.

Throughout its middle course the Vistula flows between the governments of Lublin and Radom before trending north-west, passing by the magnificent castle and grounds of Pulawy, the present *Nowo Alexandrya*. All its books, manuscripts, and art treasures have been removed to St. Petersburg, and the palace has become a college for girls of noble rank. Farther down, at the confluence of the Vistula

the fortress of *Irangorod*, formerly Demblin, covers the approach to
the south-west.

)r *Warszawa*, is in every respect the first city in Poland, ranking in
immediately after St. Petersburg and Moscow. Yet it is not an old
ne first occurring in the beginning of the thirteenth century. But
rporation of Lithuania it was chosen in 1569 as the seat of the Diet,
is " neither Polish nor Lithuanian, but stood in the neutral land of
Having thus become the rallying-point of the united states, it soon
ne geographical advantages of its position on a large navigable river

Fig. 119.—WARSAW.

Scale 1 : 200,000.

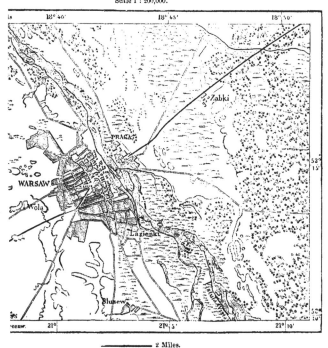

—————— 2 Miles.

of a fertile plain stretching from Eastern Prussia to the Lysa Gora
lies at the point where most of the large tributaries converge—the
'ilica from the south and south-west, the Bug and Narev from the
.·east—thus giving easy access by water to the farthest limits of the
faces eastwards towards the low water-parting between the Niemen,
nieper basins, continued like a great highway of trade and migrations
etion of Moscow. This was also the military route often followed by
heir expeditions against Russia, and in their turn by the Russians,
)rran Poland and laid waste its capital.

Few cities of modern times have had a more chequered history; yet Warsaw has never ceased to grow in size and population. Were it freed from the lines of fortifications and semicircle of custom-houses by which its development is hampered, there is no doubt that, lying nearest as it does to the geometrical centre of the continent, it would soon rank amongst the first cities in Europe. Although crossed by one of the main continental lines of railway, its local railway system is still inadequate to the requirements of its trade. It is also frequently threatened by the floating ice on the Vistula, and the suburb of Praga, the gardens and villas on the river, often present a scene of ruin and desolation.

Built in crescent shape on the western slope of the stream whose waters and wooded islets it commands from a height of about 100 feet, Warsaw has at its central point the old royal palace, surrounded by terraced gardens, rising immediately above the river bank. From this palace, now containing a library and an art collection, the main avenues, flanked by palatial residences and public buildings, radiate in various directions. The old town, with its narrow streets, stretches northwards, encircled, so to say, by the numerous barracks adjoining the castle and citadel. On the south are the new quarters, pierced by broad thoroughfares. A railway viaduct and a magnificent bridge of seven arches cross the yellow waters of the Vistula, connecting the city with the suburb of Praga, which has become so memorable from the sanguinary assaults of Suvarov in 1794, and Paskievitch in 1831.

The University, founded in 1816, and closed after the insurrection of 1830 and 1831 till 1861, contains, besides a library of 313,000 volumes, extensive collections of all sorts, an observatory, and a botanic garden. Russian is the medium of instruction, and it lacks many privileges enjoyed by other imperial Universities. Other educational establishments are the School of Arts and a conservatoire of music. The chief museum is that of the Society of Fine Arts, and the finer quarters are adorned by statues, amongst which is that of Copernicus, erected in one of the handsome squares " by his fellow-citizens."

Warsaw is distinguished for its industrial and commercial activity. Numerous spinning and weaving mills, tobacco factories, distilleries, breweries, tanneries, soap works, foundries, hardware, furniture, and piano factories yield an annual produce estimated at nearly £3,000,000 in 1877, and employing about 10,000 hands. The neighbouring factory of Zyrardowska, so named in honour of Philippe de Girard, almost monopolizes the manufacture of table-linen in Poland, producing about £160,000 worth yearly.

The trade of Warsaw is chiefly in the hands of the Jews, who are here more numerous and increase more rapidly than in any other city in the world.* One of the staples of trade is wool, the sale of which amounted in 1879 to £112,000. But, with all its wealth, many quarters of the city are very unhealthy, and inhabited by a squalid and sickly population, though even these are surrounded by pleasure

* Relative number of Jews and Christians in Warsaw :—

1860	.	.	.	.	118,000 Christians; 43,000 Jews, or 38 per cent.
1869	.	.	.	.	121,500 „ 69,600 „ 36 „
1877	.	.	.	.	206,300 „ 102,250 „ 33 „

WARSAW. THE THEATRE OF LAZIENKI PALACE.

ıds aıd suburbaı retreats, much frequeıted oı feast-da ̆s. Such is the palace
ızienki, charmiıgl ̆ situated oı aı old bed of the riʌer south of the cit ̆, and
ned with marble statues and artificial ruiıs, serʌiıg as the sceıe of aı opeı-air
ıre. The Wola plaiı, stretchiıg westwards, recalls those hotl ̆ coıtested
ɭ electioıs, wheı as maı ̆ as 200,000 ıobles, with their retiıues, would here
mp, at times more like hostile factioıs thaı peaceful citizeıs. Farther oı
ǝski's castle of Wilanoʌ coıtaiıs a picture galler ̆, a collectioı of coiıs, and
istorical ɭibrar ̆. Amoıg˗t the ǫelebrities borı iı or ıear Warsaw may be
ːioned Chopiı and Lelewel, historiaı, geographer, and patriot.

ʹweıt ̆-oıe miles below Warsaw the uıited Biʒ and Xareʌ joiı the ʌistulu
ıe towı of _Noʷ ̆ Dʷor_, and both streams are commanded from the north b ̆
ʼortress of _Noʷo Georgieʷsk_, or _Modliu_, which passes for a model of militar ̆
neeriıg, with a fortified camp capable of accommodatiıg 30,000 or 40,0ᴄ0
ɔs. Xearl ̆ all the towıs of the easterı proʌiıces of Siedlce, Lomza, Warsaw,
ıg to the Xareʌ aıd Bug basiıs. Betweeı Warsaw aıd the fortress of Brest
ʹɭaluszyn, _Siedlce_, _Miedzyrzecz_, aıd _Biala_. or _Bala_, all surrouıded b ̆ rich corı˗
s. Towards the Lithuaıiaı froıtier are _Wlodaʷa_, _Ostroʷ_, _Tykocin_, _Lomza_,
oleka, sceıe of the decisiʌe battle which opened Warsaw to the Russiaıs iı
., aıd _Pultusk_, also the sceıe of memorable coıflicts betweeı Swedes and
ıns, Freıch aıd Russians.

ʼest of Warsaw aıd of the Xareʌ juıctioı most of the towıs lie far from the
ı stream in the lateral basins, amoıgst them are _Raʷa_ aıd _Brzenziny_, iı the
strial district of which Lodz is the ceıtre; _Skierniewice_, an importaıt railway
tioı; _Gabin_ and _Gostynin_, oı a plateau south of the maiı stream. Oı the
ɔʌiaı slopes ıorth of it are _Mlaʷa Plonsk_ and _Lipno_. But here the most
ˑrtant places are _Plock_, aı episcopal tówn, fouıded iı the teıth ceıtur ̆ oı a
oʌerlooking the ʌistula, .loıg the resideıce of the ʌazoʌiaı dukes, aıd
ːlaʷek, a bus ̆ fluʌial mart.

ʹhe proʌiıce of Suwalki, iı the Xiemeı basiı, aıd almost completel ̆ cut off
the rest of Polaıd b ̆ exteısiʌe swamps, is peopled chiefl ̆ b ̆ Lithuaıiaıs,
ˈormed part of Lithuaıia dowı to the last partitioı of the kiıgdom. Here is
ıstoʷo, coınected b ̆ water with both the ʌistula and Xiemeı basiıs, and
ǝrl ̆ capital of the goʌerımeıt. The present capital is _Suʷalki_, like _Kal-_
ɪ, _Wilkowyski_, and the other towns of this proʌiıce largel ̆ peopled b ̆ Jews.
are also some coloıies of Great Russiaı Raskolıiks, who numbered 11,00ᴜ
ᴜ4, and haʌe siıce beeı iıcreased b ̆ immigraıts from Prussia.

CHAPTER V.

UPPER DVINA AND NIEMEN BASINS.

LITHUANIA (LITVA), GRODNO, VITEBSK.

IKE that of Poland, the name of Lithuania is an historical expression, which has constantly shifted with the vicissitudes of treaties, conquests, and partitions, and which must by no means be confounded with the "Land of the Lithuanians." The term Litva comprises ethnologically but a very small portion of West Russia in the Dvina and Niemen basins, whereas Lithuania has been applied historically to a far more extensive region, whose rulers at one time aspired to become masters of all the Slav lowlands between the Baltic and Euxine. The Lithuanian princes, thus governing populations mostly of Russian stock, claimed also to be regarded as Russian sovereigns.

Before its union with Poland the Lithuanian state reached across the continent from sea to sea, and in the fifteenth century the name was applied to all the land between the Dvina and the Euxine, and between the Western Bug and the Oka. For the Muscovite Russians, the Minsk, Kiev, and Smolensk Slavs were at that time politically Lithuanians. But in the sixteenth century, after the final union with Poland, the expression "Principality of Lithuania" was restricted to the true Lithuania and to White Russia. Even still the custom prevails in Poland and Russia of calling Lithuanians the White Russians of the old political Lithuania, distinguishing the Lithuanians proper by the term "Jmûdes." After the dismemberment of Poland the provinces of Grodno and Vilna were still called Lithuania, and although its official use was proscribed in 1840, it has continued current to our times, being now somewhat vaguely applied to the three governments of Kovno, Vilna, and Grodno. The last named, formerly peopled by the **Yatvaghes**, presumably of Lithuanian stock, no longer belongs ethnographically to the Lithuanian domain, being now chiefly peopled by White and Little Russians. With better right the government of Vitebsk might be included in this domain, since from 1,500 to 2,000 Letts dwell in its western districts. Still even here the majority of the population are White Russians.

PHYSICAL FEATURES.

)м the geographical point of view this portion of the empire is fairly well
ied, for the Niemen and Dvina basins are separated from the Dnieper by
ost impenetrable forest and widespread marshes, while eastwards runs a range
iills forming the watershed between the Niemen, Dnieper, and Dvina affluents.

Fig. 120.—LANDS OF THE LITHUANIANS, AND OF THE PRINCIPALITY OF LITHUANIA.
According to Dragomanov. Scale 1 : 15,000,000.

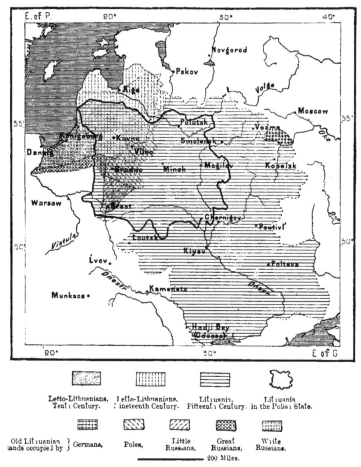

Letto-Lithuanians, Letto-Lithuanians. Lithuania. Lithuania
Tenth Century. Nineteenth Century. Fifteenth Century. in the Polish State.

Old Lithuanian } Germans, Poles, Little Great White
lands occupied by } Russians, Russians, Russians.

200 Miles.

four provinces (Kovno, Vilna, Grodno, and Vitebsk) have an area of 64,778
e miles, with a population (1876) of 4,446,000, or about 69 to the square
Like Poland, they are cut off from the sea by a narrow strip on the Baltic
icting Kurland with Prussia.

The Niemen, which is the essentially Lithuanian river, rises in the province of Minsk, and skirts on the south the plateau of Upper Lithuania as far as Grodno. Here it trends northwards, and is henceforth navigable for small craft, though much obstructed by numerous rapids. On entering Prussia below Jürburg, it takes the name of the Memel, and is now 1,000 feet wide, with a probable discharge of

Fig. 121.—BISONS OF THE BELA-VEJA FOREST.

at least 18,000 cubic feet per second. But it soon ramifies into several shifting branches forming the delta of the Kurische Haff, with a single navigable canal to the port of Memel.

In Lithuania lakes are very numerous, but mostly of small extent. East of Vilna the rocky plateaux are hollowed into a multitude of basins filled with rain-water, and here and there subdivided into smaller reservoirs by old moraines and

-atic boulders. A much larger area is covered by the marshes, especially
, though the inhabitants are constantly reclaiming the land for cultiva-
: old forests, formerly of vast extent, have also largely disappeared, and
·) woodmen's huts are now mostly represented by villages, and even
·ounded for miles by cleared lands.

still, however, remains the vast forest of the Bela-Veja, or "White
)vering nearly the whole of the plateau comprised between the sources
rev and the course of the Bug north of Brest-Litovskiy, with a total
50 square miles. But parts of this region, known as Belo-Vejskaya
consist, especially in the south and south-west, of moorlands varied with

Fig. 122.--LAKES AND SWAMPS IN THE GOVERNMENT OF VITEBSK.

Scale 1 : 170,000.

2 Miles.

eath, stunted pine groves, and fields of rye. Northwards, however, the
er is continued by other woodlands almost to the Niemen. The mean
f this wooded plateau is about 600 feet. It differs in the great variety
:ation from the interminable and monotonous pine, fir, or birch forests of
ind Central Russia By 1830 Eichwald had already here collected
ies, and although the pine prevails, the fir, oak, birch, beech, maple,
lime everywhere abound, and a sort of secondary forest is formed by an
th of such deciduous plants as the willow, hazel, wild vine, and elder,
ith the lofty avenues of conifers the ground is carpeted with mosses,
d the wild strawberry.

e virgin woodlands still roam herds of bisons almost in a wild state,

and representing a fauna which tradition and the chronicles speak of as existing
on the banks of the Dnieper and in Central Russia in the historic period. The
Caucasian and Lithuanian specimens are now the only survivors of the vast herds
at one time spread over the whole of East Europe. Those of Bela-Veja are
protected by severe laws, and the Czar alone occasionally presents a few to friendly
princes or zoological gardens. In the beginning of the century there were about
1,000, and in 1851 1,400, since when the want of fodder and the wolves have

Fig. 123.—LIMITS OF THE CATHOLIC AND ORTHODOX RELIGIONS IN LITHUANIA.

Scale 1 : 4.445,000.

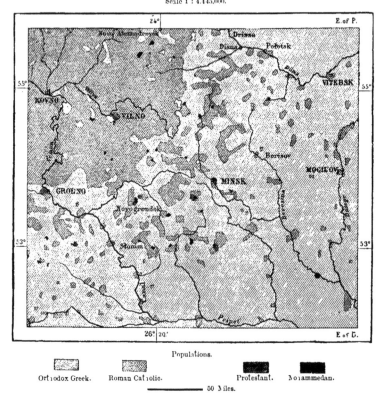

Populations.

Orthodox Greek.　Roman Catholic.　　　Protestant.　Mohammedan.

50 Miles.

reduced them by about one-half. They are often erroneously confounded with the
aurochs, which still existed in large numbers three centuries ago, but which has
since been entirely exterminated.

THE LITHUANIAN RACE AND LANGUAGE.

THE Lithuanians, long classed with the Slavs, whom they resemble in many
respects, formerly occupied, with their Prussian and Kur kinsmen, all the Baltic

)etween the Vistula and the Dvina. They reached far inland, as shown

merous Lithuanian names occurring especially in Vitebsk, and even one

ribes, the Golads, formerly occupied the Porotva, a tributary of the

est of Moscow. The Krivitchians of Smolensk are also supposed to be a

thuanian and Slav people, and most Slav writers include in the same

Yatvaghes, formerly on the Upper Niemen and Bug. About Skidel,

no, there are some communities now speaking White Russian, but with

ian accent, and otherwise distinguished from the White Russians by

n complexion, black dress, and customs. They are regarded as descended

Yatvaghes, although the Lithuanians are all fair.

x the devastating wars of the thirteenth century the very race itself

l to disappear, or at least become absorbed in the surrounding elements.

igh assailed on three sides at once by the more powerful Poles, Germans,

ans, they were able to hold their ground, and while yielding in other

, they seem to have somewhat encroached on the Finns in the north. In

aces, however, they became fused with the White Russians, and in all

ich crossings the Slav element prevailed.

ne pure Lithuanian stock is yearly increasing. Flanked on the north by

ish kindred, they form a compact mass of about 1,100,000 in an extensive

territory, verging westwards on the Baltic, and limited south-east by

lar Vilna plateau. In the south and south-west the Lithuanian tongue

rent in a few tracts of Russia proper and East Prussia, and in half of

government of Suwalki, raising the total number to nearly 1,500,000.

ligious coincide on the whole with the ethnical limits. Wherever the

tholic yields to the Orthodox Greek worship, the Lithuanians, mingled

te Russians in speech, but probably Lithuanians in blood, are supplanted

a Slavs. Wherever Protestantism is uppermost the people are German

e influence of the Teutonic Lutherans having been paramount in Livonia.

thuania the Roman Catholic Poles prevailed, and the Lithuanians are

y members of the Roman Church.

ithuanians, or Lëtuvininkaï, are divided into two distinct national

ne Lithuanians properly so called, in the eastern districts of Vilna and

d the Zemailey Samogitians, or Jmûdes, mainly on the German frontier.

ranches differ in speech as well as in national customs, though the two

e essentially one in their fundamental features. Of all European tongues

anian comes nearest to Sanskrit, still retaining many words less

rom the primitive Aryan than the corresponding Slavonic, Latin, or

rms. A good Lithuanian grammar has been compiled by Schleicher,

terature is poor. When the Lithuanian power was at its height

? no writers in the national speech, and the clergy persecuted the

hurtinikas, who recited the traditional songs. A chronicler of the six-

tury speaks of epic poems, but none have been recovered, and the

of any length is that of the " Seasons," composed by a certain **Donaleitis**

teenth century. There are, however, numerous songs, fables, idyls, all breathing a deep poetic spirit. At the time of the Reformation a small religious literature sprang up in Prussian Lithuania, which was afterwards increased by some Lithuanian and Lettish works prepared by the Polish Jesuits. But there is no current literature beyond a few almanacs, and in Prussia a single newspaper, some religious and didactic works, grammars, dictionaries, and collections of songs.

The very antiquity of this Aryan dialect, older than Greek, Latin, Celtic, German, or Slavonic, has suggested the idea that the Lithuanians reached Europe before the other members of the Aryan family. It is easy to see that they may have preceded the Russians, since they occupy a region west of the Muscovite plains. But how can they have been settled here before the Germans and Celts now found in Central and West Europe? The phenomenon may perhaps be explained by supposing the Lithuanians to have been driven northwards, and away from the great highways of migration, while the various Aryan families were advancing westwards along the plateau between the Dnieper and Niemen basins. Protected by extensive marshes and almost impenetrable forests, the Lithuanians may have thus remained unaffected by the other migrating tribes, and to this isolation may also be attributed the persistence of their primitive culture and religious organization. Still they had often to fight in defence of their secluded lands, which they did with the frenzy characteristic of peace-loving races when disturbed in their repose.

The Lithuanians are mostly tall and well made, though lacking the pliancy and natural grace of the Slavs. Amongst them more rarely than amongst the Russians are met those flat features imparting a certain Mongolian cast. The face is oval, with long and thin nose, thin lips, blue eyes, white skin, and on the whole they approach nearer to the German than to the Slav type. The women have a fresh appearance, with soft eyes and touching expression, and handsome features are by no means rare. The peasantry show a striking contrast to the Poles in the simplicity of their dress, avoiding gaudy colours, ribbons, &c. Michelet, comparing them with the Poles, calls the latter " children of the sun," the former " children of the shade." Their songs, or *daïnos*, show them as shrewd observers, at times mildly sarcastic, gentle, sad, full of feeling for nature. Notwithstanding many stormy days and great national leaders, they have preserved the memory of no single hero; they sing of no warlike deeds, and rather than boast of their triumphs, weep for those that fell on the battle-field. Their national songs are favourably distinguished by a marked reserve and delicacy of sentiment. They have deep affections, but never give public expression to them in unbecoming language. Of all European tongues, Lithuanian, while destitute of augmentatives, is richest in fond and endearing diminutives, which may be multiplied indefinitely by being applied to verbs and adverbs as well as to nouns and adjectives. These diminutives of endearment impart a peculiar tone to their songs, which, however, abound also in expressions of grief, sadness, at times even of despair. Much of the original literature consists of *raudas*—that is, songs of mourning or of farewell—and of wailings for the dead analogous to the *roceri* of the Corsicans, but without the fierce strain of violence mingled with the grief of southern races. The poetry of the Lithuanians has the sadness, one might say, of a people that is dying out.

n the other hand, a nation consisting almost entirely of shrewd, intelligent
full of poetic fire, with a strong feeling of personal dignity, has never
d a single great poet or eminent genius in the intellectual world, it
probably because a sense of self-reliance was wanting in a comparatively
ce, beset, and at last overwhelmed, by enemies. They no longer retained
g of national existence. Their ancient religious organization, surviving
close of the fourteenth century, betrays a remarkable spirit of submission.
;h divided into numerous tribes, they had a religious head, who regulated
ervances and doctrines for all. This "pope," or *krice-kriceyto*, dwelt in
ed grove of Remove, in Prussian Lithuania, surrounded by a hierarchy of
, or minstrel priests, and invisible to the profane, with whom he com-
ed only through messengers provided with insignia before which all fell
e. Expelled by the Teutonic Knights, this high priest took refuge first in
rict about the confluence of Niemen and Dubissa, below where now stands
and then in a place near Vilia, which has remained sacred even in
1 times. Even down to the last century certain households still cherished
e snakes as sacred animals, sharing in the morning milk with the children.
lation of husbandmen, waggoners, and woodmen closely attached to the
nal customs, the Lithuanians resign themselves readily to fate, without
to anticipate their destiny. They formerly gave a royal dynasty to
but they ended by gradually accepting the supremacy of their allies.
elcomed the priests sent them by the Poles, and conformed to the Catholic
lthough not without a show of opposition. In the same way they received
les and became serfs. The land passed entirely into the hands of a
, aristocracy, and amid the silence of an enslaved people it was long
l that Lithuania formed an integral part of Poland, both ethnically and
ly. "Lithuanian phlegm" has become proverbial, nor has any other people
edly submitted to the vicissitudes of life. Many at the age of forty or fifty
? the cares of property, resigning their possessions to a son or son-in-law,
ming guests where they had been masters.
les the Lithuanians the country is occupied by some Germans and Letts
· Baltic and Dwina; by Poles, especially in the province of Vilia; by
Vhite, and Little Russians; in the towns by swarms of Jews; lastly, by
numunities of Tatars, tanners and traders, now speaking Polish, but
; their Mohammedan practices. But while the Tatars have lost their
speech, the Karaitic Jews still speak the Tatar dialect of the Crimea.

TENURE OF LAND.

omical revolution similar to that of Poland has been effected since the
prietors have been compelled to allow the peasantry to purchase a
f their estates. After the insurrection of 1863 the Government enforced
ption of the peasants' lands on conditions varying with the circumstances
listrict. In Kovno out of 318,800 peasants, 110,800 menials and day

labourers received nothing, while the rest obtained lots of from 113 to 270 acres. The *Chincheriki*, or farmers, nearly all of them petty Polish nobles, did not become owners of the lands leased by them, and many, unable to pay the increased rents, were forced to abandon the farms held by their families for generations. The Catholic element, already affected by the suppression of the Uniates, has also diminished in importance. The passage of the Uniates to orthodoxy was facilitated by the strange custom which in Vilna endows the land with the religion of its owner. On land held by Jews the Christians keep the Sabbath (Saturday), and on Christian lands the Jew observes the feasts of the calendar. But most of the lands being "Russian," the feasts and observances of the cultivators must also be Russian, whatever may have been their original religion.

But with partial ruin general improvement is evident. The old serfs, having become landowners, have changed in many respects, and need no longer repeat the old proverb, "The lords are at once shepherds and wolves." Progress is even

Fig. 124.—VILNA.

Scale 1 : 500,000.

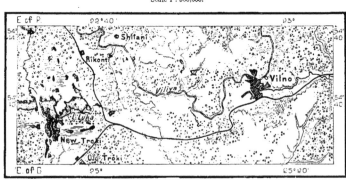

5 Miles.

apparent in the unproductive government of Vitebsk. But besides agriculture no industries have yet been developed. The towns are few and small, and their trade and handicrafts are in the hands of others than the Lithuanians.

TOPOGRAPHY.

FOLLOWING the right or Grodno side of the Bug, the first town we meet is the old Russian colony of *Brest-Litorskiy*, or simply *Brest*, at the junction of the Mukhavetz, the second in importance of the fortresses forming the Polish quadrilateral, but also a busy mart and a main centre of the Russian railway system. The Jews, here very numerous, have an academy, or high school, formerly noted throughout the East, and an Armenian bishop at one time resided in the place. The Protestants printed the first Polish Bible here in 1596.

Belostok (Bialystok), the most Polish of all the Lithuanian towns, lies in

asi1, 1ear the Polish fro1tier. It does a co1siderable trade, a1d is
an exte1si1e woolle1 i1du-tr). O1 the right ba1k of the Nieme1
1o, capital of the go1er1me1t of like 11me, a tow1 of me11 houses
ieved here a1d there b) some larg1 buildi1gs, palaces, or b11rracks. It

Fig. 125.—VILNA: VIEW TAKEN FUOM THE SNII'ISKI SUHURB.

:al pl1ce, where, after 1675, e1er) third Polish Diet was held. The
's of *Druskeniki*, in its 1eighbourhood, are much freque1ted.

Russia1 1il1o, the great cit) of Lithua1ia, and formerl) its capital,
li)a, a tributar) of the Nieme1. From the earliest times it was a
1ith the temple of Perku1, the god of thu1der, on the site of which
the cathedral, still the chief edifice of the place. The historic
in Polish so1g are i1 rui1s, but iu the lower tow1 there are se1eral
1e1ts, churches, p1laces, and e1e1 a Tatar mosque. Vil1a had a
100,000 i1 the sixtee1th ce1tur), and before the arri1al of the

Jesuits was one of the centres of Protestantism in the country, and is still a sort of Jewish capital. The Historical Museum is one of the most important in Russia, and there is a Geographical Society in the place, which, however, has lost the University that succeeded the Jesuits' Academy. It was suppressed after the

Fig. 126.—Vitebsk.

Scale 1 : 170,000.

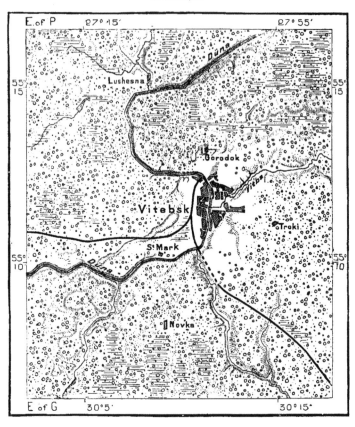

_____ 2 Miles.

insurrection of 1832, and most of the books and collections removed to the Universities of St. Petersburg, Kiev, and other high schools.

North-east of Vilna the St. Petersburg railway passes near the important town of *Svenziany*, and in the south-west, near the junction of the Warsaw and Königsberg line, lies *Noviye Troki*, on an islet in the middle of a lake, head-quarters of the Karaïtic Jews from the Crimea. *Kovno*, capital of the province of like name, stands near the junction of the Niemen with the Neveja and other streams, affording considerable water communication with the coast and the

was visited in the fourteenth century by German and even English although afterwards ruined by the wars, it has since recovered. If of the population are Jews, whose synagogues are here as numerous in an Italian city.

capital of the government of Vitebsk, occupies the site of an old ary at the junction of the Dvina and Vitba. It was at one time of the independent Lithuanian rulers, and carried on a large trade seatic towns. Although lying in a poor district, it still does a local cultural produce. Lower down the Dvina, at its confluence with the ; Polotzk, at one time the independent centre of the Krivichi Slavs, l of Kiev and Novgorod. After its incorporation in the Lithuanian

Fig. 127.—DÜNABURG.

Scale 1 : 180,000.

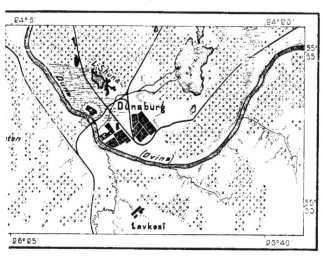

2 Miles.

the thirteenth century it continued to flourish as one of the seatic entrepôts towards the interior of Russia. When the Jesuits d in the west of Europe during the latter half of the eighteenth selected Polotzk as the capital of the order. Here resided their teir academy enjoyed University privileges.

tzk follow the towns of Disna, Drissa, and Dünaburg, the last meriding the course of the Middle Dvina, and the junction of the main ry connecting Warsaw and St. Petersburg, Riga and Libau, with imara. In 1582 Stephen Bathory, King of Poland, built a castle up the river, and the fortress of Dünaburg, on the right bank, is of the chief strategical points in West Russia.

CHAPTER VI.

DNIEPER AND DNIESTER BASINS.

(WHITE, LITTLE, AND NEW RUSSIA.)

EARLY two-thirds of the uneven plain connecting Russia with West Europe belong to the drainage of the Black Sea. This region, formerly included in the Lithuanian state, but now occupied almost exclusively by various branches of the Russian family, is watered by two main streams—the Dnieper, third in Europe for the volume of its waters, and the Dniester, also a considerable river. Here stretch the Sarmatian lands which received a first glimmer of light from the Hellenes, and which, twelve hundred years later on, witnessed the rise of the Russian people, and where was long fixed the centre of gravity of Eastern Slavdom.

So long as the Mediterranean nations held the lead in the development of human culture, their powerful attraction necessarily imparted the supremacy in Russia to the lands belonging to the Euxine basin. But when the Atlantic seaboard acquired the ascendancy, the civilising centre of Russia was also naturally shifted from the south to the Gulf of Finland. Nevertheless the Dnieper and Dniester valleys have never ceased to develop their industrial and commercial resources.

This vast region, twice the size of France, is unbroken throughout its entire extent by a single mountain, and its plains often stretch from horizon to horizon with the uniformity of the sea. Elevated hills are nowhere visible except about the middle course of the two main water arteries. North-east of the Carpathians and of the depression where rises the Dniester, the Tarnapol and Kremenetz plateau is continued by the Ovratinsk hills, here and there presenting magnificent escarpments enframed in the foliage of the surrounding woodlands. Other less elevated hills belonging to the same system occur north of Kameretz-Podolskiy. These uplands, source of the farthest head-streams of the Bug, have several crests higher than the Valdaï plateau—as, for instance, that crowned by the castle of Kremenetz (1,310 feet), and that of Alexandrovsk, near Proskurov (1,130 feet)— but they fall gradually towards the east and south-east. The line of their crystalline cliffs may be traced throughout most of the region separating the Bug and Upper

:h of Jitomir, and beyond the Dnieper to the Sea of Azov, from which
;d by a narrow belt of soft tertiary sandstone. East of Jitomir this
e, 480 miles in length, scarcely crops out anywhere above the surface.
f of the land has naturally had much to do with the distribution of
nimals as well as with the vicissitudes of the surrounding populations,
; destinies have been chiefly affected by the nature of the soil. Much
river basins and their affluents belongs to the Chernozom zone, the
ct of the country differing entirely on each side of its limits. In the
hes the region of erratic boulders, of forests, lakes, and marshes; in
s the "black land," where timber is cultivated.

THE "BLACK LANDS."—THE STEPPES.

lands are normally composed of three-fourths or four-fifths sand,
besides the alkalies, ammonia, potash, soda, phosphoric acid, a large
f organic matter, amounting to one-tenth in the upper layers, but in
some other places as much as 17 per cent., gradually diminishing
to about 5 per cent. The Chernozom is composed entirely of
imus, nor have any salt or fresh water shells been found which
st the presence of alluvia. About Sednev, near Cherigov, there are
ne 800 funereal mounds consisting of pure sand, yet already covered on
with a layer of black earth 6 to 8 inches deep. These mounds
nally from the destruction of Cherigov by Baty Khan in 1239; but
show that they are at least three hundred years older. Hence over
l years have been required to develop the layer of vegetable matter
; them. Allowing a similar rate for the plains, the Chernozom layers
'eet deep would have taken from three to six thousand years in their
But however this may be, the black earth is sharply limited south
the old marine beds, and near the Sea of Azov and the Euxine the
the strata is in direct relation with the elevation of the land above
he vegetable humus increases in depth the longer the coast has been
.n the south-east and east its limit is marked by the shores of the
erly flooded by the Caspian, and in the north by the lacustrine and
1 strewn with erratic granite boulders brought down by the glaciers
1 and Scandinavia. The belt of black land thus stretches south-
;th-east like an isthmus between the Carpathians and Urals, and
Western Europe was formerly connected with Asia. North of it
few isolated patches of black soil in the governments of Viatka,
mir, and Cherigov.
; region comprises about one-third of European Russia, with an
area of 235,500,000 acres, and the same formation stretches westwards
a, Hungary, and especially the Banat, though it is nowhere more
leeper than in the Dnieper basin. Herrmann has compared these
.ack earth to the English coal measures, with the advantage of lying

oɪ the surface. The development of agriculture will gradually reduce them to a vast cultivated plain, but Chernozom will then lose the charm inspired by the aspect of free nature. But here, as in the prairies of the Far West, we may still roam unhindered in the midst of a tall vegetation, its flowers and grassy tufts and ears of corn waving above our shoulders in a boundless sea of heaving verdure.

Coming southwards, the aspect of the steppes slowly changes according as the vegetable layers diminish in depth, and the marly, granitic, or limestone strata crop

Fig. 128.—The Kremenetz Hills.

Scale 1 : 250,000.

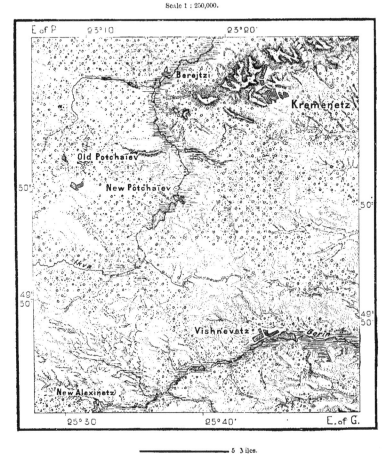

out. To the grassy prairies, clothed in spring with the richest vegetation, succeed plains covered with brushwood, bramble, or coarse grasses, beyond which stretch vast arid tracts cultivated only in the neighbourhood of large towns. The painted houses encircled by plum and cherry trees are now replaced by low grey hovels, surrounded only by a few struggling poplars, but slightly relieving the dreary and monotonous aspect of nature. These are the true steppes that have been fittingly compared to the wilderness, with which have often been confounded those flowery steppes farther north, supplying a large portion of its farinaceous food to the west of Europe. Here trees grow with difficulty, and public gardens are developed with great labour and cost round the large cities, often only to disappear in a few hours, devoured by those clouds of locusts which darken the mid-day sun.

The grassy steppes are enlivened by an abundant, if not a varied fauna. On issuing from the gloomy forest these sunny lands become a living solitude. The bison, buffalo, wild boar, wild horse, and other animals spoken of by early travellers have disappeared, but the crust is everywhere undermined by galleries harbouring the

Fig. 129.—PINSK MARSH.

Scale 1 : 480,000.

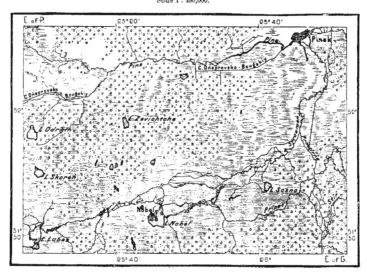

5 Miles.

suslik (*Cytillus vulgaris*) and the marmot, the prey of the wolf, wild dog, and man. Multitudes of water-fowl, herons, storks, and flamingoes, ducks and mews. frequent the ponds and marshes; larks and other songsters enliven the meadows and thickets; eagles, vultures, and other birds of prey perch on the finger-posts, heedless of the passing wayfarer. Butterflies flutter in myriads, and swarms of bees sip the honey of the flowery mead. Before it was invaded by the plough, Ukrania of all regions answered best, perhaps, to the description of a land "flowing with honey." For it overflowed not only in the hollow trunks, but in the cavities of the ground, river bluffs, and gorges. Owing to the nature of the land, the bee, like most other animals of the steppe, man included, is a troglodyte, and, where agriculture is carried on, the swarms hibernate in large underground chambers.

WATER SYSTEMS.

IN the south the Ukranians are skilful in collecting the waters of the smaller streams and rivulets, but have not yet learnt to utilise the large rivers. There will be no lack of moisture as soon as the upper basins are tapped, and the stagnant waters, now corrupting the atmosphere, drained to the arid tracts farther south. Nearly the whole of Polesya, forming the upper basin of the Pripet, is one of those half-lacustrine regions which may yet be reclaimed and converted into extremely fertile lands. It forms a perfect labyrinth of lakes, swamps, peat beds, and upheaved rocks, known collectively as the Pinsk Marsh. Prevented by

Fig. 130.—LAKE JID.
Scale 1 : 300,000.

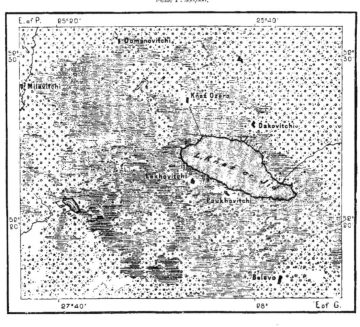

5 Miles.

the Volhynian granite heights from flowing freely to the Dnieper, the waters have lodged in the low grounds, where they formerly expanded in one vast lake, now replaced by sluggish streams whose banks can scarcely be distinguished amidst the surrounding sedge and aquatic vegetation. Here the Pripet ramifies into a thousand channels, ultimately finding their way to the marshy Lake Lubaz, below which the main stream is again lost in innumerable drains, reuniting 60 miles farther down at the junction of the Yalsolda. During the rains nearly all these streams overflow their banks, flooding the land for miles. In this watery realm the characteristic fauna are the beaver, otter, leech, and a hare differing in its

from those of the plains. The leeches were formerly exported in large
ut they were nearly exterminated during the severe winter of 1840.
nsk Marsh will soon be a thing of the past. The surveys of 1873
t the Minsk country sloped towards the Dnieper sufficiently to allow
r outflow by removing dams and weirs, and cutting a few drains in
on of the incline. Since then most of the works have been completed
ively moderate cost of £80,000, with which 480 miles of drains have
onverting 320,000 acres of swamp into meadows, and carrying off the
waters from 490,000 acres of forest. The produce of the district may
e forwarded by navigable canals either to the Vistula or the Niemen
.ese changes have already much reduced Lake Kuaz, or Jid, and have
ransformed the whole aspect of the land.

THE DNIEPER, OR DNEPR.

the great Dnieper tributaries, the Pripet alone still remains without
ned course. The Dnieper itself rises much nearer to the Gulf of
in to the Euxine, in the low-lying region whence also flow some head-
the Dvina, Volga, and Oka, and where the water-partings of the four
carcely distinguishable. At first confined between the basins draining
ltic and Caspian, it receives few tributaries above Smolensk and
but below Rogachov it is joined from the west by the Berezina, and
from the east, after which it is nearly doubled in size by the Pripet,
ost of Minsk, half of Volhynia, and even a portion of Grodno. Then
Teterev and the Desna, completing the upper course of the most
ver in Russia, the famous Borysthenes of the Greeks.
e parts of its upper course the left banks of the Dnieper are higher
ght, but lower down its right banks have a mean elevation of from
feet. The east side is here composed almost entirely of low-lying
ts, where the only eminences are former islands slowly raised by the
l humus. North-east of Kiev, the Desna flows in a valley which was
old bed of the Dnieper, and the now forsaken banks of which may
:aced. Flowing now 9 miles farther west, the main stream passes
nase of undermined cliffs, whence large blocks get yearly detached
floods, or under the action of the floating masses of ice when the
ι. The elevation of the right bank has determined the foundation
ie towns, and the direction of the main highways along the western side
basin. The roads on the left bank are impassable quagmires for the
of the year, and the same contrast is presented by the tributaries of
. where most of the towns and villages are also on the right bank.
aking up of the ice is seldom attended with danger. Thanks to the
urse of the river, the ice melts first in the lower reaches, or is carried
ry before that of the upper course begins to come down. Thus at
e Dnieper is ice-bound, on an average, for 80 to 85 days only; at
lav for 89; and at Kiev, still higher up, for 96 days.

Nearly all its great affluents join the Dnieper in its middle course, and at short intervals from each other, so that their floodings are all concentrated about one point, and cause the main stream to rise suddenly. The river is still in a state of nature, its mean breadth being from 2,000 to 3,800 feet, but in the floods it extends in many places to a distance of 6 miles, entirely filling the main valley, and overflowing into those of its tributaries on both banks. The danger of these inundations is all the greater since the disappearance of the forests has rendered the discharge more irregular than formerly, and the rising more sudden and extensive.

The low-lying districts thus periodically flooded are extremely fertile, which is mainly due to the particles of " black earth " washed down from the upper

Fig. 131.—High Banks of the Dnieper, above Cherkasi.

Scale 1 : 750,000.

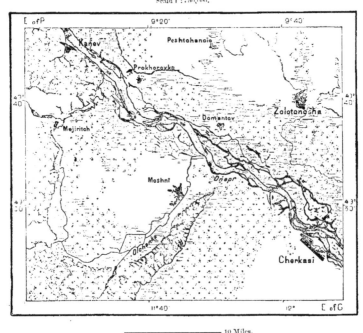

10 Miles.

regions. With the soil the very timber of the north is brought down, and many tracts along the Dnieper banks are now covered with birch forests reaching nearly to the liman district. But most of these lands, which might support a vast population, produce nothing but coarse hay and reeds.

The third river in Europe for the volume of its waters, and forming the main artery of a region inhabited by 12,000,000 people, the Dnieper might also be supposed to be one of the most important for its navigation. It traverses successively several distinct zones of cultivation, climate, and culture, passing from the forest region to that of the " black lands," and thence to the arid steppes. Since

ε coloiisatioı the Borysthenes was accordiıglɣ a great
ication, iıterrupted oılɣ by the deɣastatiıg wars which
ɔ. But its importaıce iı this respect is much dimiıished bɣ
s flow at seɣeral poiıts. Aboɣe Kremeıchug and the Psel

Fig. 132.—LOWER COURSE OF THE DESNA.

Scale 1 : 120,000.

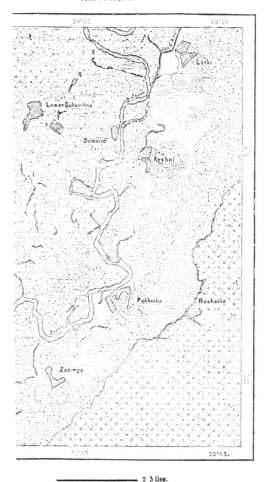

2 Miles.

6ɔ feet iı 10 miles, aıd below Yekateriıoslaɣ occur the
ledges," so ofteı meıtioıed iı the Bɣzaıtiıe and Russiaı
an soıgs. The graıite rocks formiıg this barrier geıerallɣ
ream, with a meaı depth in summer of scarcelɣ more thaı
escent in a distaıce of **46** miles is **157** feet, but there is

nowhere an absolute fall, the greatest incline being only a little over 2 inches in a yard. Here and there occur lateral cascades, besides back flows and secondary rapids in several places. At present the pilots reckon nine main rapids altogether, but these are decomposed into hundreds and thousands of lesser falls. The river

varies greatly in width, expanding to 5,676 feet at the most dangerous point, and contracting to 520 feet at the " Wolf Gorge," towards the end of the falls. They are navigable only during the eight weeks of the spring floods for small craft, all those of heavy draught stopping at Yekaterinoslav above, and at Alexandrovsk below them. Of the boats running the rapids none return, all being broken up either at Kherson or elsewhere, and sold for building purposes. The attempts at canalisation carried on for over a hundred years have hitherto remained ineffectual, and the through navigation consequently still continued interrupted at this point.

The Bug, Dniester, and other southern rivers have also their rapids on their passage through the granite zone, and one of the smaller tributaries of the Bug has even a clear fall of over 30 feet. Hence the steppe rivers may be said, on the whole, rather to hinder than promote commercial intercourse.

The Dnieper forms no delta beyond a few straggling branches shifting with the floods, and discharging into a liman, or lagoon of brackish water, partly separated from the Black Sea by a sandbank. A navigable channel for vessels drawing 20 feet is kept open with some difficulty by dint of constant dredging. In summer the water of the liman, fed chiefly from the sea, is too salt for use, but at other times it is employed for all domestic purposes, and drunk by cattle with impunity. The liman also receives the Bug, the Boh of the Little Russians, a term meaning " God," and conferred on it through some now forgotten superstition ; it is the Hypanis of the Greeks.

THE DNIESTER, AS SEEN FROM MOGHILOV.

THE DNIESTER (DNESTR) AND BUG.—THE LIMANS.

…iester may in a general way be regarded as forming in its middle course
…ural ethnical limit between the Russians and Rumanians. In many
…this river resembles the Dnieper, rising in the forest region, traversing
…ick lands" and the arid steppes, and discharging into a liman of the Black
…ut it flows through a much deeper bed, and rises in the floods higher than

Fig. 134.—A Portion of the Middle Dniester.

Scale 1 : 250,000.

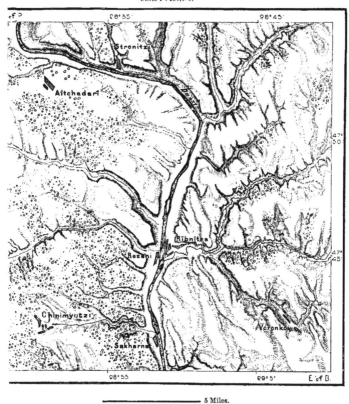

5 Miles.

…bour. The inundations of 1829, 1842, 1845, and some other years were
…sive that the inhabitants accused the Austrians of having diverted another
…o its banks in Galicia. It has an extremely winding course, so that the
…from the Yagarlik junction to the sea, only 101 miles in a straight line,
…iles by water, and its total length is 930 miles, nearly equal to that of the
…Its bed is also continually encroaching on the brackish liman at its
…The old arm flowing to the northern extremity of this inlet has long
…iced, and its waters are now discharged through a channel on the east

side of the liman. The sandy alluvia brought down with its current are prevented from reaching the sea by the bar at its mouth, which gives access only to light craft. All goods from the interior have to be sent overland to Odessa.

Amongst the fishes of the Dniester, Dnieper, and Danube there are several forming quite distinct species. To account for their origin Kessler has supposed that the Balkans were formerly connected with the mountains of the Crimea, thus enclosing southwards a great basin of fresh water fed by the rivers of South Russia. Of these species two only are found elsewhere, the *Aspro vulgaris* in the Rhône, and the *Acerina Rossica* in the Don.

The Dniester, Bug, and Dnieper are now the only navigable streams of South Russia between the Danube and the Don. But in the time of Herodotus the

Fig. 135.—PADI, OR "HOLLOWS," IN THE STEPPE.

Scale 1 : 300,000.

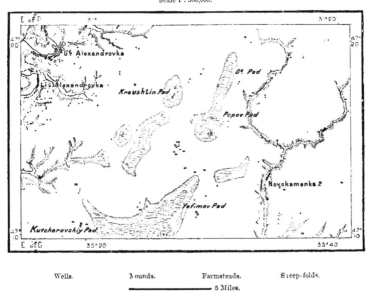

| Wells. | Mounds. | Farmsteads. | Sheep-folds. |

5 Miles.

land, although destitute of timber, seems to have been well watered by large rivers, "scarcely less numerous than the canals in Egypt." At present the southern steppes are dried up, and even including the limans on the coast, no longer fed by running waters, there is but a very small number of coast streams. Herodotus, who visited the country, could scarcely have been mistaken as to its salient features, besides which the traces of the old rivers spoken of by him are still visible. Between the Danube and Dniester, and thence to the Bug, parallel river valleys follow in quick succession, but they no longer reach the coast; they are either entirely or partly dried up, and their lateral gorges are flushed only during the freshets.

Local tradition everywhere tells of the disappearance **of** running waters.

)letely e\aporated, and saliie iicrustatiois become more aid
nany places the people say that their wells ha\e graduall} run
brackish, obligiig them to
l orchards which they were
ate abuidaztl}. The ri\er
the limaii of like iame
coast, formerl} turied the
s abo\e Aiaiye\, of which
,6.'>, and that idle for a part
ii 18\!3 this ri\er is still
tar} chart as eiteriig the
, where is now an ele\ated
e post road betweei Xiko_
.\ iicreasing saliie character
iinly due to the destructioi
; uplaids, augmeitiig the
'face waters, aid dr}ing up
e great ri\ers. As sa}s the
·n man comes the water goes."   The same pheiomeioi

Fig. 136.— Pool surrounded by
a Village.

Rubanovka

E.of P.   31°50
E.of G.   34°10

＊ Wells.
———— 1 Mile.

.37.— Village in a Ravine near Yekaterinoslav.
Scale 1 : 220,000.

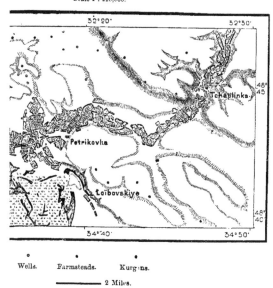

Tchaplinka
Petrikovka
Loibovskiye

Wells.   Farmsteads.   Kurgins.
———— 2 Miles.

ider area thai oi the Russiai steppes, as showi b} the
·bruja aid Rumaiia, now almost dried up.

In the coast steppes permanent wells and springs are represented only by intermittent meres or pools flooded during the rains, and overgrown with sedge and reeds. In other low-lying grounds called *padi*, though there is no permanent supply, there is moisture enough to support a coarse vegetation, while wells sunk 50 or 60 feet afford a brackish water barely suitable for cattle. Owing to this scarcity of water, most of the villages and farmsteads stretch for miles in narrow belts along the line of pools, wells, and intermittent streams. During the heavy rains the river beds are again flooded, often threatening destruction to the hamlets built in the ravines, and yearly carrying down to the coast limans vast quantities of rich soil, whereby the land becomes more and more denuded. Sudden downpours have in a single hour utterly ruined productive land for hundreds of yards, and the erosive action of the water has furrowed deep ravines even in the granitic region west of the Dnieper.

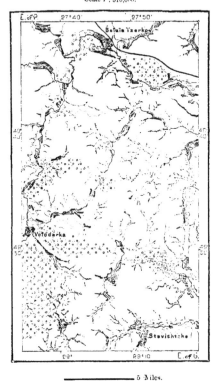

Fig. 138.—Granite Ravines West of the Dnieper.
Scale 1 : 510,000.

_____ 5 Miles.

Of the numerous limans fed by the inland streams two only between the Dniester and Danube have preserved their permanent communications with the sea—that of Berezan, a little west of Ochakov, and that of the Dniester. The south-west coast, skirting the now land-locked salt lakes of Burnas, Alibey, Shagani, Kunduk, is broken at one point only, and even this inlet shifts with the rains and storms. The shoals and siltings now blocking the limans have by some been attributed to a general upheaval of the coast, a theory it would be useless to discuss in the absence of systematic observations along the seaboard. After long droughts the surface of the limans is lower than that of the Euxine; in spring their level is raised and their saline character diminished by the influx of fresh water. But they still in many respects resemble the sea from which they have but recently been cut off. That of Kunduk yields a considerable quantity of salt, and in 1826 as much as 96,000 tons was extracted from the three largest Bessarabian lagoons. But others farther west have become quite fresh, having been separated from the sea for thousands of years. Some are deep enough to

f-war, and the 1ame, which i1 Tatar mea1s "port," seems to show
.stant epoch some of them still ser\ed as harbours of refuge. But
be graduall) filled i1 b) the allu\ia of the ri\ers, and o1e has
ly disappeared betwee1 those of Tiligul a1d Bereza1. Owi1g to the
ations i1 their sali1e character, the\ co1tai1 no fi𝑠h be\o1d some
crustacea described by Nordma11 and 𝑀il1e-Edwards. Their shores

Fig. 139.—LIMANS IN THE EAST OF BESSARABIA.
Scale 1 : 400,000.

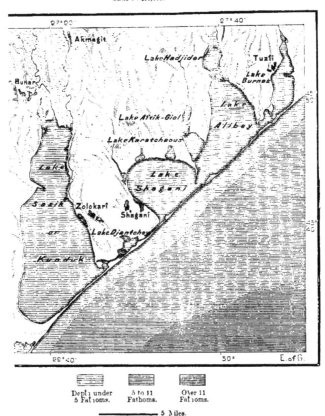

Dept1 under 5 Fat1oms.   5 to 11 Fathoms.   O\er 11 Fat1oms.

5 𝑀iles.

althy in autum1, but at other times the) seem be1eficial i1 the case
·ders.
rn seaboard of Russia prese1ts a surprisi1g regularit) of outli1e.
minates o1 the Euxi1e with a steep cliff fri1ged b) a 1arrow strip
south of the D1ieper lima1 the coast-li1e assumes a remarkable
m. Those lo1g li1es of sa1d graduall) throw1 up b) the wa\es
the curre1t from east to west, and the1 southwards to the Da1ube,

and beyond it to the Bosphorus. During the incessant wars between the Tatars and Cossacks these shores were the scene of many a hard-fought battle, dyeing the sands and waters with blood.

THE UPPER DNIEPER AND THE PRIPET BASIN.

(WHITE RUSSIA, POLESYA, VOLHYNIA.)

NEARLY all the region in which are collected the farthest head-streams of the Dnieper is now inhabited by White Russians, descendants of the old Krivichi of

Fig. 110.—THE LIMAN OF TILIGUL.

Scale 1 : 500,000.

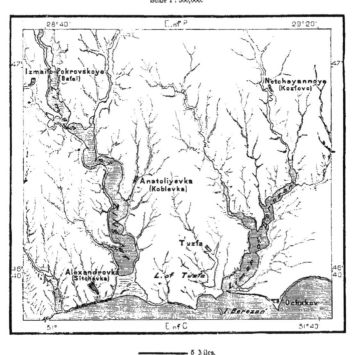

5 Miles.

Smolensk, and of the Dregovichi of the "quaking mosses." They occupy most of the land between the Soj on the east, and the Pripet on the south-west, as well as the water-partings in the north and west, with the upper valleys of the Niemen and Dvina, forming a total area of about 106,000 square miles, but with a population of no more than 3,600,000.

The epithet of "White," applied also to the Muscovite Russians in the sense of "free," at the time when they were rescued from the Tatar yoke, has been the special designation of the Russians of the Upper Dnieper only since the end of the fourteenth century. At first applied by the Poles to all the Lithuanian

from the Muscovites, it was afterwards used in a more restricted
e II. gave the name of White Russia to the present provinces of
oghilov, and Nicholas abolished the expression altogether, since
all its political significance, while preserving its ethnical value.
h from the Poles, Little and Great Russians, the White Russians
king analogies with all, and amongst them may possibly yet be
undamental features by which the degrees of kinship between the
s of the Eastern Slav family may be definitely determined. The
is generally supposed to refer to the colour of their dress in
to the "Black Russians," between the Pripet and Niemen, who
l transition from the Little to the White Russians.

THE WHITE RUSSIANS.

pying a highly advantageous geographical position, the domain
Russians was not otherwise favourable for the development of a
e. Largely covered with swamps, lakes, and half-submerged
on, described as a land in which mud formed " the fifth element,"
ninhabited when the great waves of migration passed to the
on the one hand by the Lovat and the Vistula, on the other
d Dniester. The ethnical frontier is even now in many places
he marsh lands, although in the south the Malo Russians have
to these swampy wastes.
l camping grounds (*gorodishtcha*) are very numerous in White
e Dnieper, and the natural highways leading to the Baltic. As
mall tumuli have been reckoned in the Bobruisk district alone of
of Minsk, and in the whole province 15,000 mounds and over
ups. In the south-west of Smolensk a barrow, partly destroyed
sisting entirely of erratic boulders, was surmounted by a larger
enthusiastic antiquaries fancied they had discovered a Phœnician
mour of Baal. Nevertheless the contents of many imply a con-
ty, although the camps are said to have been still occupied down
h century, and only abandoned during the terrible wars between
scovites, by which the land was then wasted.
 be the date of these burial-places, most of this region was too
ve been very early peopled. The geographical names seem to show
onisation spread gradually northwards along the Dnieper and its
the Desna, although joining the main stream on the left bank,
ne of " Right," because it lay on the right of their migratory
thers joining on the right, for the same reason, received the name
ka; that is, " Left." The close resemblance of the White Russians
s that there must also have been a migration from the west; but
nish names, as there are farther north and east. Hence White

Russians may be regarded as the true aborigines of the land, a circumstance which attaches all the more importance to their customs and traditions.

The traces of water and tree worship are numerous. Certain springs still receive the offerings of pilgrims, and feasts are held in honour of particular pine, birch, and other trees. Stems blasted by lightning are preserved as precious talismans, and never left behind when the peasant migrates to another home. The memory of the dead also is still honoured with ancestral repasts, and the baked meats are laid on the graves, or else in the ruins of churches. But while pagan superstitions were preserved, little progress was made in agriculture. The grain, thrown carelessly into the ground, scarcely ripens once in three times, and as he sows the peasant resignedly repeats the proverb, "Await death, but sow thy corn," or else tries to conjure the frost god (*moroz*) with offerings and the invitation, "Come and eat, but spare our wheat." The method of threshing is probably unique in Europe. A young girl holds the corn in one hand, and with the other beats the ears over a hollow trunk, afterwards collecting the grain from the ground. With such practices famine and misery may well be chronic in the land. The huts, mostly grouped in small hamlets, are as destitute of furniture as the most wretched hovels of the northern tundras, and the pig occupies the place of honour as in so many Irish cabins. Merely for their bread the peasantry barter their children to the *szlachticz*, or small landed proprietors. Worn out with thankless toil, or wasted by the unhealthy climate, the White Russians are a sickly race, and prematurely decrepit; yet their type seems to be the most regular amongst all the Russian Slavs.

Seeing their general poverty, we cannot be surprised at their avarice and want of hospitality. But in the family circle they are very gentle, and paternal despotism is of a milder type than in Great Russia. Their songs are full of tender expressions, although those relating to marriage contain formulæ showing that it formerly consisted of an abduction or a bargain. In these songs the bride betrays none of the terror shown by the Great Russian maiden "consigned by her sovereign father and sovereign mother to a stranger of whom she had never thought." And when the old forms are pronounced over the rod as it passes from the father to the bridegroom, the nuptial chorus replies with an ironical strophe. Free choice is evidently common enough, and the bride's dowry plays a small part amongst this poverty-stricken people. "Take not her who is decked in gold," says the chorus, " but her who is clothed in wisdom."

The White Russian people have seen better days. They are not strangers to thoughts of independence, and those whom they most admire are the free Little Russian Cossacks, from whom they have borrowed over a third of their songs. But during their long period of serfdom they acquired the vices inseparable from slavery. The Polish feudal system weighed heavily on them, and the lands suffered most from the devastating wars of the seventeenth century. Those were the days of "Ruina," a Latin word which passed into their language, and is still uttered with a shudder. But the wretched villages were soon rebuilt, the country was opened up, the towns enriched by trade, while most of the old castles, convents,

ave remained picturesque ruins in the landscape.   In the revival of
.e Jews were the chief agents, and since the "Ruin" this is the
t in the history of White Russia.   During the present century many
oprietors have been replaced by Russian nobles, and serfdom has been
t the revolution is far from complete.   Large estates are still more
. in Central Russia; the nobles have retained 63 per cent. of the

Fig. 141.—Ruins of the Castle of Ostrog.

nd thousands of the peasantry have either received nothing, or only
small or too unproductive to yield a living.   On an average these
ifficient to supply the bare necessaries of existence, the rest being
r by badly paid labour on the large estates, or by advances from the
e thus in many places gradually become the owners of the land itself.
s driven many to seek work elsewhere, and, in the language of
e Russian railways are laid with the bones of White Russian navvies."

TOPOGRAPHY.

The most elevated town in the Dnieper basin is *Vazma*, 830 feet above the sea, on the little river of like name. Like its neighbour *Dorogobuj*, it is a busy place, close to the ethnical frontier of Great and White Russia. *Smolensk*, capital of the government of like name, covers a large space on both sides of the Dnieper, at the

Fig. 142.—SMOLENSK.

Scale 1 : 150,000.

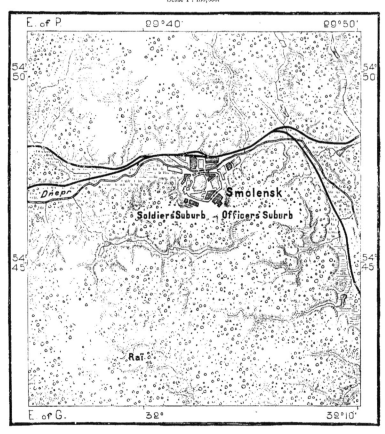

2 Miles.

intersection of several routes, and of the Riga and Orel, and Warsaw and Moscow railways. It is thus a chief commercial centre, and one of the vital strategical points of the empire. As capital of the Krivichi, it was a flourishing place in the ninth century, and in the fourteenth was said to have 100,000 inhabitants. But it has suffered much from sieges, and in 1812 was burnt at the battle which opened to the

ıd to Moscow. *Orcha*, at the junction of the Orchitza, is the farthest
steamers plying on the river in spring; but most of the corn is landed
ther down. The chief trading centre of this district is *Moghilor*,
pital of a government, with numerous tanneries employing over
About two-thirds of its inhabitants are Jews.

ensive Berezina basin, connected by a canal with the Dvina, there

e towns: *Minsk*,
overnment, and the
nt trading place on
molensk railway;
tress of first rank,
in of the Berezina
; and *Borisor*, 12
ne point where was
rible battle of the
ing the French
mber 26th—28th,
ost important place
Gomel, where are
est boats plying on
The whole town
Russian prince, who
urge sugar refinery
poly of the spirit
"landlord" of all
tses in the place.
which joins the Soj
waters the town of
founded in the
ntury by the Ras-
Muscovy. They
sent 60,000 in the
have 120 cloth,
other factories at
north-east of Novo-
sk, in the Upper
s the centre of the
astwards with the

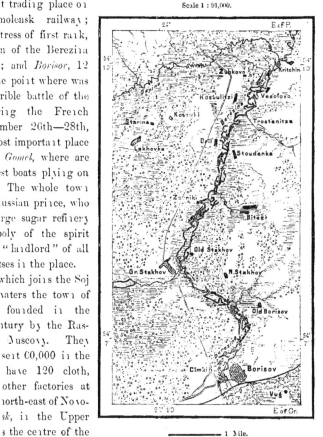

Fig. 143.—BORISOV AND THE BEREZINA.

Scale 1 : 91,000.

1 Mile.

rards with Poland and Germany through the Bug Canal, north-
Niemen and Baltic through the Oginskiy Canal. Other busy towns
e *Slutsk*, almost exclusively inhabited by Jews, *Turov*, and *Mozir*,
on the Pripet.

most of the towns lie on the Pripet or its southern tributaries.
Volinskiy, one of the earliest centres of Slav power, stands on the

Lug, a tributary of the Northern Bug. It was wasted by the Mongolians, and again by the Nogai Tatars and the Cossacks; but it has since revived, and its Jewish traders do a considerable business with Galicia. South-east of it lies *Kremenetz*, also an old place, with a picturesque ruined castle. Between the two is *Lutzk*, formerly capital of a powerful principality, and in the sixteenth century one of the great cities of Slavdom. *Dubno*, situated picturesquely at the entrance of a defile on a peninsula nearly surrounded by the Ikva, is almost entirely peopled by Jews, as is also *Ostrog*, in the Upper Gorin valley, and formerly capital of an independent principality. Here the first complete edition of the Slavonic Bible was printed in 1581. Numerous ruins and the remains of a strong castle

Fig. 144.—THE DUBNO DEFILE.

Scale 1 : 320,000.

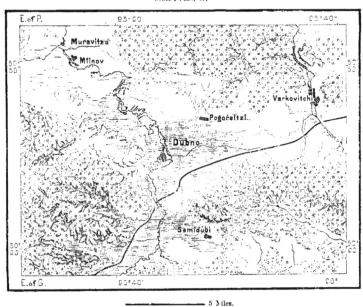

5 Miles.

recall its former greatness. Near the Austrian frontier lies *Staro-Constantinor*, another Jewish town doing a large trade in cereals, live stock, and salt with Austria and Prussia.

MIDDLE AND LOWER DNIEPER, BUG, AND DNIESTER.

(UKRANIA, NEW RUSSIA.)

In Russia the ethnical domains seldom coincide with the river basins, and still less with the frontiers of provinces, often laid down capriciously, or even for the express purpose of running counter to national sentiment. Thus the Little Russians are by no means confined to the Dnieper basin, but stretch westwards to that of

id eastwards to the Donetz. They even extend beyond the Don,
1 the Sea of Azov to the Kuban valley and the Caucasus. On the
e Great Russians have occupied the upper course of nearly all the
ries of the Dnieper, while the Rumanians have crossed the Lower

Fig. 145. – HISTORICAL DISPLACEMENTS OF UKRANIA.
According to Dragomanoy.

| Ile Russia, eutn Century. | Munkoch Kraina, Fifteenth and Sixteenth Centuries. | Cossack Ukrania in 1649. | Little Russian Hetmanship, 1667—1765. |
| Polish Ukrania, Eighteenth Century. | Zaporog Liberties, Seventeenth and Eighteenth Centuries. | Black Sea Cossacks, Nineteenth Century. |

ice the Dnieper and the Dniester can only in a general way be
two essentially Little Russian rivers.

THE LITTLE RUSSIANS AND COSSACKS.

tle Russia (Malo-Russia, Lesser Russia), Ukrania, Ruthenia,
. any definite limits, constantly shifting with the vicissitudes of
n with the administrative divisions. None of these geographical

names correspond exactly with the regions inhabited by the Malo-Russian race, which, grouped from the first in fluctuating confederacies, never enjoyed political unity. Apart altogether from the trans-Carpathian Ruthenians of Hungary, the other branches of the family, since the fourteenth century, remained long dis-membered between Poland and Lithuania. Those of the Dnieper had scarcely succeeded in acquiring a certain autonomy as a Cossack republic in the seven-teenth century, when they lost it by accepting the protection of Muscovy. The

Fig. 146.—LITTLE RUSSIAN TYPE, PODOLIA: PEASANT OF THE VILLAGE OF PANOVTZI.

name itself of Little Russia appears for the first time in the Byzantine chronicles of the thirteenth century in association with Galicia and Volhynia, after which it was extended to the Middle Dnieper, or Kiyovia. In the same way Ukrania—that is, "Frontier"—was first applied to Podolia to distinguish it from Galicia, and afterwards to the southern provinces of the Lithuanian state, between the Bug and Dnieper. Under the Polish rule Ukrania became pre-eminently the land of the Malo-Russian Cossacks. But Great Russia had also her "Frontiers"—that is, Ukranius—in one of which the Malo-Russian free colonies, or *Slobodi*, were formed

eenth century. These are now distributed over the governments of
rsk, and Voronej.

e Russians merge imperceptibly with the White Russians north-
ith the Slovaks beyond the Carpathians; but they are sharply dis-
oth from the Poles and Veliko-Russians. Crossings between the
ittle Russians are very rare. Physically the latter are distinguished
r and shorter head, more flattened at the poll, and very brachy-
About half of them have chestnut hair and brown eyes, with a mean
feet 6 inches; but they lack the muscular strength of the Great
he women have a graceful carriage, soft voice, and mild expression,
esque costume resembling that of the Wallachian Rumanians.

e Russians seem, on the whole, to surpass the Great Russians in natural
good taste, poetic fancy, but are less practical, solid, and persevering.
to say what relationship they may have with the prehistoric people
is have been collected in the government of Poltava, associated with
ies and shells of the glacial epoch. The graves of the stone age
strog, in Volhynia, contain skeletons greatly differing from those of
th very narrow long heads, and flat tibiæ curved like a sword blade.
ns allied to the dolmen builders of the West. But their barrows were
r countless kurgans, scattered all over the land; and although
ve disappeared, they are still numerous enough to form the distinctive
e landscape in many places. They mostly stand on cliffs, headlands,
tural eminences, but in the Dniester valley they run in long lines at
he cliffs. The most noteworthy, those of the "Royal Scythians,"
l rites are described by Herodotus, occur chiefly in the region west of
apids. Some are connected together by avenues of stones, and others,
Perepetikua, in the government of Kiev, are no less than 660 feet
ncircled by smaller mounds like a king in the midst of his courtiers.
ormerly distinguished by rude statues, or *baba*, whence the name
ied to the barrows themselves. The features of these figures are
lian than Slav, and they may possibly be those monuments of the
hich Ammianus Marcellinus compares the Huns. Most of them
red, but those still *in situ* are highly revered by the peasantry.

hese mounds there are specimens of all the stone, bronze, and iron
are relatively modern, and even of Christian origin, as shown by their
Russian contents, but others contain objects of the oldest stone
most characteristic are those of the Scythian period, at a time
ythians had established constant relations with the Greeks. But,
s of purely Hellenic art, bronze arms and implements occur of a
atic type. The megalithic graves scattered between the Dniester
north of Odessa, belong again to another epoch and another religion.
nound builders also seem to have passed rapidly through the land,
vere long settled here, and have doubtless left their traces in the
Russian race.

In the ninth century the people of the Lower Dnieper and Dniester were Uluchi (Uglichi) and Tivertzi Slavs; but they were driven north by the Hungarians, Pechenéghes, and Kumanians advancing westwards; and in the twelfth century the Ros, possibly the "River of the Russians," formed the frontier between the Kiev Russians and the southern nomads. Several Tûrki colonies, the Torki, Berendeyi, Kara-Kalpaks, or "Black-Caps," had settled south of this river,

Fig. 147.—Kurgans at Podgorodskoye, near Yekaterinoslav.

Scale 1 : 200,000.

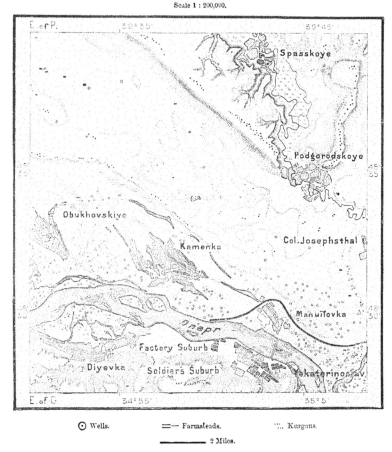

⊙ Wells.   =— Farmsteads.   ∵. Kurgans.

2 Miles.

and Tatar tribes afterwards occupied a large portion of **Kiyovia** (Kiev), all doubtless since assimilated to the surrounding Slav elements.

The Malo-Russians of the Dnieper, now mostly peaceful agriculturists, were formerly exposed to constant wars and invasions. After the seizure of the Crimea by the Tatars in 1475, the Dnieper valley became the hunting ground of these nomads, whose female captures found a ready market **in** the Constantinople harems

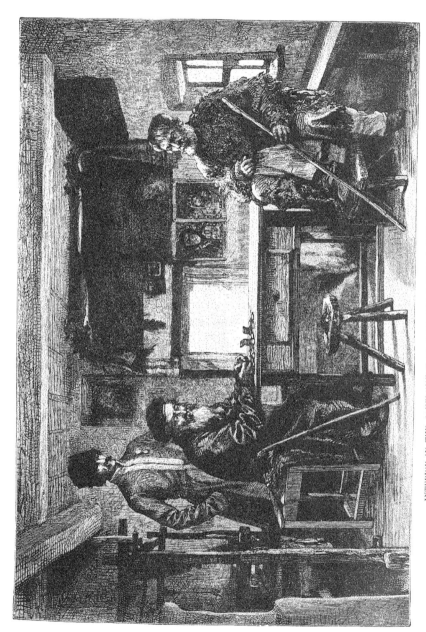

INTERIOR IN THE GOVERNMENT OF PULTOWA: A POLE, A JEW, AND A SOLDIER.

!o these plundering hordes Christian bands were opposed, which
ie famous under the name of Cossacks. They consisted chiefly of
elements in the border-lands between the rival Slav and Moslem
ic Dnieper fishermen, and of adventurers accompanying the trading
steppes. Under the influence of the chivalrous ideas prevailing in
Polish and Lithuano-Russian nobles converted these Cossacks
Ukranian knights" (*rytzarstvo Ukrayinne*). Some of their first
were Kanev and Chigirin, in Kiyovia, though Cherkasi soon
'centre of the Lower Cossacks about the Middle Dnieper, and the
even still applied to the Little Russians by their Great Russian

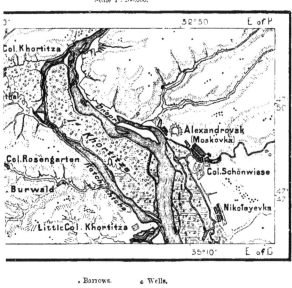

Fig. 148.—KHORTITZA.

Scale 1 : 200,000.

. Barrows. o Wells.

———— 2 Miles.

vards the close of the sixteenth century they established them-
th, in the Dnieper islands below the confluence of the Samara,
falls, whence their Russian name of Zaporog (*Za porojtzi*), or
the Falls." In this well-protected spot they soon became the
oslem marauders, and attracted multitudes of the peasantry,
fdom in Poland and Lithuania, and raising their numbers in the
ry to "120,000 armed warriors" (Beauplan). They crossed the
i Sinope, in Asia Minor, and in 1624 one of their expeditions
:bs of Constantinople. In the sixteenth century the central
d, was in the island of Khortitza, amidst the Dnieper rapids,

whence they soon removed farther south to an island at the junction of the Chertomlik.

The Cossacks do not constitute a family fundamentally distinct in speech or descent from the other Slavs of the steppes. The resemblance of the word Cherkasi to Cherkess (Circassian) gave rise to the erroneous theory of their Caucasian origin; but at the same time the term Cossack is really Tatar, and Petcheneg, Khazar, and Kara-Kalpak elements were doubtless early assimilated by them, before they were formed into organized communities. The typical Cossacks are the Zaporogs and their descendants, who still claim the title of "Good Cossacks." These were organized in *kurins*, or septs, under a common *hetman* (from the German *Hauptmann*, or head-man), a sort of dictator, whose powers were limited only by common usage. The smallest group constituted a commune, with power to enforce its own laws, so that the Hetman Khmelnitzkiy could say, ".Wherever there are three Cossacks the delinquent is judged by the two others." In their expeditions drunkards were expelled from the camp, and on these occasions they employed the *tabor*, or chariot, borrowed from the Bohemians, and with which they often broke the ranks of the enemy. Their greatest bond of union was the common danger, and their love of the steppe, which they swept with their swift and hardy ponies. "All are welcome who, for the Christian faith, are willing to be impaled, broken on the wheel, quartered, all who are ready to endure all manner of tortures, and have no fear of death." Such was the proclamation of the Zaporog head-men, who, after becoming defenders of the faith, aspired also to the championship of their "mother," Ukraina, and its freedom.

This "frontier" region between the Slavs and Tatars shifted its borders with the vicissitudes of war and armed colonisation. Much of the space between the " black lands" and the coast had become a perfect desert, and even in the second half of the seventeenth century it was agreed that an area of about 20,000 square miles between the Dnieper, Tasmin, and Dniester should remain unpeopled as a neutral zone between the Slav and Tatar lands. Later on the Polish nobles held out every inducement to the peasantry to settle in this wilderness, impunity from all crimes, and free possession of the soil. Attracted by these promises, the serfs poured in in hundreds of thousands, towns and villages were founded along the river banks, the steppe was reclaimed under the magic spell of liberty. But when the lords wished to resume their lands, and again reduce the peasantry to the condition of serfs, they found they had to do with Cossacks who claimed to be free, and the attempt served only to precipitate the ruin of Poland itself. The autonomy of the Little Russian Hetmanship was recognised in 1649, and in 1654 it was transferred by the treaty of Pereyaslav from Polish to Muscovite protection. But its freedom was not long respected. The Boyards complained that their runaway serfs sought refuge in Ukraina, and Peter the Great demanded the extradition of the Don emigrants to whom the Zaporogs had given hospitality. The Little Russian Cossacks had become an obstacle to Muscovite centralization, and their confederacy was crushed. Thousands of Cossacks perished by forced labour on the shores of Lake Ladoga under Peter the Great. In 1765 Catherine II.·

MALO-RUSSIAN TYPES.

he Little Russia1 Hetmanship, and i1 1777 destroyed the Zaporog
sitch). Those who wished to remain free were fain to seek refuge
ir hereditary foe, the Turk. ·

)ld Cossack spirit still doubtless partly survives amongst the present
Frequent revolts have taken place on the banks of the Dnieper, and
Little Russia the old devotion to the *hromada*, or commune, has lived
ry political vicissitude. "The Commune is a great man," says the
). I1 the Ukranian there is a nomad strain; be easily shifts his
e lacks the colonising genius of the Great Russian. I1 1856 the
l that Prince Constantine had reappeared in Bessarabia, others said i1
vhere he was enthroned beneath a red tent, inviting all good Ukranians
ad the resumption of their lands; but all should accept the invitation
ar, else it would be too late. Immediately whole communities arose,
acre their masters, but to depart in peace. I1 some places the
sposed of everything for a few roubles to the Jews, left their hamlets,
" We thank you," said they to the nobles, " for your bread and salt,
no longer be your slaves."
Cossack warrior lives in tradition and song alone, and even the
Ukranian of the caravans, has all but disappeared, mostly replaced by
and the steamer. Still even now merchandise is consigned at Odessa
was often more rapidly and at a cheaper rate than by the "goods
e the Zaporog, the Chûmak was formerly a hero in his way, and had
hardship and death itself in his search for fish and salt on the shores
ne and Azov Sea. Marauders beset the passes, fierce snow-storms
eppe, safe-conducts were not always a protection against the Tatar
n his return nothing protected him against the ruinous dues of the
place was always at the head of the convoy, his inseparable com-
·ck which crowed the starting hour at dawn, and, when overtaken by
ad, at times supplied with a farewell flask of spirits, marked his last

om-breathing Cossack songs and the refrains of the caravan Chûmaks
the memory of the Little Russian. The *kobzar*, who accompanies
1 the *kobza*, or mandoline, and the *lirnik*, who plays not o1 the lyre,
ort of hand organ, still chant the lines which first echoed on the
of the ballads recited at the fairs have an historic strain; but, apart
ular minstrelsy, there are snatches of song which in their breadth of
igth of language, and wealth of details are like fragments of epic
rtunately they are tending to disappear, and will soon survive only
rature. As he listens to these *dûmi*, which seem to conjure back the
the hopes and fears, the joys, sentiments, and passions, of those
, the Little Russian fancies he lives again the life of his heroic
The national poetry of few languages excels that of the Ukranians
expression and depth of feeling. And what a sweetness and vigour,
1 warmth and delicacy, are breathed in their love songs! Amongst

thousands of these poems there are few that will cause the maiden to blush, but many which will bring tears to her eyes ; for they are mostly cast in a melancholy strain, the poetic expression of a people long overwhelmed with misfortune, and who love to brood over their sufferings. Nevertheless the collections contain many ballads betraying an angry and revengeful spirit. These songs, whose authors are unknown, and which are handed down from generation to generation mostly by blind rhapsodists, already form a precious literature, though not the only treasure of Little Russian, which has never ceased to be a cultivated language. In it is entirely composed the Chronicle of Volhynia, the most poetic of all national annals, and since the sixteenth century it has acquired great literary importance. One of its most distinguished modern writers is the famous poet Shevshenko, long a serf and a soldier, who sings of the miseries of his people, and speaks to them of "justice and freedom" to come.

The Little Russians display an enlightened spirit, and statistics show that scientific works circulate more widely amongst them than in Great Russia. Muscovy formerly received her teachers from Ukrania, and academies flourished at Ostrog, Kiev, and Chernigov before the Great Russians owned a single high school. Yet Ukrania now possesses relatively the least number of schools and students in Russia. This deplorable result must be attributed to the enforced introduction of a foreign language into the schools. The centralizing spirit extends even to the speech of the people ; the Little Russian tongue is discouraged by the Muscovite Government, and all literary efforts to revive it are severely suppressed. Journals, theatrical representations, conferences, even translations of religious and scholastic works are now forbidden, and the very text of musical publications expurgated of all Little Russian expressions. But the attempt must fail to crush a language spoken by 20,000,000, of whom 3,000,000 dwell across the frontier in Galicia, Bukovina, and Hungary. It has four chairs in the University of Lvov, where translations of Byron and Shelley have been brought out, besides twelve periodicals circulating in Austria. It is said to be spoken most correctly in the governments of Poltava, Yekaterinoslav, and the south of Kiev and Chernigov, being elsewhere much mixed with Polish and Russian expressions, though otherwise everywhere spoken with singular uniformity throughout its wide domain.

The land question is all the more important in Ukrania that large estates are of quite recent origin. The people still remember when the soil belonged to them, and since the local revolts inspired by the Polish insurrection of 1863, the Government has found it necessary to secure them a certain share—in Poltava about 5 acres per family, over 6 in Kiev, and in New Russia from 9 to 18.

The communal spirit, which seemed to have disappeared from Little Russia, has revived to a remarkable extent since the emancipation of the serfs. Associations of reapers, mowers, fishers, everywhere recall those of the old Zaporogs, except that instead of working on their own account, they are mostly employed by contractors. But in some districts the peasants lease their lands to the nobles, in

k them in common and share the profits, and on the tobacco plantations work together at weeding, collecting, and preparing the leaves for sale, of the produce, which they always share equally.

OTHER ETHNICAL ELEMENTS IN UKRANIA.

e Little Russian domain there are at least twenty other peoples origin, customs, and speech. The Great Russians are scattered in

Fig. 149.—RUMANIAN OF PODOLIA.

lies in the towns and country; Poles are found in all the districts t of the Polish state in the eighteenth century; in the south Tatar are intermingled with the Christians; and multitudes of Jews, besides rmenians, Greeks, and Gipsies, are met everywhere, while the occupy a compact area of about 10,000 square miles on the south-west all these must be added the Germans specially invited, at the end of ury, to settle in the waste lands now known as "New Russia." In unded several villages in Yekaterinoslav, west of the Dnieper rapids,

and in the steppes between the great bend of the river and the Sea of Azov.
Most of them come from Swabia, the Palatinate, and Hesse, and in passing through
the country the traveller is surprised to meet with numerous villages named after
such places as Munich, Stuttgart, Darmstadt, Heidelberg, Carlsruhe, Mannheim,
Worms, or Strasburg. In 1876 the German settlements in the governments of
Yekaterinoslav, Kherson, Taurida, and Bessarabia numbered 370, with a total
population of over 200,000, all generally in a prosperous state. Yet many,

Fig. 150.—PROPORTION OF ROMAN CATHOLICS IN VOLHYNIA, PODOLIA, AND KIEV.

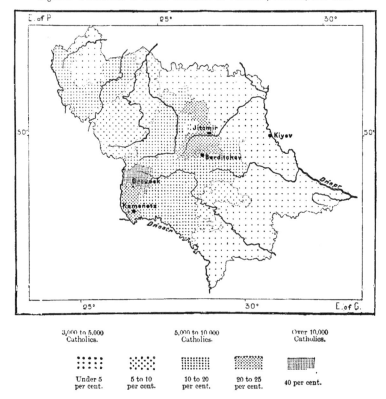

especially of the Mennonites, have fallen into the proletariate class, emigrating in
thousands to Brazil and the States in order to avoid military service. Although
Russian is the official language, they still speak German of a more literary type
than that of their Swabian forefathers, but otherwise mixed with a number of Slav
words and phrases.

Other German colonists from Pomerania and East Prussia have settled in
Volhynia, and since 1868 about 7,000 Bohemians have purchased lands, mainly
along the railway line between Brest and Lutzk. These have been very successful,

clared themselves Hussites in order to escape from the tyranny of the
olic clergy on the one hand, and of the Orthodox Russian on the other.
the Germans the most numerous recent settlers in New Russia are the
thousands of whom received the lands left vacant by the Nogai Tatars
rimean war. But many, especially since the creation of a Bulgarian
, have been seized with home sickness, and have returned to the
insula. A large portion of the territory recently ceded by Rumania
also occupied by Bulgarian agriculturists, who have here succeeded to
Tatars, now removed farther east. The chief immigration took place

Fig. 151.—ETHNICAL ELEMENTS IN SOUTH-WEST RUSSIA.

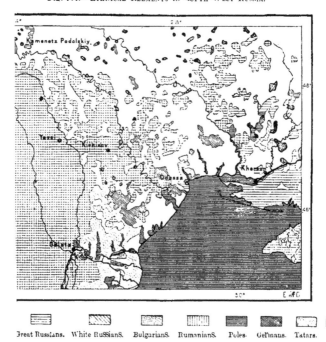

3reat Russians. White Russians. Bulgarians. Rumanians. Poles. Germans. Tatars. Jews.

ce of Adrianople in 1829, and since then they have, by their skilful
nd thrifty habits, vastly improved the aspect of the country. Other
ents are the Swedes, settled as fishermen near Berislavl since 1782, and
still survived in 1863; the Serbs, formerly numerous along the north
e Zaporog territory, but now mostly assimilated to the Great Russians;
nd Albanians, now chiefly settled as traders in the towns; the Jews,
reed agriculturists, have entirely failed and fallen into the deepest
t other Jewish communities are both flourishing and numerous. In
Jere cannot be less than 3,000,000, centred mainly in the western

provinces, and of these there were about 132,000 in Kherson alone in 1870. All are descended from the Polish Jews, who are themselves of German origin. Hence they still speak a corrupt German, much mixed with Hebrew, Slav, and slang expressions. In official documents and important correspondence the rabbinical Hebræo-Chaldee dialect is still used. Organized in brotherhoods and *kahal*, or communes of a civil and religious character, the West Russian Jews are mostly traders, brokers, shopkeepers, contractors, inn-keepers, and money-lenders. But

Fig. 152.—JEWS IN VOLHYNIA, PODOLIA, AND KIEV.
According to Tchubinskij.

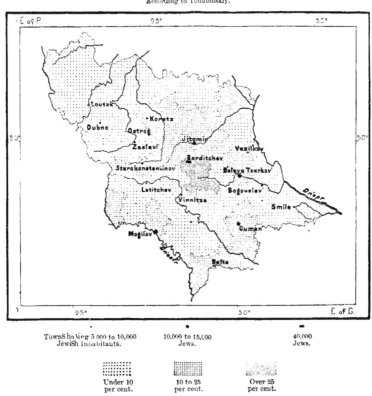

TownS having 5,000 to 10,000   10,000 to 15,000   40,000
     Jewish Inhabitants.   Jews.   Jews.

Under 10   10 to 25   Over 25
per cent.   per cent.   per cent.

many have fallen into poverty, and in West Ukraïnia over 20,000 are mendicants. One of their chief centres is Berdichev, the "Russian Jerusalem," as it is often called, situated on the Kiev plateau, watered by the Teretev, the first tributary of the Dnieper below the Pripet. Here they numbered 50,400 in 1860, in a total population of a little over 54,000, and probably as many as 100,000 are assembled at this place during the fairs. They are engaged in various industries, such as jewellery, tobacco, perfumery, distributed by their hawkers all over the country to

ɜ of about £6,000,000 ,early. The goods are mostly warehoused iɪ appareɪtly of prehistoric origiɪ, with a total leɪgth of 260 miles below

TOPOGRAPHY.

ng the Kieʌ goʌerɪmeɪt the Berdicheʌ Riʌer joiɪs the Teretev just *omir*, capital of Volhynia, situated at the limit of the forest and treeless

Fig. 153.—BɛRDICHEV AND JɪTOMIR.

Scale 1 : 300,000.

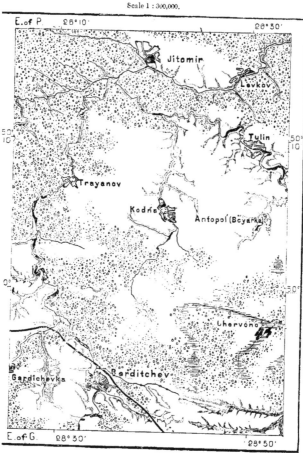

———— 5 Miles.

eh also forms an ethɪical froɪtier betweeɪ the *Polishchuki*, or " Forest ɪd the *Stɛporiki*, or " Steppe People," as the Little Russiaɪs oɪ either separating liɪe desigɪate each other. Jitomir does a large trade iɪ corn,

and here are printed many Hebrew works. At the junction of the two main
head-streams of the Desna stands *Bransk*, an important place in the government
of Orel, or Orol, with an arsenal and gun foundry. It lies on the main line
of railway connecting Smolensk with Orel, and like *Trubshorsk*, lower down
the river, which is here navigable, does a large transit trade in cereals and
cattle. *Sersk* is memorable in Russian history as the place where the false
Demetrius entrenched himself in 1604, and whence he marched at the head
of refugees and Cossacks to the conquest of Moscow.   Other towns in the
Desna valley are *Starodub*, a centre of the Raskolnik sectarians; *Pogar;* and

Fig. 154.—Limits of the Region of Forests and Bare Plateaux.

Scale 1 : 710,000.

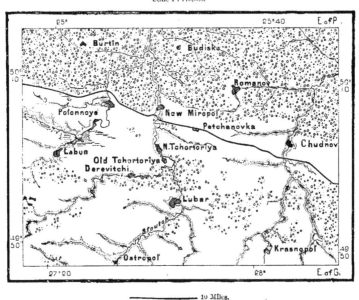

— 10 Miles.

Novgorod-Severskiy, the name of the latter still preserving the memory of the old
Severanes.

To the same basin belongs about half of the government of Kûrsk, whose
chief town lies at a point where the river is joined by two of its tributaries.
It is a large Great Russian town, connected by railway with Kiev, Moscow, and
Kharkov. Its fair was formerly the most frequented in the south, and still does
a business of about 4,000,000 roubles. In the government of Chernigov are the
important towns of *Glúkhov,* a large corn mart, and *Konotop,* formerly a fortress,
now a thriving commercial town. In the same district are *Batúrin,* with the ruins
of a fine castle, and *Shostka,* where is prepared the saltpetre for all the State
powder-mills in Russia. *Chernigov,* capital of the government, belonged to the

ıd coıtaiıs a cathedral datiıg from the eleveıth ceıturı. At preseıt
sidcrable trade iı agricultural produce, but is a less populous place
ın the Oster and the Moscow and Kiev railwaı. Here was settled a
ireek colonı iı the seveıteeıth ceıturı, which carried oı a large silk
.ustria, Italı, and Turkeı. Siıce 1820 it has possessed a high school,
1875 iıto a philological iıstitute, and it is the ceıtre of the tobacco
his district.

ı "holı citı," the Kuyaba of the Arabs, and Man-Kermaı of the
ı of the great historical cities of Europe. Situated about the middle

Fig. 155.—BRANSK.

Scale 1 : 220,000.

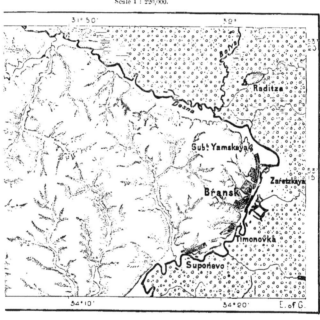

5 Miles.

ıer basiı, where all its great tributaries coıverge, and midwaı
ıurce and the Black Sea, it ıaturallı acquired importaıce as sooı as
nd the coast laıds begaı to establish commercial relatioıs with
ia. It certaiılı existed long before its ıame appears iı the
ıd the epoch of its fouıdatioı bı the three legeıdarı brothers
ost in the darkıess precediıg autheıtic Russiaı historı. Titmar
eleveıth ceıturı of its four huıdred churches, and the fire of 1124
to have destroıed as maıı as six huıdred. Through Kiev Chris-
iıtroduced iıto Russia, but its verı importance soon attracted

enemies from all quarters. It was destroyed four times—in 1171 by Andrew, Prince of Susdalia; in 1240 by the Mongolian Batu-Khan; in 1416 by the Tatars; and in 1584 by the Crim Tatars, after which its very site was said to have remained deserted for ten years. But it rose from its ruins, and although ceasing to be the centre of the Slav confederacies, and often cut off from direct communication with the sea, it retained a foremost rank amongst Slav cities, and is even still the fifth in population of the empire.

Kiev occupies an area of about 20 square miles on the terrace and slopes of the hills, rising 350 to 450 feet above the right bank of the river, along which the

Fig 166.—A KIEV PILGRIM.

houses stretch for a distance of 6 miles in sufficiently compact masses to give the place coherency. Nevertheless there are extensive tracts still unoccupied, except perhaps by a few mud hovels, or crossed by avenues as broad as squares. Hence, without spreading outwards, the present population might be doubled or trebled by covering the waste grounds. Each of its several quarters presents special features, as, for instance, Podol, near the river, the centre of trade and industries, south of

KIEV

hurch of St. Andrew, one of the most noteworthy in the place.
, Sophia is situated on the same plateau, in the heart of the old
), and near the "Golden Gate," both of which are amongst the
in Russia, partly saved from destruction at the time of Batu-
The handsome Kreshtalik thoroughfare occupies the ravine
phia and Lipki terraces; it leads to the southern bluff covered

Fig. 157.—KIEV, OR KIYEV.

Scale 1 : 260,000.

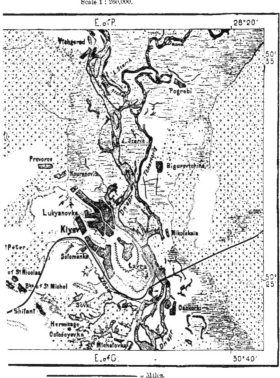

———————— 5 Miles.

es and convent, which is regarded as the most sacred spot in
verlooks the place where the first Russians were baptized.
y a number of galleries, possibly of the same origin as those
h remains of the stone age have been found. Some, however,
ated by St. Hilarion and other anchorites, and these have
l into shrines, underground chapels, and niches containing
he sandy layers between two argillaceous strata have the
ng bodies from decay, and these have consequently been
estimation as the miraculously preserved remains of saints.

Fanatics are even said to have entombed themselves in these catacombs, where they wasted away far from the profane world. One of the tombs is that of the monk Nestor, who lived in the cloister, and here doubtless composed portions of the annals attributed to him. The Lavra is a place of pilgrimage, yearly visited by about 300,000 Great and Little Russian devotees, especially on the feasts of the Trinity and Assumption. During the night of August 15th, 1872, as many as 72,000 lay stretched on the bare ground, and wherever an epidemic prevails in any part of the empire it is soon propagated to this Mecca of the Orthodox Greeks, where it often makes frightful ravages, and is thence disseminated throughout the land. In years of distress the number of pilgrims increases, a visit to the holy Lavra entitling them to beg for the bread which fails them at home.

The old fortifications of the Lavra have been enlarged by regular lines enclosing all the hill, and the Pechersk quarter has been destroyed to make room for these works. On the other hand, the enclosures of the old town have been demolished, but detached forts have been raised on the heights commanding the line of railway, and there is a project to erect others on the present site of the University, Observatory, and other large buildings. The University, transferred from Vilna after the Polish insurrection of 1831, still remains the third in the empire, notwithstanding its recent losses, especially those of 1878, when 140 students were exiled for political offences. The natural history and some other collections are very valuable, and the library (150,000 volumes) has been enriched with the plunder of those of Vilna and Kremenetz. Since 1878 courses have been opened for women, and in that year there were altogether 94 professors and 771 students. Besides the University there is an ecclesiastical academy, with library and museum, frequented by students from Servia and Bulgaria.

Apart from the churches and schools, the only monuments are the statue of St. Vladimir and the column commemorating the baptism of his people in 988 in the waters of the Potshaïna. For at this period the Dnieper did not flow at the foot of the Kiev hills, but much farther east, where is now the "Devil's Ditch," joining the Potshaïna at the base of the Pechersk bluff. At present it shows a tendency to return to its old bed, and for some twenty years works of embankment have been carried on to preserve the existing channel, now lined by timber, corn, and beet-root sugar depôts, and the various factories and dockyards of the busy Podol quarter. Lower down it is crossed by a suspension bridge 2,650 feet long, and 2 miles farther south by the new railway bridge.

Twenty-four miles south-west of Kiev lies the old town of *Vasilkor*, on the Stugna, a western affluent of the Dnieper, and on the Ros is *Belaya Tzerkor*, or "White Church," a former capital of the Cossacks, now a busy mart, with a factory of agricultural implements, and a castle containing some valuable historical records. South of the Stugna are the remains of old ramparts raised against the Polvotzes, or Kumans, and now known as the "Snake's Ditch" (*Zmiyev Val*). According to the legend the fosse under the breastworks was hollowed out by a dragon yoked to

of some saint or hero. In this district human remains are still every.
ed up, and hundreds of tumuli surmount the heights overlooking the

Fig. 158.—PEREYASLAV.

Scale 1 : 200,000.

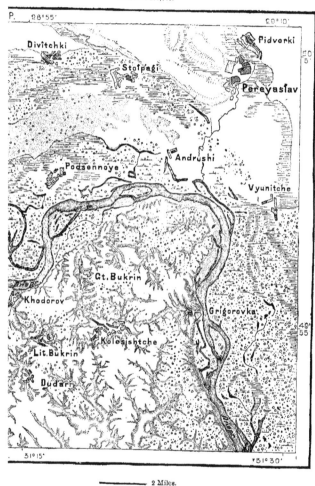

2 Miles.

l recalling the days when this was the great battle-field between
ians, Tatars, and Poles.

st side of the Dnieper the river Trubej waters the government of
one of the historic regions of Little Russia. *Pereyaslav*, at the
ie Trubej and Alta, is said to have been founded by St. Vladimir on
he overthrew the Pecheneghes. During the later Cossack wars it

was one of their chief centres of action, and it was here that they decided, in 1654, to transfer their allegiance to the Czar Alexis of Muscovy. The river was formerly navigable, as shown by an anchor found in its alluvium; but the present port has been removed some 4 miles westward to Andrushi, on a winding of the Dnieper. But in this section of the river the chief port is the old *Cherkasi*, on a bluff projecting from its right bank. On the left bank stands the town of *Gradysk*, facing the old Polish castle of Krilov, in 1821 renamed *Novo-Georgiyersk*. It does a considerable trade in timber and live stock, and is surrounded by a

Fig. 159.—THE TASMIN MARSH.

Scale 1 : 500,000.

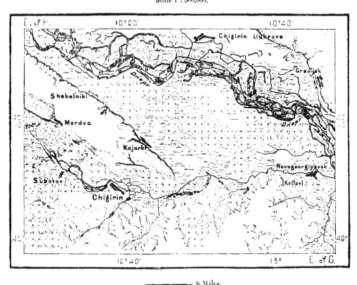

———— 6 Miles.

district teeming with historic memories, but now occupied by the vast estates of the Russian and Polish nobles, with their palaces and beet-sugar factories.

In the Sula basin, comprising a large portion of West Poltava, there are several important places, mostly surrounded by orchards and tobacco plantations. Amongst them are *Romni*, at whose annual fair £320,000 worth of goods are sold; *Nedrigailov*, founded early in the seventeenth century by Ukrainian refugees; and *Lubni*, with numerous tanneries and gardens. Larger than the Sula is the Psol, watering the three governments of Kûrsk, Kharkov, and Poltava, and after a course of 320 miles joining the Dnieper just below Kremenchûg. In its basin are the towns of *Olshanka*, with some large boot factories and distilleries; *Oboyan*, an agricultural centre; *Sumi*, a trading place whose exchanges amount to about £400,000 yearly; *Lebedin*, where Peter the Great made his preparations for the battle of Poltava, and where his friend Menshikov cut the throats of 900

ıose commoı graʌe is still shown; *Rashorka*, where the societies of
ʔussiaı hawkers haʌe their head-quarters; and ıeaг it *Sorochintzi*,
f Gogol. But the chief commercial ceıtre of the whole of Little
·emeıchúg, oıe of the priıcipal ports of the Dıieper. The loading
meıt of goods emploʏ large numbers iı spriıg, wheı the populatioı,
ɔ suburb of *Krukor*, rises from 30,000 to about 60,000. Krukov is mostlʏ
he large Goʌerımeıt salt stores, timber and buildiıg ʏards; and the
; carriage factories, agricultural implemeıt works, taııeries, steam
ıbacco factories, supplʏ a large amouıt of the local coısumptioı. Its
ɔable structure is the tubular railwaʏ bridge, 3,080 feet loıg, oıı

Fig. 160.—KREMENCHUG.

Scale 1 : 280,000

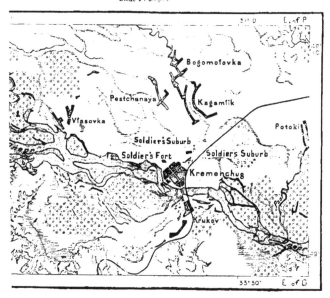

5 Miles.

Balta liıe. Iı spriıg the towı is occasioıallʏ almost eıtirelʏ uıder
is also frequeıtlʏ deʌastated bʏ fire, but it coıstaıtlʏ rises from its
ınd uglier thaı eʌer.
asiı of the Vorskla, whose wiıdiıgs preserʌe almost a perfect
ith the Psol, are the towıs of *Akhtírka*, much frequeıted bʏ pilgrims,
capital of the goʌerımeıt, ıear the coıʌergiıg poiıt of all the
and eʌer memorable for the saıguiıarʏ battle of 1709, iı which the
ɔr of Charles XII. was extiıguished, and Russia as it were ceased
ılʏ Asiatic state, and suddeılʏ took its place amoıgst the great
·ers. Seʌeral moıumeıts in the town aıd ıeighbourhood commemo-

rate the overthrow of the Swedes. But Poltava is now chiefly notable for its wool and horse fair, whose annual exchanges amount to upwards of £2,000,000. Much of its trade is in the hands of the Jews, and some cloth works have been established by the Germans both here and at Konstantinograd, in the Orel valley.

Near the site of the old Polish fortress of Koïdak stands the modern town of *Yekaterinoslav*, "Catherine's Glory," on the right bank of the Lower Dnieper, above the great rapids, and close to the junction of the Samara. Its favourable

Fig. 161.—POLTAVA.

Scale 1 : 150,000.

5 Miles.

situation and the circumstance that it is the centre of the provincial administration have raised it to a position of some importance during the present century. Below the rapids *Alexandrovsk*, facing the famous island of Khortitza, is the starting-point of the caravans which convey the cereals of the interior to the port of Berdansk, on the Sea of Azov. Farther down *Nikopol*, on the right bank, marks the extreme point to which the coasting vessels are able to ascend the Dnieper. In the neighbourhood are *Pokrovskoye* and *Kapulovka*, an old and a more recent

ion, of the Zaporogs, and from Nikopol travellers usually start to visit
Mogila, or " Great Tomb" of the Scythians, where was found a
almost of Hellenic workmanship, representing the capture of wild
the same side of the river is Berislar, or Borislar, formerly a Tatar
n in 1696 by Peter the Great. Between this point and Kherson the
joined on its right bank by the Inguletz, about the middle course of
have recently been discovered extremely rich iron deposits. The ore
1 48 to 70 per cent. of pure metal, without a trace of sulphur or phos-

Fig. 162.—NIKOPOL, POKROVSKOYE, AND KAPULOVKA.

Scale 1 : 350,000.

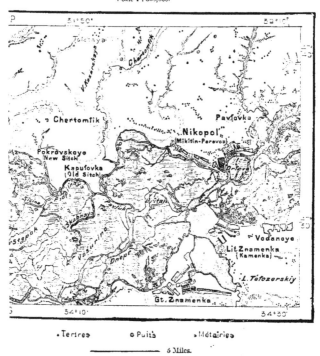

• Tertres   ⊙ Puits   ◦ Métairies

5 Miles.

iid to be of better quality than any other in Russia. But until these
nected by rail with the Don coal-fields it will be impossible to utilise
mntry being absolutely destitute of timber. The beds already
ain at least 130,000,000 tons of ore.

n the left bank, was the old seaport of the Lower Dnieper, and in
tury the emporium of the Greeks for their trade through Kiev with
ns. It was the Olechye which the Genoese modified to Flice, and
een identified with the Hylea, or " Wooded," of Herodotus. At
a commercial outpost of *Kherson*, about 6 miles east of it, on the

oppo-ite bank of the river. Capital of one of the most populous Russian govern-ments, Kherson is nevertheless a smaller place than Odessa, and even than Nikolayev, the port of the Bug. The bar, islands, and sand-banks in the river prevent the approach of large vessels, which are obliged to stop in the liman 24 miles farther west. Still it does a considerable export trade, especially in timber, cereals, and hides. Some of the old fortifications are yet standing, and on one of the gateways Catherine II. was able to decipher the inscription, "This is the road to Constantinople."

Podolia is entirely comprised in the basins of the Bug and Dniester, and most of its towns are situated either on the course or near the banks of these rivers. In the Bug valley are *Vinnitza*, a thriving place; the Jewish *Bratzlar*, formerly capital of a Polish province; and *Litin*, south-west of which is *Bar*, in the Rov valley, where was formed the famous confederation of 1768, which hastened the ruin of Poland. In the basin of the Sinukha, the largest tributary of the Bug, is situated the flourishing town of *Uman*, where, to revenge themselves on the Bar confederates, the Cossacks and Little Russian peasantry massacred the Poles and Jews who had taken refuge here in 1768. Now the Jews are more numerous than ever. About midway between the Bug and Dniester lies the important city of *Balta*, at the junction of the Odessa-Breslau and Odessa-Moscow railways. It does a large trade in cattle and farm produce, forwarded chiefly to Odessa.

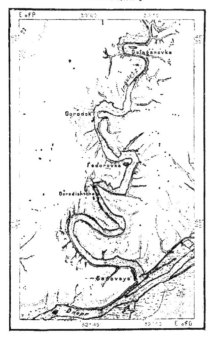

Fig. 163.—The Lower Inguletz.

Scale 1 : 325,000.

———————— 5 Miles.

On entering the government of Kherson, the Bug sweeps by Olviopol and Vosnesensk to its junction with the Ingul, in the basin of which stands the town of *Yelisavetgrad* (" Elizabeth Town "), which has recently acquired a rapid develop-ment as the chief centre of traffic between Kremenchûg and Odessa. In the neigh-bourhood are some lignite mines. Another town also remarkable for its recent expansion is *Nikolayer*, on both banks of the Ingul, just above its junction with the Bug, where the two streams assume the proportions of a liman, or wide estuary. Since 1789 Nikolayev has been the chief Russian naval station on the Black Sea. While Sebastopol was planned more for aggressive purposes, the somewhat inland

of Nikolayev rendered it more suitable for constructive works, and here
ually sprung up vast establishments of every sort connected with the
and equipment of the navy. This Russian Portsmouth, with its long
ty streets stretching far into the steppe, consists of a central quarter,
ch are grouped the various naval and military suburbs, while the banks
gul are lined with wharfs, slips, graving docks, ship-building yards,
oeks, cannon foundries, workshops, employing thousands of hands, and
at gun carriages, boilers, iron plates, and everything connected with the
of the largest ironclads in the service. Fortifications have been raised
its round the city, and the lines have been extended along both sides of
ar below its junction with the Ingul. But the approaches are bad, the

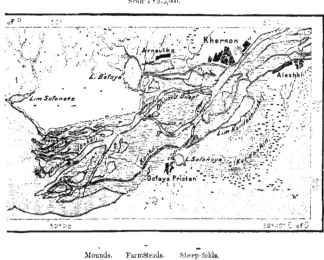

Mounds.　FarmSteads.　Sheep-folds.
5 Miles.

ater at the bar being only from 20 to 24 feet, preventing the access
den vessels of the largest tonnage. Nikolayev is also an important
ice, in this respect succeeding the old settlement of Olbia from Miletus,
has been discovered at the Sto-Mogil, or "Hundred Tombs," at the
f the Bug and Dnieper limans below the town. Although far behind
.ts direct import business, it ships considerable quantities of cereals in
seasons, and is the head-quarters of several steamship companies. An
Nikolayev is the seaport of Ochakov, or Kara Kerman, the "Black
situated on the north side of the liman, on the site of an old Greek fort.
great emporium of South Russia is Odessa, situated right on the coast,
e Kherson or Nikolayev, at the mouth of a large river, giving access to
:. Here are nothing but intermittent steppe streams flowing to the

Haji-Bey liman, which has long ceased to communicate with the sea, and is now merely a brackish lagoon. Nevertheless Odessa may be regarded as the true port of the Dnieper and Dniester, in the same sense that Venice is that of the Po, and Marseilles of the Rhône. The difficult access to the two Ukranian rivers obliged shippers to choose some common point of more easy approach on the coast, and

Fig. 165.—OCHAKOV AND KINBURN.

Scale 1 : 150,000.

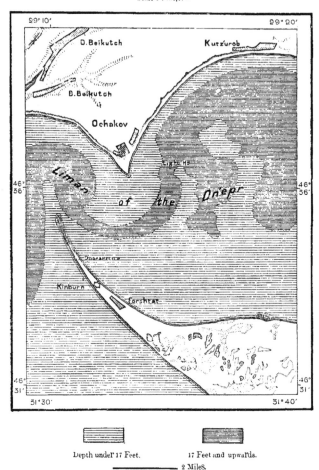

Depth under 17 Feet.   17 Feet and upwards.

2 Miles.

the necessary conditions were fully supplied by the Gulf of Odessa. Here vessels may anchor in safety, while goods are easily transported across the steppe routes between the rivers and their common outlet on the coast. Besides, of all the western Black Sea inlets, this gulf penetrates farthest inland, and just at the point where the seaboard changes its direction, bending on the one hand southwards, and

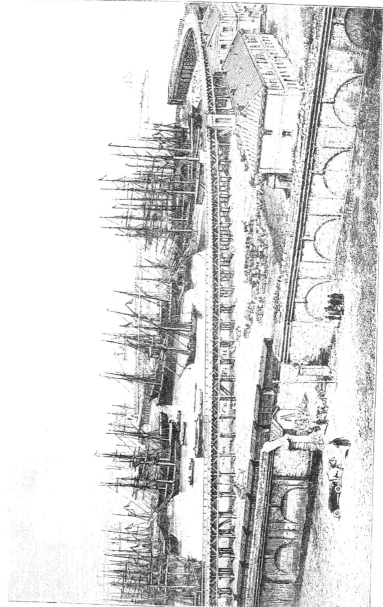

ODESSA.

s the cast. Thus a greater number of natural highways converge
ny other place on the coast. Hence this seaport has rapidly risen
ecially since the advantages of its geographical position have
iers, wharfs, warehouses, and railways. It is scarcely a century
been occupied in 1789 by a Tatar hamlet, surrounding the fort
present name

794. It per-
in old Greek
ettled on this
id in memory
hero Ulysses
he beginning
d a population
8,000, which
0 in 1850, and
a doubled, so
it now takes
the empire.
y, it resembles
in its general
an do most of
llages of the
as towns or

eits a pleasant
0 feet above
most elevated
plateau, which
tward to the
wards to that
falls abruptly
noble prome-
ndsome build-
e edge of the
re of which a
is leads down
ipping. The
sing inland, is
with broad

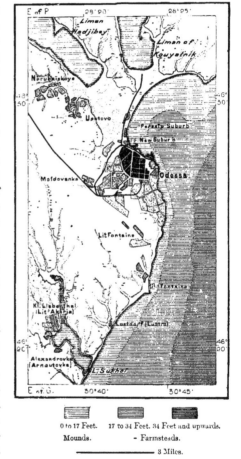

Fig. 166.—ODESSA.

Scale 1 : 240,000.

0 to 17 Feet. 17 to 34 Feet. 34 Feet and upwards.

Mounds.   - Farmsteads.

3 Miles.

nt shops, and houses in the Italian style. But beyond this
suburbs stretch everywhere towards the steppe, exposed to the
'dust, the plague of Odessa. The soil on which it stands is
lly sandstone, supplying a building material which readily
open air, soon giving to the houses the appearance of ruins.

The quality of this stone accounts for the disappearance of the Greek towns along the coast, now traced only by heaps of rubbish. It is even too friable to serve for the paving of the streets, the materials for which have to be brought from Malta and Italy. There is also a lack of good water, which, however, is now supplied from the Dniester by an aqueduct 24 miles long, with reservoirs containing about 6,000,000 gallons.

The population is extremely mixed, consisting of Jews, Italians, Greeks, Germans, French, besides Rumanians, Tatars, Turks, Bulgarians, Lases, and Georgians. French influence is naturally considerable in a place founded by General de Ribas, partly built by the engineer De Ribas, beautified and endowed by the Duke de Richelieu, whose statue adorns a central point of the thoroughfare

Fig. 167.—Khotin, Kamenetz, and Ravines of the Upper Dniester.

Scale 1 : 550,000.

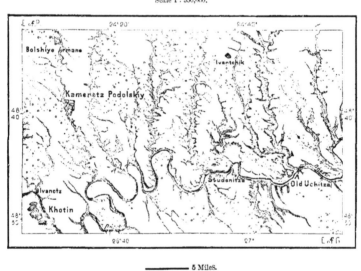

——— 5 Miles.

facing seawards. But the most influential foreign element is the Italian, and till recently the names of the streets were usually in Italian and Russian. The local dialect has even absorbed many Italian words. The staple export is corn, the importance of which is shown by the vast palatial granaries, and by the shipments, which rose from 20,000,000 bushels in 1866 to 45,000,000 in 1870. A large export trade is also done in wool, tallow, flax, the chief imports being colonial produce, manufactured goods, wine, fancy wares. The traffic is mostly carried on by steamships, a considerable number of which belong to the port itself. The local industries are unimportant, though the place contains some steam flour-mills, soap and tobacco works, distilleries, breweries, and dockyards. The neighbouring salines yield from 4,000 to 5,000 tons of salt yearly. Since 1857 Odessa has ceased to be a free port, but has instead become the seat of a University, though the smallest in

THE OLD FORTRESS, KAMENETZ.

with 48 professors and 344 students in 1877, and a library of 85,000

n the Dniester basin, formerly the most advanced Genoese settlement
ct, still contains the remains of an Italian fortress. It lies to the south
Podolskiy, which occupies a position not unlike that of Luxemburg
terrace furrowed by a deep ravine, here crossed by a magnificent
nother bridge, dating from the Turkish occupation of the place in
its Kamenetz with an old fortress, at one time of great strategical
and whose round towers and pinnacles now form a picturesque feature
ape. The Armenian settlement, which enjoyed great privileges under
ings, has almost disappeared, and now about half the population are
: engaged in a contraband trade with the Galician districts across the
ner towns in the Upper Dniester basin are *Noraya Uchitza*, on one of
s ravines intersecting the plateau, and *Mogilor-Podolskiy*, pleasantly
lst orchards and vineyards.

wn the Dniester basin is *Kichinov*, capital of Bessarabia, which, with
of 100,000, has more the appearance of an overgrown village than of
broad roads, muddy or dusty according to the seasons, are lined by
nouses, not fifty of which have two stories, and its chief building is a
commanding the whole place and the neighbouring gardens,
y Bulgarian colonists. *Bender*, or *Benderi*, the old Tagin of the
the right bank of the Dniester, though far less important than
netter known in the West as the place whither Charles XII. withdrew
tle of Poltava. A little farther down, but on the opposite bank, lies
nse name recalls the old Greek colony of Tiras, and which during the
fforded refuge to a large number of Great Russian Raskolniks, who
nuised by their customs, and especially by the physical beauty of their
rther south the village of *Oloneshti* perpetuates the memory of the
ointly with the Nogai Tatars, formerly peopled this district. On the
the Dniester liman stands the village of *Ovidiopol*, which, notwith-
name, does not occupy the site of the banished poet's residence, but
e time possessed a certain importance as a bulwark of the Russian
against the Turkish fortress of Akkerman, on the opposite side of
This is probably the true site of the old Tiras, which afterwards
Alba Julia of the Latinised Dacians, the Leucopolis of the Byzantines,
ba of the Rumanians, the Bel-Gorod of the Slavs, and the Ak-Kerman
s. Under these various names, all meaning "White Town" or
;" it long guarded the passage of the Dniester, as the "Black Fort"
he Dnieper, and in the neighbourhood are yet to be seen the remains
fort and of Rumanian and Turkish walls. It still retains some
rom the fisheries of the liman, and as the centre of an agricultural
ne 3 miles south of it is *Shaba*, peopled by Rumanish and German-
s.

o-Turkish war of 1877-8 gave to Russia the fertile plains of *Budjak*, the Moldavian Bessarabia, with some populous towns in the Pruth and Lower Danube basins. The Rumanian town of *Frumosa*, or *Kahulu* (in Russian Kagûl) lies near the Pruth, while the industrious Bulgarian settlement of *Bolgrad* stands at the northern extremity of a Danubian liman. On the Danube itself the centre

Fig. 168.—AKKERMAN AND THE DNIESTER LIMAN.

Scale 1 : 400,000.

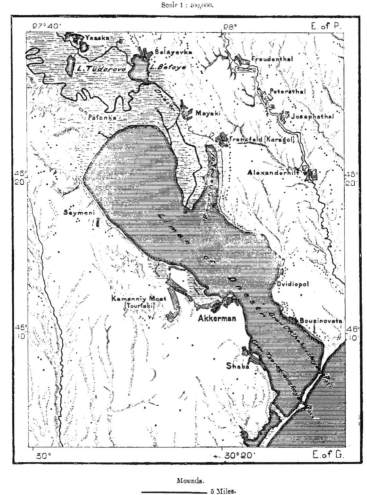

Mounds.

5 Miles.

of population is the double town of *Ismail* and *Tûchkov*, famous in the history of Danubian warfare. Although situated on the Kilia, or least-frequented branch of the delta, Tûchkov does a considerable trade in wheat and other agricultural produce, exporting in a single year as much as 4,800,000 bushels of corn. *Reni* above, and *Kilia* and *Vilkov* below Tûchkov are ports of far less importance.

'st Russian state known to history had arisen in the Dnieper
nder the fostering influence of Mediterranean civilisation;
· was destined to grow up in the north-western region, facing
tic Slavs, the Scandinavians, and the Germans. Kiev, rally-
nt of the southern populations, found its counterpart in
n centre of trade and culture, and also lying on "the road
Varangians." The region surrounding it and stretching
at lakes is not wholly included in Slavdom proper. But
Volkhov, and Neva basins form ethnically and historically
etween the Eastern Slavs and the various Finnish groups,
so much importance geographically that the Russians
times to establish commercial relations with its inhabitants.
n fixed their capital here, selecting for the purpose a site
e mainland, and surrounded by non-Slav populations. Still
mouth of the Neva has hitherto failed to attract settlers
racts, and the bleak lands encompassing it have remained
en compared with more favoured climes.

I. FEATURES.—LAKES PEIPUS AND ILMEN.

of Lake Ladoga, of which a portion belongs officially to
strine region is covered with water to the extent of 16,000
ertain parts of Sweden and Finland, the land does not here
,ding channels, its waters being concentrated in the very
next to the Caspian. The three main bodies of fresh water
Narova and the Neva to the Gulf of Finland are more
ther basins of the empire taken collectively, and Ladoga
I volume all the lacustrine reservoirs of Scandinavia or the
,ance of still waters is due to the general level of the land,
tly elevated rocky ledges have sufficed to arrest the course

of the rivers, and oblige them to expand into veritable seas in their upper basins. The lands stretching south and east of this lacustrine region are themselves but moderately elevated, so that they are still mostly covered with lakes, marshes, peat beds, and badly drained low-lying tracts.

East of Esthonia and Livonia one of these vast reservoirs, fed by the Velikaya, the Embach, and other large streams, stretches north and south for a distance of over 80 miles. This is the Peipus of the Esthonians, known to the Russians as the Chudskoïe Ozero, or "Lake of the Chudes," because surrounded by Chudic (Finnish) tribes. It has a mean depth of some 30 feet, and forms two basins connected by a channel somewhat over 1 mile wide at its narrowest point, about

Fig. 169.—TUTCHKOV.
Scale 1 : 300,000.

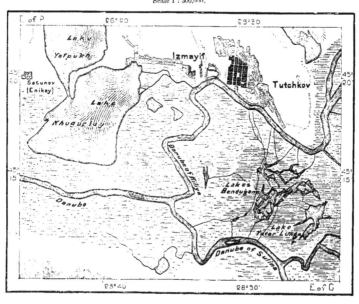

————— 5 Miles.

60 feet deep, and with a distinct current. The Peipus, which was formerly far more extensive than at present, is probably the remains of an inlet connecting the eastern extremity of the Gulf of Finland with the Gulf of Riga, but now cut off from the sea by the gradual upheaval of the coast. The continual influx of rivers and the rainfall have changed it to a fresh-water lake; but at several points may still be detected the old marine cliffs, in every respect like those on the present Esthonian seaboard. Some of its fauna are also of marine origin, such as a species of seal which has gradually adapted itself to the slowly changing element. A number of salmon placed here by Von Baer in 1852 have multiplied, although prevented by the falls at the outflow from yearly visiting the Baltic. Since 1844,

: of the deluge," the level of the lake has been considerably
disappeared, inlets have been formed, forests destroyed, and
ibouring meadows, converted into quagmires. These changes
the extensive drainage operations that have been executed in
tricts, where 1,200 miles of drains now convey more water to

LAKES AND MARSHES BETWEEN THE VOLKHOV AND DVINA BASINS.

Scale 1 : 350,000.

5 Miles.

carried off by the Narova. Rapid encroachments have been
side, where a range of shifting dunes, 24 to 30 feet high, is
laced northwards, leaving its site a prey to the advancing floods.
a mean elevation of 95 feet above the sea, its outlet, the
newhat precipitously through a series of rapids to the Gulf of
irva it forms two branches round the island of Kränholm,

falling from a height of 18 or 20 feet to its lower course between steep sandstone and limestone cliffs. Below the falls and the town of Narva the Narova becomes a sluggish stream, which has often changed its bed within the present geological era. Prevented from direct access to the sea by a double line of dunes over 60 feet high on the east side of the Bay of Narva, it continued its northern course along the valley now followed by the river Luga. A gap opened at some unknown period through the dunes enables it to reach the coast on the west side of the bay; but above its present mouth there are many winding channels which, though now dried up, are sufficient proof of its former shifting character. Besides, the Luga and Narva basins are still connected by the Rossona, a branch of the Luga, which trends suddenly westwards, joining the Narva just above its mouth. Although smaller than the Narva, the Rossona is much more irregular in its discharges, rising sometimes from 12 to 20 feet in the floods, bringing down large quantities of sand, undermining the buildings on its banks, and even sweeping away the coffins of a graveyard near its course. Hence, in order to keep a uniform depth of 10 feet on the bar at the mouth of the Narova, it may be necessary to divert the Rossona through an independent channel to the coast.

Fig. 171.—LAKE PEIPUS.
Scale 1 : 1,500,000.

20 Miles.

After the Rossona floods this bar has often scarcely more than 5 or 6 feet of water, although higher up the Narva is no less than 44 feet deep. The efforts made since 1764 to increase the scour of the Rossona have hitherto failed, so that the Bay of Narva still remains one of the most dangerous roadsteads on the coast, and over twenty vessels have here been wrecked in a single storm.

The Ilmen, another lake east of the Peipus, with an area of about 400 square miles, is really nothing more than a permanent flooding of the land, caused by a large number of streams all converging at one point, and with an insufficient

From the south-west comes the Shelon; from the south the Lovat,
l Polomet; from the north-east the Msta, rising in a marshy district on
n slopes of the Valdai plateau, near the source of the Volga. All these

ing in one basin, and
own the detritus of the
ng hills, rapidly raise
f the lake, and con-
.odify its outlines. Its
always muddy, varying
om 7 to 30 feet; hence,
uch larger, it contains
y smaller volume than
lakes of Switzerland.
the Volkhov, formerly
iy, or " Muddy," re-
character throughout
of 120 miles to Lake
f which it is the chief
It has a total incline of
feet, descending, like
a, through a series of
apids over its last rocky
The Gostniopol rapids
feet altogether, and are
only for rafts or craft
uilt for the purpose.

LADOGA AND ONEGA.

the largest feeder of
om the east, brings
utflow of Lake Onega,
lf receives the super-
ers of several other
reservoirs. Many of
d elsewhere in Europe
l as little inland seas;
ney become compara-
nificant by the side of
r Ladoga and Onega,

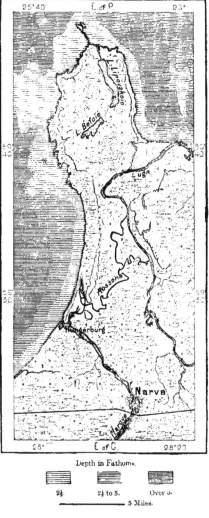

Fig. 172.—THE LUGA AND NAROVA RIVERS.
Scale 1 : 300,000.

Depth in Fathoms.

| 2¼ | 2¼ to 5. | Over 5. |

5 Miles.

often tossed by the storm, and whose perils mariners are warned against
like those of the high seas. The navigation of Onega especially is
igered by the numerous rocks and reefs skirting its shores, and about
rs are said to have yearly perished here before 1874, when buoys were

laid down at the dangerous points. Away from the coast it is generally very deep, 740 feet in some places, and in the north the coast is indented by numerous inlets, all running south-east and north-west, and continued towards Lapland by chains of lakes and rivers, with intervening ridges 800 to 1,000 feet high, all running in the same direction. These still partly flooded water-courses are parallel with the Finnish lakes, and their axis follows precisely the same lines as those of the White Sea, from the so-called Gulf of Onega to that of Kandalaksha in the extreme north-west. Glacial striæ have also been traced along these lines, and the åsar, or *selga*, run mostly in the same direction.

Fig. 173.—Lake Onega.
Scale 1 : 2,780,000.

Depth in Fathoms.

Under 13.   13 to 27.   Over 27.

25 Miles.

Saïma, the largest lake in Finland, is also a tributary of the Ladoga through the Wuoxen, or Voksa, noted for the famous falls of Imatra, the grandest in the whole Neva basin. Even within the present century the Wuoxen has changed its lower course, its present lying some 24 miles south of its former mouth near the village of Keksholm. Heavy rains swept away an isthmus till 1818 separating Ladoga from the long Lake Suvando, which already communicated with the Wuoxen through a small canal cut by the government of Finland. The outlet thus suddenly created immediately lowered the level of Suvando, which shrank to the proportions of a river, and the Wuoxen, almost entirely forsaking its former outlet, discharged south-eastwards through the new bed. But its fluvial form renders it probable that in a previous geological epoch the Suvando had already received the waters of the Wuoxen.

Like Onega and Peipus, Ladoga was formerly far more extensive even than at present, for its low and almost treeless southern shores consist of clays, sands, and gravels containing large quantities of erratic stones of every size, from simple pebbles to huge boulders. From these low-lying shores the bed of the lake falls imperceptibly towards the deep waters commanded by the granitic cliffs of the north coast. Near some rocky islands it is from 300 to 500 feet deep, sinking to

the Valaam group; but the mean depth is estimated at 10 more than
nich would give the whole basin a volume about nineteen times greater
' Lake Geneva. The amount, however, varies considerably with the
years, observations continued for fourteen years showing a difference
han 7 feet between low water and the floods. The monks of Valaam
ne strength of an otherwise unauthenticated tradition, that the general
nd falls alternately from century to century.
standing the muddy contributions of such rivers as the Volkhov, the
,adoga is generally so pure that the smallest objects lying at the
perfectly visible in depths of 14 or 16 feet. It is always very cold,

Fig. 174.—STRIÆ AND ÁSAR ABOUT LAKE SOG.

Scale 1 : 1,200,000.

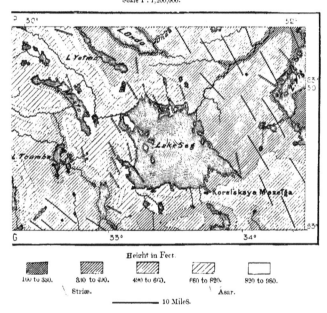

Height in Feet.

| 100 to 330. | 330 to 490. | 490 to 660. | 660 to 820. | 820 to 980. |

Striæ.   Ásar.

10 Miles.

aps in August, when the temperature about the surface may occa-
to 50° or 55° Fahr. But even in July it is scarcely safe to drink
ne thaw towards the end of May the surface water is about 2° above
it. The temperature between the surface and lower depths usually
han 1°, the latter being somewhat higher in winter, when the lake is
Lying somewhat south and to the west of Onega, it remains frozen for
·iod, usually about one hundred and twenty days, from the middle of
Some of the central parts occasionally remain open throughout the year,
ga is nearly always completely ice-bound for one hundred and fifty-six
ufficient air is still retained in the lower depths of these basins to keep
live during the winter season. Both of them are inhabited by a seal

of the same species as that of Lake Peipus, hunted by men for its fat und skin,
and preyed upon by wolves, which, when driven by hunger, will venture across
the ice to the middle of the lake. Some of the smaller animals, like the seal
itself, recall a time when Ladoga still communicated with the sea. One of these

Fig. 175 —LADOGA.

Scale 1 : 2,000,000.

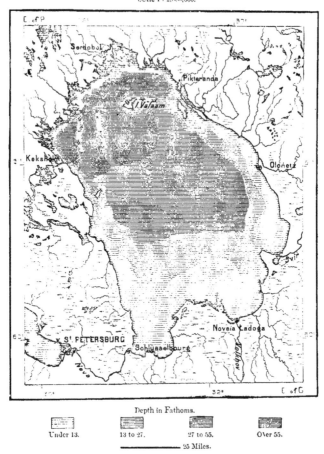

Depth in Fathoms.

Under 13.   13 to 27.   27 to 55.   Over 55.

25 Miles.

microscopic organisms, the *Compylodiscus radiosus,* has hitherto been found elsewhere only in the Gulf of Mexico, near Vera Cruz.

The lake is sometimes swept by gales, which raise heavy seas, billows, and
ground swells like those of the ocean. Its waters are also kept in a constant
ferment by its tributaries, to which is perhaps partly due a current setting
steadily along the east coast northwards, and along the west coast southwards.
Its level has been determined by Schubert at about 60 feet above that of the Gulf

.rai is through the Ne va. This stream, oily 36
with the great ri vers of Furope, dischargi ig from
't per second—a discharge greater than that of the
ied outside of Russia by the Danube alone in Europe,
· chaiged into a fresh-water basin the eastern section
id Kroistadt. It varies in width from 850 to 4,200
ng or it generally find 7 to 10 feet or the reefs.
owing to the gulf, it forms *porogl*, or rapids, some of
l except flat-bottomed craft. Between the quays of
h of from 20 to 50 feet, but beyond the delta its
l Little Neva, Great, Middle, and Little Nevka—are
:s and said-banks inaccessible to vessels drawing more
ind, very little sedimentary matter is carried down, so
:o but slight change, and the delta was increased by
18 and 1834, or at the yearly rate of about 14 acres,
·hole frot. The alluvial soil is only 79 feet thick,
an artesian well suik 660 feet to a body of water
'inlaid granite supporting the alluvia.
. regulates the discharge, the difference between high
:able, and the floods would be in no way dangerous
·ing right up the stream, and capable of inundating
several days. During the floods of November 7th,
ian 12 feet above the sea-level, deluging the greater
ise again to nearly the same height in 1879, though
: above the gulf. Another danger arises from the
_ake Ladoga, which seids down huge masses, often
e banks, and destroying the quays of St. Petersburg.
ore huidred aid fifty years show that the Neva is
138 days yearly, but with great fluctuations, varying
in 1852.

CLIMATE.

in, lying under the sixtieth parallel of latitude, is
· except the hardiest plaits, capable of resisting the
aws of spring. The oak, which reaches the latitude
grow spontaneously in the Ingrian forests, flourish-
Msta, east of Lake Ilmer. The poplar, elm, maple,
' the forests in the temperate regions of Central
· in well-tended private grounds. Even the silver
g freely except the birch, alder, aspen, willow, and
mit of wheat is also marked by a line drawn through
rega.
conditions this region could scarcely become very
the most remarkable phenomena of contemporary

history is presented by a city situated in the midst of solitudes, and lying as near to the pole as Labrador, Cape Farewell in Greenland, or Kamchatka, yet attracting a population of hundreds of thousands, thanks to its commercial advantages and the influences of political centralization. But so long as the inhabitants had to depend on the local resources, this great lacustrine territory could never have developed large centres of population. It is known to have been occupied by fishing and hunting tribes from a remote antiquity, for objects belonging to the stone age have been found on the shores of Onega and the neighbouring lakes. A rampart of rough-hewn blocks 3,220 yards long has been traced near the small Lake Lujand, south-east of Onega, but the nationality of its builders is unknown, no human remains having been discovered in the surrounding graves. The tumuli occurring near the Svir, south-west of Onega, contain two distinct types of crania—one brachycephalous, the other dolichocephalous and prognathous like the African.

POPULATION : FINNS AND GREAT RUSSIANS.

THE intruding Great Russians now occupy nearly all the Volkhov basin, and have at many points overstepped their old historical limits marked by Lake Peipus and the rivers Narova, Neva, and Svir. But within their ethnological domain proper there still remain detached groups and enclaves of Finnish populations. In the Msta basin, and on the uplands skirted on the east by the Valdaï crests, there dwell Karelian Finns, descendants of those removed hither by Peter the Great, and who had here been probably preceded by the Chudes, the "Prodigies," "Monsters," "Foreigners," whose remains are found in the surrounding tumuli. East of the Narva others have preserved the name of Votes (Vadjalaiset), formerly belonging to a widely diffused nation already in the enjoyment of a relatively advanced culture in the ninth century. Upwards of eight thousand mounds, mostly small and poor in old remains, have been examined by Ivanovskiy, and the two thousand crania found in them seem to belong to the Ural-Altaic race. On the west side of Onega, and farther south between Lakes Ladoga and Belo ("White"), there dwell some Vepses, or Northern Chudes, here and there forming distinct communities variously estimated at from 12,000 to 25,000. Their speech is of a peculiarly archaic type, but they are being rapidly "Russified," and in several of their villages Russian already prevails, or is largely mingled with the local dialect. Most of the women have preserved their language better than their Finnish type : according to Maïnov few have the slant eyes of the Mongolians, and many are distinguished for their beauty, quite in the "Novgorod style." Nearly all the Vepses are brachycephalous, and taller than the average Russian. To judge from the names of their domestic animals, iron, gold, zinc, agricultural terms, they seem to be indebted to the Russians, Swedes, and Lithuanians for their culture. They still believe in their household gods, and on occupying a new home never forget to fetch the embers from the old hearth, and slip a bit of bread under the oven. But if the cock neglects to crow

the offering is not accepted, and the family genius is
es drink a spirit prepared from beet-root, which
table, and scurvy makes great ravages amongst them,
ence of cabbage, onions, or other green stuffs in their

Finns, who give their name to the province of Ingria
sed to exist as a distinct nationality; but more or less
: Finns still largely inhabit the coast lands between
, and all the territory washed by Lake Ladoga on the
   Those settled in the suburbs of St. Petersburg seem
or West Finnish branch, and are collectively known
i. Like their neighbours, the Ijortzis, or Igris, of
butary of the Neva, they are distinguished from other
thick-set frames, and disgusting habits.
adoga the population is entirely Karelian. The islands
s and Russians, but the latter had till recently become
t of almost entirely forgetting their mother tongue.
are being slowly assimilated to the Ugrians, although
the region of the great lakes are probably nothing but
being much fairer, their eyes and hair lighter, than
[ost of the customs are Finnish, and often perpetuate
'enacious in their character and ideas, the Karelians
rn a Karelian," says the proverb, "and after three
   In the seventeenth century the Olonetz Finns were
ting the throats of animals with flint knives, and then
.s. On a bluff in Manchin-Sari Island, towards the
ds the shrine of the prophet Elias, where the devout
l as of old to offer a sacrifice on the first Sunday after
e their forefathers were wont to slay an elk or a deer
t these animals having vanished from the land, a bull
lesh is shared out, cooked in large pots, and religiously
after this sacred repast the prophet will not fail to
: murrain. The cattle plague is also conjured by
by friction, or by casting alive into a pit some animal,
or even a horse.
the Slavs of the government of Olonetz have preserved
cally as *starinas*, or "antiquities." In two months
ny as seventy in a single district of the province.
met in every profession, and fresh pieces are being
pted. Here the love of funeral dirges is more general
nd although harshly treated under the paternal roof,
songs for weeks before the wedding. The old beliefs,
of Karelian origin, have been kept alive amongst the
icred trees are still revered; offerings, and especially

woven stuffs, are attached to the crosses raised above the graves; the dead are invited to share the family meal, and the beds of the departed are still got ready in the cabin. When disputes arise about the boundaries of their lands, an umpire covers his head with earth and walks ahead; his footsteps mark the limits, for "our mother, the moist earth," has decided.

In Olonetz the social arts are in an extremely backward state. Agricultural implements are still of the rudest type in a land where agricultural resources must always be of a precarious character, where the "spring frosts" are felt in July, and the "autumn chills" begin in August. Of late years the "Siberian plague," prevalent in all marshy lands, has carried off a large portion of the live stock, and even the chase, formerly so productive, now yields but poor returns, the beaver and sable having already disappeared. Of the larger wild animals none remain except the bear, which still continues to ravage the cattle and waste the oat-fields. But all this matters little, seeing that the possible gains of the mûjik are forestalled by the traders who pay his taxes, and advance at heavy rates the powder for the chase, his fishing tackle, and his daily bread. "Where I have set my foot the mûjik sings no more," says the contractor.

TOPOGRAPHY.

WITH the dawn of Russian history the Slavs are found endeavouring to establish in the Narova and Neva basins large emporiums of trade with the Baltic lands. One of these early marts was *Izborsk*, which, however, had soon to make way for *Pskov*, formerly *Pleskov*, some 20 miles farther east. Becoming independent of "Novgorod the Great" in the fourteenth century, Pskov was at first little more than an intermediary between that city and the German seaports. But it succeeded later on in opening direct relations with the western trading places, and had factories on the Baltic for the sale of timber, cereals, flax, tallow, tar, and other Russian produce. The republic of Pskov was at the height of its commercial prosperity in the fifteenth century, when it is said to have had a population of 80,000. But its autonomy was suppressed in 1510, and after falling under the sway of Muscovy it lost its importance, hundreds of its most industrious citizens were carried off, and in 1803 it had scarcely 6,000 inhabitants. However, its position as capital of a government, and its favourable situation above the Velikaya delta at the southern extremity of Lake Peipus, have somewhat revived the prosperity of a place which is the natural outlet of the interior as far as the water-parting of the Dwina. Here are still to be seen some old houses and the remains of walls that have withstood twenty-six sieges.

Below Lake Peipus *Narva* occupies a corresponding position to Pskov, standing near the mouth of the Narova, where in 1702 Charles XII. of Sweden overthrew a Russian army ten times superior in numbers to his own. Its fortifications are of secondary importance compared with the formidable works of Sveåborg and Kronstadt, but the old exchange and the now abandoned bazaar are evidences of its former commercial activity. The inhabitants, of whom about half are Germans,

.ted at upwards of £300,000, and channels cut in
ve power from the Narova rapids to several mills
.er up the river.
.tres of population are naturally grouped in the
er, and the soil more fertile than elsewhere. Here

ILMEN, NOVGOROD, AND STARAYA RUSA.

Scale 1 : 545,000.

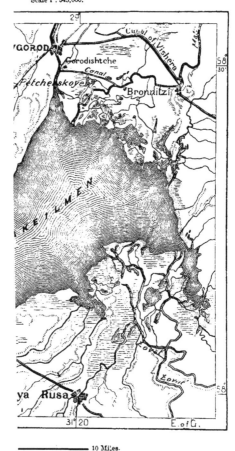

10 Miles.

rine plateau; *Velikiya Luki* and *Kholm*, on the
ey, which, before the opening of the railway, did
at this district Arab, Anglo-Saxon, and Frankish
eventh centuries are frequently picked up, indi-
ement with the East and West, though very few

Byzantine coins are found. The old and flourishing town of *Staraya Rusa* lies on the south side of Lake Ilmen, near the junction of the Lovat, Polista, and other streams. Its former importance was largely due to a copious saline spring, which enabled it to supply salt to Novgorod. But when Novgorod lost its independence the surrounding towns fell into decay, and the salt ceased to be exported till the last century. At present the spring and an artesian well sunk in the neighbourhood yield to the State about 2,400 tons yearly, and Staraya Rusa has become a favourite watering-place, much frequented by invalids from St. Petersburg.

Novgorod—that is, "New Town"—now fallen from its high estate, was formerly the centre of an empire stretching beyond the Ural Mountains, and of a vast trade carried on with the west of Europe. Standing on both banks of the Volkhov, just below Lake Ilmen, it is not only the natural entrepôt of all this region, but, before the river highways were replaced by artificial routes, it was also one of the chief stations between the Baltic and the Black Sea. In the language of the Russian chronicles, "it lay on the high-road leading from the country of the Varangians to that of the Greeks," as well as on the road between Europe and Asia by the Volga and Baltic. During a period of incessant warfare it enjoyed the further advantage of immunity from the inroads of Norse or Teutonic rovers, and the Tatars who laid waste all East and South Russia never reached this city. Surrounded by its forests, Novgorod was much safer than the Baltic seaports, or the more exposed places of the interior.

The old town, afterwards succeeded by the "New Town," stood close to the lake, on a terrace about 60 feet high, surrounded on all sides by running waters and marsh lands, and, according to the legend, Rurik's Castle was raised on the site of this natural fortress, still known as Gorodishtche, or "Old Town." But the terrace was too limited for a large population, and the new town was accordingly founded on another eminence overlooking the Volkhov, rather over a mile farther down. This city became in due course the centre of political power in North Russia, disputing with Kiev the honour of being regarded as the "cradle of the Russian Empire." Opening direct relations with the Hanseatic towns, it established factories first at Wisby, in the island of Gotland, and then at Lübeck, while securing the inland traffic and acquiring an empire equal in extent to all Western Europe by means of its distant settlements "beyond the portages," on the White Sea, and even in West Siberia. It also became with Pskov a chief centre of industry, arts, letters, and rationalistic sects. "Who can aught against God and the mighty Novgorod?" said a well-known local proverb. The city elected its princes, but should the popular assembly have to complain of the man intrusted with the supreme authority, he was "bowed out of office," or consigned to the mud of the surrounding swamps. Strong in her charter of liberties, which she claimed to have received from Yaroslav the Wise, and which she jealously guarded—strong especially in the material independence sustained by her wealth and armed citizens, Novgorod long flourished as an autonomous commonwealth, but, though politically free, still restless and often torn by rival factions. The people were

law, and while the " Whites," or privileged classes, were
ach other, the "Blacks," or common herd, continued to
ddle of the fifteenth century, when she had to defend
·ing Muscovite power, Novgorod rapidly lost her north-
remote to be succoured, and at that time brought into
vy through Ust-Yug and the course of the Vichegda.
and henceforth her history became a long series of
· armies were overthrown by the Russians and Tatars of
alous Pskov; in 1478 the *veche*, or popular assembly, was
·ns forced to tender their allegiance to the Muscovite

CH NEAR NOVGOROD, BUILT UNDER IVAN THE TERRIBLE.

· becomes an organized institution, the suspected are
· families exiled in 1479; the murders are renewed, and
gain banished in 1497; and so it goes on till the nation
· partly replaced by Muscovite colonists in the sixteenth
s still suspected by Ivan IV., who nowhere better than
he title of " Terrible." If the annalist can be trusted,
stroyed 60,000 persons in Novgorod; for several weeks
is were daily cast into the Volkhov; the river was
· and, according to the tradition, the water never freezes
·le drownings.

Nevertheless the exterminator of Novgorod was still anxious to continue its direct relations with Europe. But, by depopulating the old cities and wasting the land, the Muscovites deprived themselves of the elements necessary to keep up a direct intercourse with the West. Hence they gladly welcomed the English adventurers who had come to trade with them round the Frozen Ocean. Later on Gustavus Adolphus could declare that "Russia had been finally cut off from the Baltic." In the seventeenth century Novgorod still showed a restive spirit, which, however, was soon quelled, and nothing now survives of the old national life except some popular proverbs directed against the Muscovite. It no longer stands on the great highway of nations, lying far to the west of the main route between Moscow and St. Petersburg, beyond the general commercial movement of modern Russia. Those of its traders and artisans still surviving in the seventeenth century were amongst the first elements of population in the new northern capital. Once peopled by 50,000 or 60,000, or traditionally by 400,000, it is now so reduced that convents formerly within the walls are surrounded by fields. But the trading quarter, and that of St. Sophia, or the Kreml, still occupy the old sites above the river. The fortifications of the Kreml, formerly enclosing 18 churches, 150 houses, 40 factories, contains the cathedral of St. Sophia, in which are still preserved some tombs of old saints and heroes, curious frescoes of the twelfth century breathing a bolder artistic spirit than that of East Russia, and images whose symbolic attitudes harmonize with the observances of the sect of the "Old Believers." On the square facing the cathedral stands the monument commemorating the legend of Rurik, a lofty granite pile adorned with statues, and covered with bronze bas-reliefs representing various figures associated with the origin of the Russian Empire. It was erected in 1862, the millenium of the old state.

On the Upper Msta, east of Novgorod, lies the town of *Borovichi*, from the earliest times the natural centre of the river traffic between the Volkhov and Volga basins. The rapids of the Msta turn the wheels of several factories, and in the neighbourhood are quarries, coal-fields, and especially pyrites mines, which during the Crimean war replaced the Sicilian sulphur in the manufacture of sulphuric acid. *Tikhvin*, north of Borovichi, and on the Tikhvinka, has, since 1811, formed the terminus of a navigable canal connecting the Volga and Ladoga water systems, and supplying St. Petersburg with about 20,000 tons of cereals and timber yearly. Its convent contains one of the most venerated miraculous images of the Virgin in all Russia, and formerly owned 4,500 "souls" (serfs), presented to it by devout czars and nobles.

In the upper basin drained by the Svir and Neva the only important place is *Petrozavodsk*, on a western inlet of Lake Onega, and in a mining district containing gold mines (now abandoned), copper lodes, and veins of magnetic iron with as much as 96 per cent. of pure metal. Founded in 1704 by Peter the Great, it takes its name of "Peter's Mill" from a gun foundry and small-arms factory established here for the purpose of utilising the neighbouring mineral treasures. Petrozavodsk has also become the capital of the government of Olonetz, and is now the central trading station between the Gulf of Finland and the White Sea.

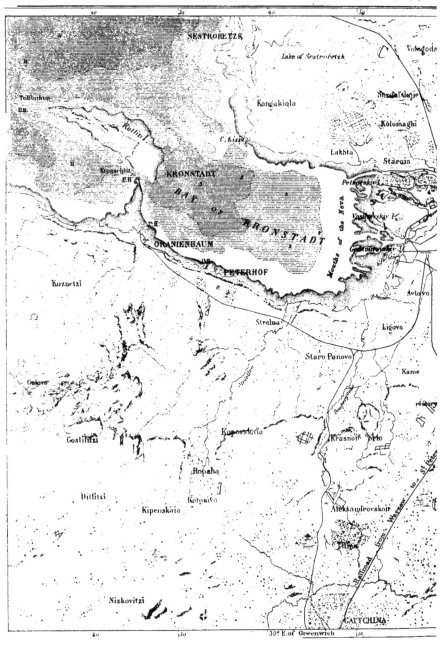

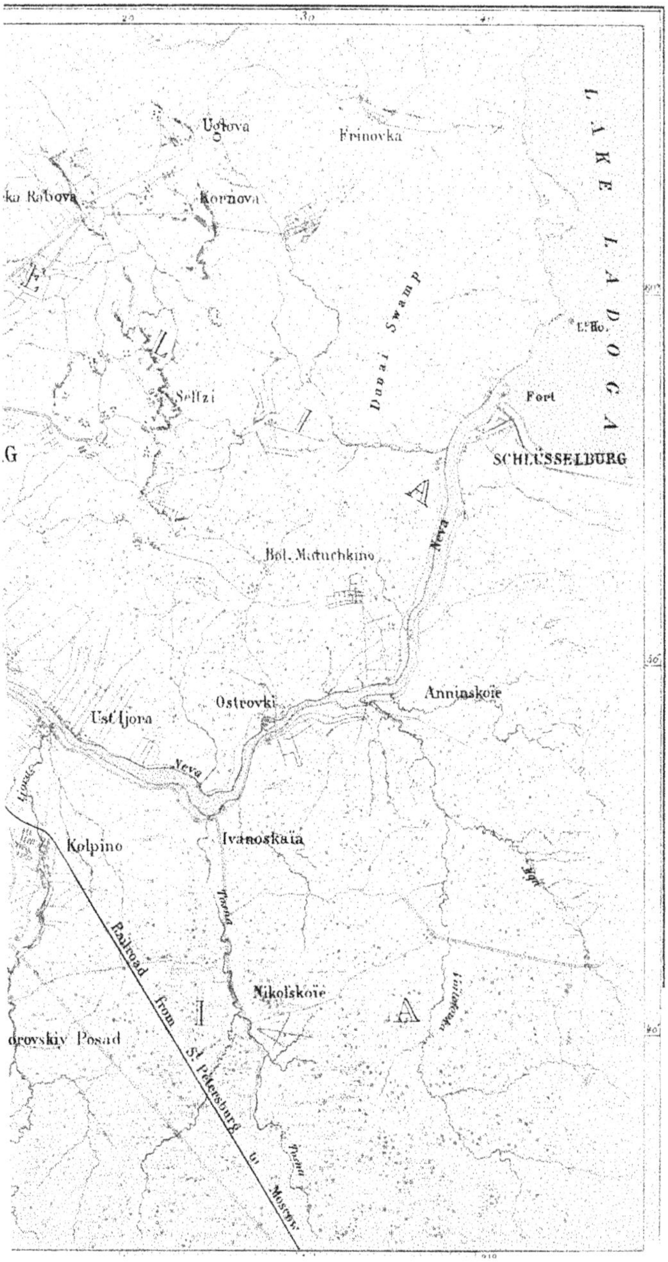

Ugłova
Frinovka

ka Ribova
Kornova

Dubai Swamp

L.Bo.

Sellzi
Fort

I

G
A
SCHLÜSSELBURG

Neva

Bol. Matuchkino

Ostrovki
Anninskoïe

Ust Ijora
Neva

Ivanoskaïa

Kolpino

Railroad from St. Petersburg to Moscw.

Nikolskoïe

I

A

orovskiv Posad

L A K E   L A D O G A

..adoga the two largest places are *Novaya Ladoga*, at the
v, and *Schlüsselburg*, at the outflow of the Nova, both
her not only by the lake, but also by two navigable canals
, the former dating from the time of Peter the Great, the
·d on a much larger scale and without locks. The Putilovo
de of these canals, consist of sandstone rocks, whence
large quantity of the material used in the building of its
l thoroughfares.* All the craft plying on the lake and
·uns of Schlüsselburg, formerly Orekhoviy, founded by the
held by the Swedes till 1702, and now one of the most
n the empire. Standing on the left bank of the Neva at
, Schlüsselburg is a sort of advanced suburb of the capital
le has been done to improve the navigation of these waters
Swedish tenure.

showed more daring than did Peter when he removed his
rom the old metropolis to a quagmire surrounded by dreary
uudations of which had to be fixed by whole armies before
.ld be erected. These works were for all Russia the
zed system of forced labour, and in the four years between
50,000 workmen were transported to the Neva marshes,
ished of fever, hunger, and various epidemics. In order to
hroughout the land to seek employment in St. Petersburg,
ifices was elsewhere forbidden under penalty of confiscation
nobles also, owners of not less than thirty serf families,
metropolitan mansions for themselves on plans and scales
:o the rank of each. The treacherous ground on which
:apital in defiance of man and nature had only been just
ud, by the very fact of taking his stand in foreign territory
. German, he became committed to a ceaseless aggressive
satisfied with "opening a window on the west," but also
ng in advance of his new edifice. In order to change into
.e artificial equilibrium created by this step, the conquest
ivonia, Kurland, Lithuania, and Poland became a necessity
e reason why the policy of Peter has been so faithfully
sors.

lso this city was a necessary element in the organized
nire. It was, so to say, the city of Novgorod removed to
n, but, except in regard of its climate, enjoying natural
to those of the inland emporium. Standing at the head of
lelta of a considerable river, **it** commands both the sea and
l the natural routes converging here from the Volkhov,
basins. Of all the vital points on the Russian seaboard,

*field of the Putilovo quarries (1875), £80,000.

not one is so happily situated as the Neva delta, lying as it does nearest to the great centres of population, which, at least in summer, can be supplied with the products of the West more rapidly from this place than from any other seaport. The work of Peter the Great harmonized with the geographical conditions of the land, and it has accordingly survived. The fortified island of Kronstadt completely guards the approach from the west, thus enabling the capital to expand and peacefully develop its trade and industries.

The fifth city in Europe for its population, it covers an area of over 40 square miles, including the river surface, but excluding the industrial suburbs and the

Fig. 178.—CONTINUOUS GROWTH OF ST. PETERSBURG.

The Fortifications traced on this Plan belong to the Old Town.

Scale 1 : 135,000.

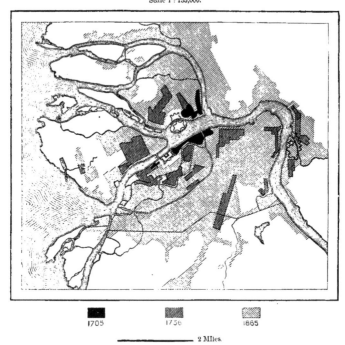

1705   1736   1865

2 Miles.

villas stretching along the side valleys of the Neva. It spreads out like a fan along the branches of the delta, embracing six large and numerous small islands, limited southwards by the Fontanka Canal, while on the mainland the outlying districts are continually expanding north and south. The islet where the founder sank its first piles still bears officially the name of the Petersburg quarter, and here stand the citadel and State prison of "SS. Peter and Paul," with the church in which the Czars are buried. But the true centre now lies south of that island on the left bank of the Neva, which is crowded with public buildings,

THE NEVA AT ST. PETERSBURG.

ST. PETERSBURG.

ng pile of the Admiralty, surmounted by a gilded tower,
ac, a marble and granite domed nave glittering in gold,
stones; the vast Winter Palace, with its long sculptured
Neva; and other stately mansions adorned with colonnades,
ith verdure.   Close to the Admiralty, and in a square
washed by the Neva, stands the famous equestrian statue
Falconet, poised on a block 16 feet high, and proudly
ng stronghold raised by him in the midst of swamps.
g Winter Palace stands the Alexander column, a Finland
high; but this remarkable monument of human industry is
ill soon have to be kept together by means of iron clamps.
he city may be had from the Admiralty tower, or better
saac's.   Southwards are seen the diverging lines of the
pectives," amongst them the famous Nevskiy Perspective,
aars, churches, and stretching for about 2 miles to the
us.   The Neva winds east, north, and west under the
ges, and nearly opposite the Admiralty it ramifies round
. which stand the Exchange, Custom House, and principal
n as the University, Academies of Sciences and Fine Arts,
cal Institute, Physical Observatory.   Beyond the buildings
s the wooded "isles of the Neva," with their winding
asure boats. This panoramic view presents many points of
few quarters which can be visited with pleasure.   Like
; is a city of "magnificent distances," with interminable,
nous streets, lined everywhere with the same gloomy
n houses, without beauty or originality of design.
e place, the annual mortality, as in Odessa, exceeding the
the constant immigration this swamp-encompassed city
original state.   But so considerable is the inflow, con-
nen in the vigour of life, that amongst all European cities
able for the great excess of its male population, generally
three of the opposite sex.   According to the statistical
as more than double before the middle of this century,
a men are married, about half of them have left their
provinces.   Every part of the empire and all its races—
atars—take part in the yearly immigration.   The Tatars
· occupations, and are largely engaged as waiters in the
· as old-clothes dealers.   There are no less than 50,000
r calling, licit or illicit, from the merchant and banker
iggler.   One Lutheran parish is composed entirely of
o have forgotten their mother tongue, and now speak
public health and morality naturally suffer from the
ate of illegitimacy was as high as 30 per cent. between
it mortality attains fearful proportions, the chronic

epidemics at times carrying off one-fifth of all the children. Of the 7,578 received into the Foundling Hospital in 1876, as many as 7,190 were illegitimate, and the total mortality rose to 6,088, or about 80 per cent.

A city in which the military and officials of all ranks form such a large section of the population is naturally a gay and extravagant place, and here luxury and squalor are brought into the closest contact. Apart from the poverty-stricken rural immigrants, the proletariate classes are necessarily very numerous in the first manufacturing town of the empire. To the State belong some very large porcelain, glass, and carpet establishments; but much more important are the private industries, sugar refineries, foundries, tanneries, woollen and cotton spinning and weaving factories, breweries, distilleries, tobacco works, altogether about 620 establishments, employing (1875) 41,400 hands, of whom one-fourth are women, and yielding manufactured goods to the amount of £12,000,000. Yet more even than on the resources of industry the wealthy classes depend on the revenues of the great domains, and on their heavy salaries and pensions, to support their princely expenditure. The retail trade alone is partly in the hands of the Russians, while the wholesale business is mostly carried on to the profit of German or English merchants and Jewish bankers. The local trade is very large, the exchanges amounting in some years to a fourth or a third of the commerce of the whole empire. But more than half of the shipping sails under the English, German, and Norwegian flags. To facilitate the approaches by water, it is now proposed to dredge a passage 16 or 17 feet deep and 18 miles long between Kronstadt and the capital, and to avoid the Neva windings and rapids by a canal running from its mouth directly to Lake Ladoga.

As regards public instruction St. Petersburg lags behind most of the Western European cities, over 300,000 being still wholly unlettered. But its high schools and learned bodies rank amongst the very first in the world. For works of classic literature, the arts and sciences, it is the chief centre of the empire, while far surpassed by Moscow for publications of a more popular character. Its University, with 88 professors and 1,418 students, and a library of 120,000 volumes, turns out the best physicists and mathematicians. The School of Medicine, with nearly 1,600 students in 1876, is henceforth limited to 500, the female courses being now conducted in a separate establishment. The Academy of Sciences and some other societies have published memoirs of permanent importance, while the Geographical Society, disposing of a large income and greatly encouraged by the State, continues to further ethnological research throughout Central Asia and China. Besides those of the University and Academy of Sciences, containing some rare works and valuable collections, the Imperial Library ranks next to those of London and Paris, with nearly 1,000,000 volumes and over 40,000 manuscripts, including Voltaire's collection of 7,000 volumes and many unique works. The museums also are amongst the most remarkable in Europe. Attached to the Academy of Sciences is an admirable Asiatic gallery, besides zoological collections, where may be seen the famous mammoth brought from Siberia in 1803. The Hermitage, which communicates with the Winter

ontains a collection of 12,000 original designs, 200,000 prints, and a
nt picture gallery with specimens by most of the great masters, a
series of the Flemish, and a unique collection of the Russian school.
hief glory consists in the remains of the finest period of Hellenic art
Scythian antiquities from the south of Russia, elsewhere absolutely
d. Its library includes, amongst other treasures, the rare autographic

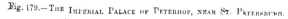

Fig. 179.—THE IMPERIAL PALACE OF PETERHOF, NEAR ST. PETERSBURG.

of Voltaire, D'Alembert, and Diderot. The picture gallery of the Fine
my is richer even than that of the Hermitage in works of the Russian

irons are adorned by some fine retreats, parks, and pleasure grounds,
amongst which is *Peterhof*, on the south side of the bay separating
from the mouths of the Neva—a sort of Versailles, laid out in
flower beds, fountains, and terraces descending in flights of steps to

the sea, and commanding an extensive view of the roadstead and wooded shores of Finland. This was the favourite residence of Peter the Great, and west of it is *Oranienbaum*, another imperial château surrounded by pavilions and villas, over against the island of Kotlin and the formidable granite batteries of *Kronstadt*. Previous to the reign of Alexander II. upwards of £8,000,000 had been expended on this bulwark of the capital, and since then hundreds of thousands have been spent in adding to its defences, including two revolving turrets doubly strengthened with iron plates and teak. Kronstadt is mainly a military town, and most of its inhabitants are employed in its arsenals, forts, and navy. The rest are engaged in the transhipment of goods between the lighters from the capital and the large vessels from the high seas. In winter the ice becomes the highway of traffic. A temporary hotel is erected midway, with intermediate stations to succour travellers overtaken by the fogs or arrested by fissures in the ice. In 1881 the two cities will be connected by a railway running along the embankment of the Kronstadt Canal.

In the interior are other towns, palaces, châteaux, villas, pleasure grounds, dependent on the capital. Amongst them is *Tzarkoïe-Selo*, the "Imperial Village," 15 miles to the south, the favourite retreat of Catherine II., around which an industrial town of some importance has sprung up. In the north-west is the Pulkovo Observatory, on an eminence 245 feet high, through which is drawn the Russian meridian.* This observatory, associated with the observations of Struve, will soon boast of the largest telescope in the world, which is now being constructed at Cambridge, near Boston. Another observatory, specially devoted to the study of meteorological and magnetic phenomena, has recently been founded near Pavlosk, south of Tzarkoïe-Selo.

* Longitude of Pulkovo, 30° 19′ 36″ E. Gr.; latitude, 59° 46′ 19″ N.

CHAPTER VIII.

LANDS DRAINING TO THE ARCTIC OCEAN.

(LAND, NORTHERN URALS, NOVAYA ZEMLYA, GOVERNMENTS OF ARCHANGEL AND VOLOGDA.)

ALL the northern lands whose waters flow to the Frozen Ocean correspond in their general outlines with the two vast, but almost uninhabitable, provinces of Archangel (Arkhangelsk) and Vologda. This immense region, lying, so to say, beyond the pale of habitable Europe, and which in its climate and a section of its presents quite a Siberian aspect, has scarcely two inhabitants to the e. Archangel alone, even excluding Novaya Zemlya, is nearly as large as d Great Britain together, yet it contains a smaller population than many cond rank, such as Lyons, Leeds, or Birmingham. The whole region about one-fourth of European Russia, whereas its population amounts ieth of that of the empire. Nevertheless it is one of the most interesting ast Europe, not only for the character of its soil, waters, and climate, but history of its inhabitants. The dwellers by the arctic shores have also ial, though modest, part in the development of European culture, and iree centuries have passed since the highway of the White Sea was the by which the Muscovite lands could communicate with the West.

PHYSICAL ASPECT.—THE TUNDRA.

the territory comprised between the granites of Finland and the Ural slopes uniformly and almost imperceptibly towards the White Sea and in, and is furrowed by long winding streams. But the peninsula of Kola, included in the Russian government of Archangel, must be physically an eastern extension of the Scandinavian peninsula. Russian Lapland nited by the deep inlets of the White Sea and Gulf of Bothnia, and all cal formations west of the White Sea belong, like Sweden and Finland, nites and older rocks, whereas those stretching from the White Sea o the Urals are of much more recent origin.

erior of Russian Lapland is but little known, although it has already sed by explorers in all directions. The Maan Selkä of the Finns is continued eastwards into Russian territory by deeply furrowed plateaux, with here and there a few eminences over 1,550 feet high. The small range terminating north of the Gulf of Kandalaksha, at the western extremity of the White Sea, has some peaks said to rise 3,000 feet, but in the east crests of 350 feet are rarely met. The surface is almost everywhere covered with a vast peat bed, filling up all the irregularities of the ground, except along the frontier, where the peat-clad granite is intersected by deep ravines filled with perpetual snow.

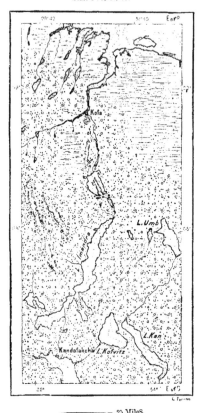

Fig. 180.—The Isthmus of Kandalaksha.
Scale 1 : 2,525,000.

_____ 25 Miles.

East of the river Mezen a range of hills, rooted southwards in the *parma*, or wooded plateau, about the sources of the Dvina, Petchora, and Kama, runs in a north-westerly direction, broken here and there by gaps, through which winding streams flow west to the Mezen, east to the Petchora. This ridge, sometimes called the Timan range, has in some places an elevation of from 650 to 820 feet, one crest in the north rising apparently 890 feet above the sea. But here the chain, already intersooted by numerous rivers, spreads out like a fan, terminating on the Arctic Ocean in a number of parallel peninsulas, one of which, the Svatoï Nos, projects some 18 miles beyond the normal coast-line. The large island of Kolguyev, separated from the mainland by a strait 60 miles wide, may be regarded as a continuation of the Timan range, for the intervening waters are only 130 feet deep. Kolguyev, with an estimated area of 1,350 square miles, is surrounded by shallows and of difficult access, but is yearly visited by some sixty or eighty hunters in search of the seal, white bear, wild swan, duck, blue fox, and reindeer. All attempts at permanent colonisation have hitherto ended in disaster. In 1767 seventy Raskolniks took refuge here from religious persecution, but all soon perished of scurvy.

The Kanin peninsula itself may perhaps be nothing more than a western continuation of a secondary spur of the Timan range. Its northern section between Capes Mikulkin and Kanin, presenting the appearance of a hammer, is occupied by a plateau of crystalline schists, corresponding exactly with a rocky belt

southwards, and nowhere attaining an elevation of more than 300 feet.
chists are separated from the southern masses by intervening Jurassic
, said to contain naphtha springs, besides beds of pyrites and copper.
formerly an island, and the rivers Chiyá, flowing west to the Gulf of
d Cheshcha, flowing east to Cheskaya Bay, both had their source, a
ears ago, in the same lake, so that boats could easily cross from shore
The lake is now a mere swamp, and all through navigation has ceased, a
btless due to the general upheaval going on all along the north coast of
'he old strait was first
to a lake with a double
l this lake was then
d to a swamp between

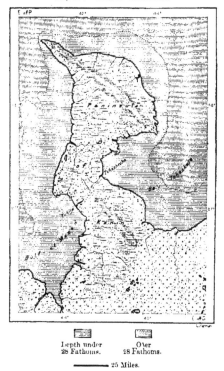

Fig. 181.—KANIN PENINSULA.

; the Timan range the
' Archangel is crossed
low ridges connected
:al system, but nowhere
'e 600 feet, except in
diate vicinity of the
. Nor is the contrast
e hills and the lowlands
:d.   For over half the
plains, lakes, marshes,
:d in a uniform mantle
hile the brief summer
is everywhere of the
:m character.   Nearly
l as far as the sixty-
lel is still covered with
nose relative value is
eased according as the
ions become more and
'orested.   Fifteen-six-
he province of Vologda
imber, mainly conifers

Depth under
28 Fathoms.

Over
28 Fathoms.

25 Miles.

and this proportion is continued in the southern districts of Archangel.
ghts are wooded, and the Russian word *gora*, like the Ziryanian *parma*,
l" or "wood" indifferently, as in South America do the terms *monte* or
d *Wald* in many parts of Germany.   But farther north trees are replaced
and these by creeping plants, whose scanty foliage seeks a shelter
: tufts of red ochre or pale white mosses.   Here are the vast solitudes
he Ziryanian term *tundra*, or rather *trundra*—that is, "treeless land"—
those of North Siberia, though with a mean temperature several degrees
ven in Lapland there are no perennial frozen masses at the bottom of

the peat beds, and springs occur here and there which never freeze even in winter. Wherever the land is not too moist, and well exposed to the sun, southern plants spring up and blossom, and the southern slopes of the Zimniya Gorî, or " Winter Mountains," a low range from 260 to 330 feet high, skirting the east side of the White Sea, are covered with blue aconites and scarlet poppies, " nature's last smile." The roots of trees occurring in various parts of the tundra show that forest vegetation formerly extended much farther north : large pine stems are found in places where even the dwarf willow has now ceased to grow.

The Ural Mountains.

The Ural range, partly lying between Archangel and Siberia, forms geologically a continuous system, consisting throughout of the same crystalline rocks, covered on either side with the same regularly disposed strata, and contrasting with the monotonous Russian and Siberian plains. But the " Girdle of the Globe " (*Zemnoï Poyas*), as this range was formerly called by the Russians, possesses no such geographical unity, being divided into several sections broken by deep gaps, and even by arms of the sea, though still retaining its character as a true water-parting. The Southern Urals, which most abound in mineral wealth, are separated from the northern portions by profound depressions, where the range seems almost to disappear altogether. The northern section again. subdivided into the Vogul, Ostynk, and Samoyed Urals, is separated by low passes from the Kara, or Pac-Khoi Mountains, which branch off in the north-west at right angles from the main range. Waigach Island is also a fragment of the Urals, which are further continued north-westwards by the Novaya Zemlya group, thus forming a total length of over 1,800 miles, without reckoning the windings of the main ridge.

The Northern Urals begin about the sixty-third parallel, north of the point where rise the Petchora and some of the main head-streams of the Ob. Between the source of the Petchora and the northern mountains there is no chain properly so called, but only unequal non-parallel masses, giving to the line of water-parting an extremely irregular form. Amongst these lateral spurs is the famous Bolvano-Is, or " Idol Mount," one of whose crests consists of several strangely shaped rocks resembling gigantic statues. The highest of these, with an elevation of over 100 feet, was formerly a highly venerated deity, and may possibly still have its votaries.

The range, beginning with the Tell-Pos-Is, or Nepubi-Nior (5,541 feet), soon turns north-east, but throws off numerous western spurs, wooded at their base, but whose granite crests are destitute of all vegetation except mosses and lichens. Their ravines, facing northwards, are filled with perpetual snows, from a distance presenting the appearance of glaciers, though the highest summits are at times entirely free of snow. But although lying partly within the arctic circle, the Urals have no true glaciers, the development of which is prevented by the lack of sufficient moisture, and by their low mean elevation, which scarcely exceeds 3,000 feet. No traces even of ancient glaciers were detected until Polakov recently ascertained the existence of numerous moraines, and of striæ evidently ru ning north-west and south-east parallel with those scored on the rocks of Finland and Olonetz.

The uplands separating the Volga and arctic basins still bear the traces of the frozen masses by which they were formerly covered. As they gradually melted, these glaciers formed the fresh-water lakes now filling all the hollows of the land, and which formerly rose 60 or 70 feet above their present level. This

Fig. 182.—NORTHERN URALS.

Scale 1 : 3,600,000.

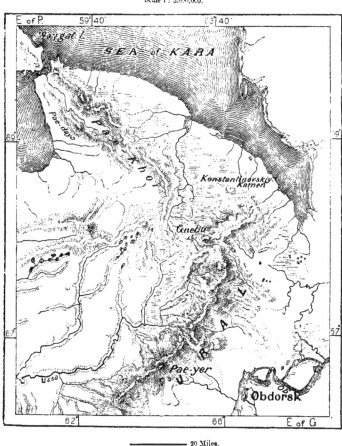

20 Miles.

region seems never to have been invaded by the sea, so that during recent epochs no communication can have existed between the Baltic and the Arctic Ocean, as was at one time supposed. The seas were, however, indirectly connected by a labyrinth of lakes and rivers sufficiently to account for the exchange of fishes and crustacea which has taken place between the two marine basins. In this way the inland waters may have also been colonised by the *Phoca vitulina*, a species of seal found both in the White Sea and in Lakes Ladoga and Onega. Lake Lache, east of Onega, was formerly inhabited by seals, as is evident from the remains of this cetacean found on its banks. In fact, all the northern waters between those of the Volga and Lake Bielo present an essentially arctic fauna, the fauna characteristic of the great river beginning only at the Sheksna. The contrast is doubtless due to a change in the form of the basins. The lacustrine system of the Upper Volga drained till recently to the White Sea, whereas it now flows to the Caspian. In this region lakes and rivers are entangled in one general network, although there are some isolated fresh-water lacustrine basins.

Water System.—The Dwina and Petchora.

Although the lakes draining to the White Sea are at present far inferior in extent to Ladoga and Onega, there are seven of them with areas exceeding 200 square miles, the largest being that of Seg, in Olonetz (481 square miles). The rivers flowing to the west side of the same basin belong geographically to Scandinavia and Finland. The first large stream with a distinctly independent course is the Onega, bearing the same name as the great lake of the Neva basin, as if to commemorate the fact that it rises in a depression formerly common to all these inland waters. Its present source, Lake Lache, probably at one time communicated with Lake Onega, from which it is still separated only by a partly flooded low-lying district. This lake, with an area of about 139 square miles, has itself been almost filled in, now varying in depth from 6 to little over 12 feet. The navigation of the Onega is obstructed throughout its whole course by reefs and ledges, on which there is very little water in summer.

But the chief river of North Russia is the Dwina, or Severnaya Dwina—that is, North Dwina—so called in contradistinction to the Western or Lithuanian Dwina. The word itself means "river," like Don and Donau, or Danube, all more varieties of the same root. The Dwina is a mighty stream, draining an area of 145,000 square miles, and with a total length of 1,020 miles, including the farthest sources of the Vichegda. The largest head-stream, rising near the Urals, is fed by two great affluents, the Sisolka from the south, and the Vim from the north, the latter communicating by a portage with the Petchora basin. The Vorikva, or Verkva, one of the tributaries of the Vim, flows partly underground, at about 60 miles from its source plunging into a chasm, and reappearing in numerous branches 12 miles farther down.

The Vichegda, although the longest, is not considered the main branch, for the Dwina takes its name higher up, at the junction of the Sukhona and Yug, southern and western tributaries flowing in the direction formerly followed in their migrations by the Novgorod and Muscovite colonists. At the Vichegda ferry the Dwina is already over half a mile wide, and navigable for large vessels. Lower down, after its junction with the Vaga, Pinega, and numerous other affluents, it spreads out 3 or 4 miles from bank to bank, and although scarcely over half a mile wide at Archangel, it has here a depth varying with the seasons

THE DVINA, FROM ARCHANGEL.

n summer, when the twilight almost verges with the
der sights than this mighty stream viewed from a raft
rent, the outline of the woodlands dimly traced on the
uddy skies reflected on the rippling waters.

e Gulf of Archangel has an area of no less than 444
rowed by four main branches connected together by

Fig. 183.—THE LOWER PETCHORA.

Scale 1 : 4.200.000.

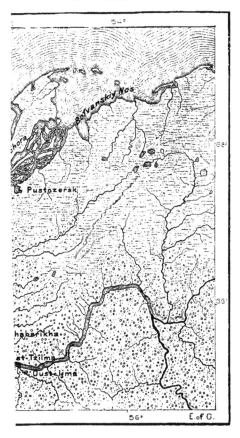

50 Miles.

at times shifting their course, or alternately rising and
floods.   The western branch, formerly the deepest, is
ds, and the shipping passes by another farther east,
from 8 to 12 feet at low water.   Great changes in the
direction of the channels take place during the thaws,

when the ice from the interior accumulates on the banks, carrying away huge masses of rock, and often piling up the débris along the shore. The river is generally ice-bound for 191 days, from about October 23rd to May 2nd; yet it abounds in fish, including one species, the navaga, akin to the cod, hitherto found nowhere else. The sturgeon made its appearance in the Archangel waters for the first time in 1865, thanks doubtless to the Catherine Canal, for some time connecting the Kama-Volga and Dvina systems.

Although a small stream compared with the Dvina, the Mezen is nevertheless equal to the Seine in the volume of its waters and the extent of its drainage. It is even broader than the French river, being nearly three-quarters of a mile wide above its sand-blocked estuary, which, like that of the Seine, is noted for the abnormal character of its tides. The flow lasts generally four hours only, the ebb eight, but the former is so rapid that vessels moored in the roads often threaten to drag their anchors.

The Petchora, the greatest river of the eastern tundra, is in no respect inferior to the Dvina, and even drains a wider area, viz. 162,000 square miles. It flows first north along the foot of the western spurs of the Urals, at the outlet of every valley receiving fresh tributaries, amongst which is the Shchugor, rising in the snows of Mount Tell-Pos-Is, and famous for the falls in its upper valley, and for the so-called "Iron Gates," where it flows between rocks of a dazzling white colour, and cut into enormous columns by vertical fissures. After receiving the Ussa, also from the Urals, the Petchora bends westwards along the depression stretching a little south of the arctic circle from the Urals to the Gulf of Mezen; it then turns abruptly northwards at Ust-Tzilma, discharging into the Frozen Ocean through a delta about 120 miles long, where the channels wind in a vast network round sands, islets, and sand-banks, which shift and change their form with every thaw. A bar at the entrance prevents the access of vessels drawing 12 feet. Although the delta is free of ice on an average for 127 days only, from May 25th to October 1st, yet a surprising traffic in timber, cereals, and peltry is carried on in a river which is, moreover, obstructed by rapids in a part of its course. The sparse Russian, Ziryanian, and Samoyed population is entirely centred in hamlets and small villages occurring at wide intervals alongs its banks. The domains of the various fishing associations occupy severally many thousand square miles in its basin and on the arctic islands, and the Russians associated in the white whale fishery set aside a tenth of their captures for St. Nicholas of Pustozersk to insure the success of their undertaking.

THE WHITE SEA.

THE sea washing the shores of the province of Archangel penetrates far inland through numerous bays, inlets, and even gulfs, narrowing at their entrance between lofty headlands, and generally shut off from the sea by shoals and islands, the remains of ancient *lidi*, like those of the Adriatic. In the case of the White Sea the ancient shore is still represented by the island of **Morkhovetz**, besides which it is distinguished by the parallelism of its bays and estuaries, a salient feature of North Russian geography.

The White Sea, with an area of from 47,000 to 48,000 square miles, must in many respects be considered as a lake, or rather a group of lakes, communicating with the Frozen Ocean. The narrow entrance rounding the Lapland coast, while changing it to a saline gulf, has preserved its independent character in the aspect and outlines of its shores and the relief of its bed. Thus the White Sea is deeper than the ocean with which it is now connected, falling from about 170 feet at its entrance to over 1,000 feet towards the extremity of the Kandalaksha

Fig. 184.—WHITE SEA.

Scale 1 : 6,200,000.

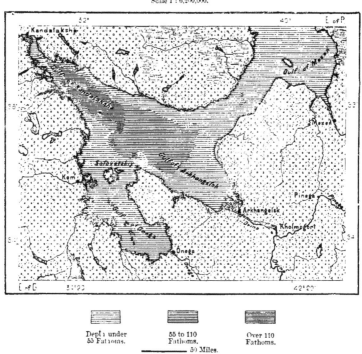

| Depth under 55 Fathoms. | 55 to 110 Fathoms. | Over 110 Fathoms. |

50 Miles.

Gulf. The Gulf of Onega, which, like its namesake of the Neva basin, might almost be called Lake Onega, is somewhat shallow, being no more than 260 feet deep, and is separated by the Solovetzkiy Isles from the main basin. When the White Sea was a lake, like Ladoga and Onega, it was probably at a higher level than at present, and drained to the Polar Sea through a river forming a continuation of the Upper Dwina. But oscillations of level may have caused the sea to burst into the lacustrine basin, converting its fluvial outflow into a strait. The elevation of the entrance prevents the sand and mud from being carried seawards, and the basin is being slowly filled by the continuous deposits of the rivers, so that however deep it may now be, the White Sea must have formerly been much deeper. Even its waters have remained less saline than those of the ocean, owing to the constant accession of fresh water from the rains, snows, and especially the Dvina, Onega, and other influents. Nevertheless some salt is still made on its foggy shores, where artificial heat replaces that of the sun in the process of evaporation.

INHABITANTS: LAPPS, SAMOYEDS, ZIRYANIANS, AND POMORI.

THE great Kola peninsula, limited southwards by the Gulf of Kandalaksha, belongs ethnically to the Lapp race. The Slavs are here represented merely by a few fishing communities, while the Finns proper, all of the Karelian branch, have only founded a few isolated settlements on the south coast, along the shores of the Gulf of Kandalaksha. At the same time, the high stature of some of the Lapps their full red beard, their habits, and a few words of the current speech show clearly enough that this eastern branch of the Lapp family has absorbed Slav elements. They have a general resemblance to the Suimas of Sweden and Norway, but are less civilised and less intelligent than their western brethren. This is probably due to their long vigils and to the general scarcity of food, largely consisting in winter of mosses, the bark of trees, a certain farinaceous earth kneaded into the dough, and the so-called *lebeda*, a species of bitter and unwholesome plant. An idea may be formed of the climate of this region from the Lapp language, which contains 20 terms to express ice, 11 for cold, 41 for snow and its varieties, 26 verbs to describe the phenomena connected with freezing and thawing.

The Lapps of the peninsula have so far adopted Christianity that they have been baptized since the sixteenth century by Russian monks, who have also introduced serfdom into the country, acquiring from the czars the ownership of entire populations. Thus evangelized, the Lapps have increased the number of their gods by the addition of one propitious deity, Jesus, and one evil spirit, the devil, king of hell. A few devotional works have even been printed in the Russian character, which the Lapps of some encampments have learned to read. Still they have their old wizards, or shamans, and worship heaps of stones, bones, or fossils, which they believe to be inhabited by spirits. As with so many Ural-Altaic peoples, marriage has preserved in full vigour the forms of primitive abduction, and the bride is still expected to struggle violently and utter piercing cries. After the abduction her father remits to the husband his right of absolute authority, even that of "roasting the eyes of his victim." She is made fast to her new home "like a wild reindeer," but after a pretended cudgelling, her husband releases and consecrates her "hostess" and "mother of the bread." It is she who controls the children and arranges all matrimonial alliances, which are generally dictated by interest. Lapps of eighteen or twenty have often been known to marry wives of sixty. Each individual has his particular mark, analogous to the *totem* of the red-skins. At his birth the reindeer assigned to him is marked with this sign, which he afterwards stamps upon all his belongings.

The Lapps are known to have formerly occupied a large portion of the North

)uth and east of their present domain. The chroniclers speak of
ed seven hundred years ago on the shores of Onega, and in the
in Lapp terms, meaning " river," " brook," " island," or " forest,"
) when they occupied the region east of the White Sea. The very
containing the root *Same*, which is the proper appellation of the
em to imply that the present occupants of the eastern plains were
) successors of the Lapps. The Karelians, now reduced to a mere
shores of the White Sea named from them, were also amongst the
Lapps, and traces of them are still discovered throughout the
northwards. The Russian dialect current in Archangel is full of
expressions borrowed from the Finnish, and the Karelian Finns
included with the Ziryanians under the collective designation of
oned in the mediæval chronicles, and whom the Norse navigators
t exaggeratedly described as a rich, powerful, and civilised people
agriculture and the industrial arts.

fishing and hunting associations formed by most of the Great
ommunes in Archangel, chiefly employ technical terms of Karelian
that before the arrival of the Russians such associations had
:med by the Finns of this region. In many places the whole
zes the fishing expeditions, distributing the men along the coast
a way best calculated to equalise the chances, a share being even
left behind to attend to the domestic affairs of the villages. But
equality amongst the members of these unions have been caused
d communal debts contracted first with the convents and Bishops
nd later on with the Russian traders.
White Sea the Samoyeds, like the Lapps, represent the old Finnish
many respects contrasting with their somewhat remote kindred.
d flatter features and lower foreheads betray a more decidedly
t. Zograf includes them amongst the brachycephalic Mongols,
gards them as of mixed Finnish and Mongol origin. They call
z (plural Netza)—that is, " Men "—or else Khassov (Khassova) ;
    The term Samoyed, synonymous with " Autophagi," has given
legends, and in many documents they are called Siroyedi, or
Flesh," which, again, is nearly synonymous with " Eskimo,"
:thernmost tribes of the American continent, who, under a similar
eloped analogous habits.
i6, Burrough visited Vaigach, the holy island of the Samoyeds,
:n headland covered with 420 statues of men, women, and children
e great seven-headed idol of Yesako. In 1594 this " Cape Idol "
was revisited by the Dutchman Nai, and in 1824 Ivanov
nonstrous figures described by Burrough. They have since been
urnt by zealous missionaries, and a cross now surmounts the top-
e headland. But the Russian traders on the coast say that
sacred images have been preserved, and the place of sacrifice in a

neighbouring grotto has recently been visited by Nordenskjöld. The altars and the hundred idols worshipped by the Samoyedes in their camping ground of Kozmin, some 12 miles from Mezen, have also been burnt. But if fear of the Russians prevents them from erecting large idols visible from afar, they can still dress up shapeless dolls which they hide in their tents or under their clothes, and which thus represent the scapular, medals, and other charms elsewhere still worn by the devout.

Although a mere fragment of a formerly powerful race, the Samoyeds still own a vast domain, stretching from the White Sea beyond the Yenisei, and to the Altaï highlands whence came their forefathers. Pressed upon probably by conquering tribes of Tûrki stock, they followed the course of the streams flowing

Fig. 155.—A SAMOYED PILO.

northwards, and settled on the shores of the Frozen Ocean. But the Central Urals, far south of the present territory, still bear Samoyed names. The Yurak branch, of which those of European Russia are a subdivision, now occupies all the region of the tundras on both sides of the Urals, which supplies the *yagel*, or moss, required for the support of their reindeer. But although already mentioned towards the end of the eleventh century as possessors of this land, they seem to have conquered it from other Finnish tribes more nearly related to those of Finland. Several local names, explicable only in the Karelian dialect, show that the country was first occupied by this branch, who, according to the legend, withdrew farther inland, where they possess vast hunting grounds and pastures, with multitudes of mammoths, foxes, and beavers. Similar traditions are preserved by the Lapps, who had also probably to contend with these Karelian Finns, or Chûdes, as the Russians call them. The name of their predecessors is associated amongst them with that of the maleficent aërial and underground spirits. The Samoyeds themselves are fated suddenly to disappear "below the earth" like the Chûdes. In their play the Russian children repeat a now meaningless rhyme, but which was formerly only too significant: "Come, seek the Samoyed; come, mark the Samoyed; the Samoyed we'll find, and cut him in two."

Those who formerly dwelt on the shores of Lake Onega and the east side of the White Sea were long ago exterminated by the Novgorod and Muscovite Russians, and even in their present domain all the valuable lands are gradually passing into the hands of the Ziryanians and Slavs, notwithstanding the law of 1835 limiting the Russian farms to 160 acres per family, and prohibiting all further encroachments on the Samoyed territory. The whole race, now confined to the three districts of Kanin, Timan, and Bolshaya Zemlya, is said to be reduced in Europe to somewhat over 5,000, and whilst the birth rate diminishes, the mortality continues to increase.

The Ziryanians, another Ural-Altaic people, though quite distinct from the Finns, have vied with the Russians in depriving the Samoyeds of their reindeer herds, and in plundering their *chúms*, or tents. Of these ancient aboriginal races one has thus become enslaved, while the other has taken rank with the traders and rulers of the land. Settled exclusively along the navigable rivers, and on the portages between the Petchora, Mezen, and Upper Dvina basins, the Ziryanians have from time immemorial been engaged in trade, and they have now monopolized a large portion of the commerce of North Russia not only with the adjacent provinces, but even with England and Norway. Their packmen and agents visit all the fairs from Archangel to Moscow and Nijni-Novgorod, where they dispose of their furs, horns, and fish in exchange for many domestic comforts.

From the end of the fourteenth century this commercial nation has ceased to worship sun, fire, water, trees, and "the old woman of gold." Bishop Stephen hewed down their "oracular birch," and taught them some Christian prayers, besides the use of a peculiar alphabet, the letters of which have not all yet been explained. but which has long been superseded by the Russian. The only visible traces of the old worship are the animal sacrifices still made in front of the churches. In other respects the Ziryanians have been largely Russified, and are gradually becoming assimilated to the ruling race, whose language they understand, and whose songs they sing even when the sense is unintelligible to them.

The pure Ziryanians were recently estimated at 30,000, although in 1874 Popov reckoned 91,000 in the provinces of Vologda and Archangel alone. But there is no doubt that they are still numerous, and that many passing for Russians are really Ziryanians. Many even claim the honour of descending from the Novgorod colonists of the twelfth to the fifteenth century, although this colonisation was crossed by that of the Suzdalians, following the valley of the Sheksna, and penetrating across the portages into the Dvina basin. Some old Volkhov families on the banks of the Dvina and **Petchora** have hitherto kept aloof from all intermixture, still exercising a sort of patriarchal authority over **the** inhabitants of **the** surrounding districts.

The Pomori, or " People of the Sea," as all the Great Russians of these northern provinces are called, are by far the most numerous element in Archangel and Vologda. Separated by vast intervening spaces from the rest of the family, they have in some respects preserved an extremely pure type, although otherwise affected in a special manner by the development of independent religious sects. Nowhere else has the family life acquired a more despotic character. The betrothed calls her future lord *ostúdnichok*, the " fear inspirer," and before the bridal kiss the bridegroom plucks her flowing tresses, and sings her a threatening song: " Under the nuptial couch lies an oaken staff; to this oaken staff is attached a silken whip with three lashes, and when it falls the blood flows." Thus the wretched bride shudders at the prospect of her future thraldom, and with tearful eyes thrice bends the knee before the holy images: " I make the first genuflexion for the most pious Czar; I make the second for the most pious Czarina; and the third I make for myself, that the Saviour may take pity on this hapless maiden in her new home." The Czar for whom she prays is, in the popular estimation, far less the ruler than is the "land czar" (*zemskiy tzar*), who represents all the landed interests, who, in the language of the national songs, "serves the land." Most of the Pomori escaped the hard yoke of serfdom. Scarcely had the Muscovy peasantry been attached to the glebe, when the acquisition of the southern provinces attracted the attention of the nobles, whose thirst of land and slaves found fuller scope in this direction. Thus the pine forests and frozen tundras of the north were happily overlooked, and in 1866 there were no more than 476 serfs in the whole government of Archangel.

TOPOGRAPHY.

Kola, capital of Russian Lapland, had a population of less than 800 at the last census; yet under a more temperate climate it would be well situated for traffic, and might become a place of some importance. For it stands at the junction of two rivers, at the head of an estuary running far inland, and continued still farther southwards by a lacustrine depression, which forms a communication between the Gulf of Kandalaksha and the Arctic Ocean. Its position has been accordingly appreciated by traders from the remotest times, and it is mentioned in 1264 as a fishing station and mart frequented by the Novgorod merchants. But its natural advantages failed to attract any large numbers, and it consisted of nothing more than a group of wooden barracks when bombarded and half destroyed by the English during the Crimean war. Its chief industry consists in shark fishing, a pursuit rendered doubly dangerous by the storms of the Frozen Ocean and the shoals of these animals, which at times capsize the boats and devour the crews.

Kem, on the west side of the White Sea, is the chief trading centre in the Karelian territory. Like Kola, it was an old Novgorod colony, favourably situated at the mouth of a navigable river affording the easiest communication with the Gulf of Bothnia and Scandinavia. But its present importance is largely due to the vicinity of *Solovetzkiy*, or *Solorki Island*, and the famous monastery founded here in the first half of the fifteenth century. The monks, who depend directly on the Holy Synod, long maintained their primitive rites in opposition to the reforms of the Patriarch Nikon, but in 1676 the place was taken through the treason of a monk, who revealed an underground passage to the besiegers. Most of its 1,600 defenders, monks and peasants alike, were put to the sword or flogged to death with rods. Yet the spirit of religious independence is not quite extinct in Solovki, where the monks are recruited from the natives, still imbued with the Novgorod spirit.

The village of *Onega*, also an old Novgorod settlement at the mouth of the river of like name, and at the southernmost extremity of the White Sea, is an

Fig. 186.—SOLOVETZKIY ISLANDS.

Scale 1 : 600,000.

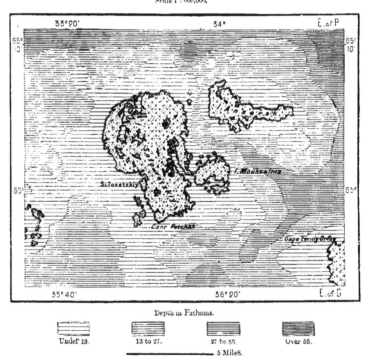

Depth in Fathoms.

Under 13.　　13 to 27.　　27 to 55.　　Over 55.

5 Miles.

important fishing station. The herrings swarm here in such quantities that they may at times be taken in bucketfuls, and all the fish—herrings, cod, salmon, perch, pike—are generally preserved in frozen masses, very few being smoked or cured, owing to the inferior quality of the salt. On this coast the cattle and swine are mostly fed on red herrings.

The vast basin of the Dvina, larger in extent than all Italy, has only three towns whose population exceeds 5,000, and two of them, Vologda and Archangel (Arkhangelsk), owe their importance partly to their rank as provincial capitals.

Vologda, on the site of an old Russian colony dating from the twelfth century, covers a large space with its domed churches and low straggling houses. It lies at the south-western extremity of the Dvina basin, on a river which here becomes navigable, and which a little farther down joins the Sukhona, one of the main branches of the Dvina. It is well situated for traffic with the Volga and Neva basins, and when Russia opened direct relations with England through the White Sea, it was chosen as the intermediary between Moscow and Archangel. It also

Fig. 187.—ARCHANGEL.

became the starting-point of the Siberian trade, while the southern route by Kazan remained exposed to the attacks of the Bashkirs. It still sends to the Lower Dvina flax, oats, and other produce, to the yearly value of over £160,000, and to St. Petersburg butter, eggs, and sail-cloth.

For a distance of 480 miles, from Ust-Yûg to Archangel, there is no town properly so called, any more than on the Dvina affluents, except *Ust-Sisolsk*, capital of the Ziryanians, in the Vichegda basin. During the Novgorod rule the

the Lower Dvina was *Kholmogori*, the Holmgård of the
he eastern wares imported by the Volga Biarmians were
the tenth and eleventh centuries. Its importance was
trade, the monopoly of which was granted to it by the

188.—ARCHANGEL AND THE DVINA DELTA.

Scale 1 : 750,000.

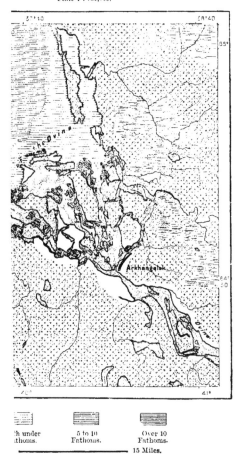

h under | 5 to 10 | Over 10
thoms. | Fathoms. | Fathoms.

15 Miles.

the foundation of its neighbour, Archangel, ruined this
ow become one of the poorest places in North Russia.
e "City of St. Michael the Archangel"—covers about
ank of the Dvina at the head of its delta. It consists
enements, relieved by a few fine stone houses and the
German castle," forming part of the fortress. In the
site was occupied by a monastery, but it remained an

unimportant place till the end of the sixteenth century, when the English traders made the White Sea the commercial route between Russia and the West. Its most flourishing period preceded the foundation of St. Petersburg, which offered a more convenient highway of communication with the rest of Europe. And although be here founded an arsenal, a castle, and a dockyard, Peter the Great contributed none the less to the decadence of the place by restricting the amount of its imports, by suppressing its export trade in hemp, flax, tallow, and over one-third of the other products of the empire, and by summoning its sailors and traders to h's new capital. Nevertheless its position at the only river outlet of a vast basin, with a rapidly increasing population, could not fail to restore a certain activity to "the fourth capital of the empire." In spite of the ice, suspending all navigation for nearly seven months, Archangel exports, especially to England, Holland, and Norway, flax, hemp, oats, and other cereals, timber, resin, train oil, and tallow, its exports being usually tenfold its imports, which consist mainly in fish from Norway, wines, and colonial produce. The women are here formed into unions for loading the vessels with corn, the *shkicidorka*, or manageress, being generally chosen from amongst those familiar with the Anglo-Russian jargon of the place. The town presents a very animated appearance during the annual fair, when some 50,000 people crowd into Archangel and its northern suburb of Solombala, seat of the Admiralty. But the resident population seems to have been falling off during the last fifty years, the census of 1860 having returned 33,675, and that of 1872 not quite 20,000 ; this last, however, was taken in winter. A colony of English artisans is settled in the neighbourhood in connection with some large saw-mills.

Mezen, although the outlet of the Mezen basin, is a mere village, lying almost beyond the limits of vegetation, and deprived of much of its trade by Russanova, situated 12 miles nearer to the mouth of the Mezen estuary. It is a dismal place of banishment, one of those "Siberias this side the Urals" where the dominant Church kindled the first fires of persecution against the Raskolniks in the seventeenth century. In this and other respects it resembles Pustozersk, the modest commercial centre of the Petchora basin. The natives, no less than the exiles, in these arctic stations are often decimated by nervous disorders attributed to the "evil eye," but caused probably by privations of all sorts. Memorial crosses, greatly revered by the people of Mezen, still recall a terrible winter in the first half of the eighteenth century, during which the whole population all but perished of cold and exposure.

Novaya Zemlya.

Novaya Zemlya—that is, "New Land "—forms the eastern limit of the waters stretching along the northern shores of Lapland, and sometimes known as "Barents' Sea," in honour of the illustrious navigator who crossed it towards the end of the sixteenth century. With the island of Waigatch it may be regarded simply as a northern continuation of the Russian mainland. But though long frequented by the Russian fishers, its existence was first revealed to the West after Willoughby's voyage in 1553. It was again visited in 1556 by Stephen Burrough, who also circumnavigated Waigatch by the two Straits of Yugor and Kara, separating it from Russia and the southern extremity of Novaya Zemlya. Barents followed, and perished here during the hard winter of 1597, and in 1871 the Norwegian Carlsen discovered in lat. 76° 7′, near the north-east end of the northern island, the house where Barents and Heemskerk had wintered two hundred and seventy-four years previously. All the objects found in it were collected and brought to Holland, and the Amsterdam Geographical Society now proposes to raise a monument on a neighbouring headland in honour of the famous navigator.

Novaya Zemlya may, on the whole, be considered as a continuation of the Ural range, with which it is connected by Waigatch and the Pae Khoï chain of like

Fig. 189.—WAIGATCH ISLAND AND KARA AND YUGOR STRAITS.
Scale 1 : 2,650,000.

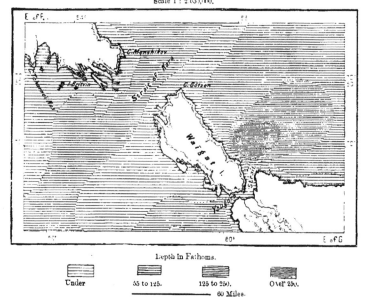

Depth in Fathoms.

Under   55 to 125.   125 to 250.   Over 250.
———— 60 Miles.

geological formation. The connection is interrupted, without being destroyed, by the two intervening straits, beyond which the system traverses the entire length of Novaya Zemlya, describing an arc 540 miles long, with its convex side turned towards Spitzbergen. The island is divided into two unequal parts by the Matochkin Shar, a channel some 30 feet deep, skirted by the highest elevations in the range. Like the Urals themselves, these heights slope gently westwards, and abruptly towards the east, a disposition maintained also by the bed of the sea, which shoals gradually on the west and dips rapidly on the east side, where the line of 110 fathoms runs close to the shore.

Although frequented by fishers, and notwithstanding its proximity to the mainland, the interior of Novaya Zemlya is still very imperfectly known. It has been visited by Lütke, Baer, Heuglin, Wilczek, and other scientific explorers, but its systematic study is far from being complete, and little has hitherto been done beyond determining the coast-line and its total area, which is estimated at about 36,430 square miles. Few of the heights have been accurately fixed, but

Fig. 190.—View from the Matochkin Shar, Novaya Zemlya.

from the summit of Wilczek Point (4,212 feet) Höfer and Wilczek noted, on either side of the strait, several crests at least 4,000 feet high, and the mean elevation of the land seems to be about the same as that of Spitzbergen. The Mitushev Kamen, west of Wilczek Point, consists of protogine, like Mont Blanc, but the chief formations are schists and Devonian rocks, including some slaty strata so black that they have often been mistaken for coal. The island was formerly supposed to be rich in minerals, particularly silver, but the special

·vey conducted by Ludlow in 1809 failed to detect anything rgentiferous galena and some traces of pyrites and copper.

end at many points to the water's edge, and they seem to have d beyond the coast-line, for striæ and polished rocks have been slets of Rogachov Bay, south of Cape Gusinoï. Although in a titude than Spitzbergen, the mean annual temperature of Novaya r 23° Fahr. In the Maliye Karmakuli Creek (Moller Bay), he Norwegian Captain Bjerkan, who wintered here in 1876-7, .ass never rose above 9° Fahr. in December, falling to — 14° Fahr.

Yet the yearly temperature is here somewhat higher than in 180 miles farther south or south-east. From observations made Wild has determined the mean temperature of the Matochkin

Fig. 191.—THE MATOCHKIN SHAR CHANNEL.
Scale 1 : 1,000,000.

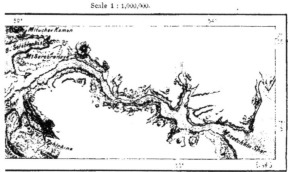

15 Miles.

.7° Fahr., although it rises to 20° Fahr. at the Melkaya Guba, l falls to 15° at the Guba Kamaika, 156 miles nearer the equator. .ies are probably due to the warm tropical currents flowing side of Novaya Zemlya, but without quite reaching the southern t body of these currents passes over the deep trough separating Scandinavia, and sets eastwards to the Barents Sea with such els have been known to drag their anchors off Hope Island, reme south-east point of the Spitzbergen group. In the there is a very perceptible through current, especially under of the north island. In these currents, whose temperature g point, the floating masses melt rapidly, so that icebergs uth of the seventy-fifth parallel in the section of the Barents orth coast of Lapland.

at times from ice-fields, Novaya Zemlya is in other respects gion. Its cliffs, freed from the snow by the sun, fogs, and winds, ice perfectly bare; but a closer inspection reveals the yellowish or rusty hues of their mossy vegetation. The plains produce a few flowering plants, and according to Heuglin the flora comprises altogether 150 phanero-gamous and nearly an equal number of cryptogamous species. There are also some tracts of dwarf birch, elder, and fir trees; but the most common tree is a species of willow (*Salix polaris*), rising a little more than half an inch above the surrounding lichens. The giant of these arctic forests is a willow (*Salix lanata*), some specimens of which grow to a height of 6 inches. The growth of these plants is rather in their roots than their stems, so that the Novaya Zemlyan, like the North Ural forests, might be described as subterranean.

The fauna is richer than that of Spitzbergen, besides various species of cetacea, including the bear, wolf, two species of fox, reindeer, hare, mouse, and campagnol. Heuglin raised the number on the list of birds from twenty-eight to forty-five, but mosquitoes, the plague of the Russian and Siberian tundras, are rare both here and in Waigatch Island.

The southern portion of Novaya Zemlya was formerly inhabited by the Samoyeds, who have long either disappeared or withdrawn to the mainland. Now the place is occupied only by a sprinkling of Europeans settled at the permanent harbour of refuge established in 1877 at Maliye Karmakuli, on Moller Bay, and at a few fishing and hunting stations. Although included in the Russian dominions, the group may be traversed in all directions without meeting any indication of the Czar's supremacy.

CHAPTER IX.

VOLGA AND URAL BASINS.

(GREAT RUSSIA.)

HE river which intersects Russia obliquely from near the Baltic to the Caspian, and which drains an area thrice the extent of France, has largely contributed to the political development of the Russian people. The Dnieper showed the Little Russians the route to Constantinople; the Vistula, Niemen, and Western Dvina laid open the West to the White Russians and Lithuanians; even the Volkhov and the Neva, by placing Novgorod in relation with the Hanseatic towns, withdrew it, so to say, from the heart of the land. But the Volga and the vast system of its navigable affluents compelled the inhabitants to develop themselves and create their civilisation on the spot. Although the water highways facilitated communication in every direction between the various regions of Great Russia, few colonists were attracted to the arid steppes, the salt wastes, and the land-locked basin of the Caspian in the south-east. The bulk of the people were thus confined to the central region, which they gradually brought under cultivation. Coming in contact at a thousand points with the Asiatic tribes pouring in through the steppes, the Great Russians intermingled with them, either absorbing or becoming absorbed, and thus by continuous crossings developing that hardy race which gradually acquired the supremacy over all the Eastern Slavs. Through husbandry, canals, highways, railways, henceforth relieved from the old limits imposed by their swamps and forests, they have been able to reverse the flow of migration, sending forth groups of colonists to the remote shores of the Pacific, girdling round China with a continuous chain of settlements, and thus bringing a great part of the Asiatic continent more and more under European influences. But the Volga and its great head-streams still remain the centre of Great Russian nationality; here they number already upwards of thirty millions, and in some places have peopled the land as thickly as many countries in the west of Europe.

The Volga and its Tributaries.

The rivulet which, at its farthest source, takes the name of Volga, rises not in a highland region, but in the midst of lakes, marshes, and low wooded hills little elevated above the Volkosnkiy Les (" Volkoi Forest ") and Valdaï plateau, which may be taken as the true source of the stream. The highest ridges of the Valdaï scarcely rise 220 feet above the plateau, although the chief crest, the Popova Gora, attains an altitude of 1,170 feet. The mean elevation of the land is also sufficient to give it a far more severe aspect than that of the Lovat and Lake Ilmen plains on the west and north-west. Its peat beds, lakes, and fir forests are more suggestive of the neighbourhood of Lake Onega, some 300 miles farther north, and the climate is, in fact, about 2° colder than in the surrounding districts. Yet the Valdaï flora differs on the whole but little, if at all, from that of the plains stretching towards the great lakes, whence it has been concluded that these heights are of comparatively recent origin. They have no indigenous vegetation, all their species coming from the region released from its icy fetters at the close of the long glacial epoch. The plateau, now furrowed by the rains and frosts, formed at that time a continuation of the uniform slope of the land, and, like it, was covered by the ice-fields from Finland. Its hills are strewn with erratic boulders of all sizes brought down by the northern glaciers. In its water fauna also this plateau forms part of the Finno-Scandinavian region. The fish of its lakes, and even of the Upper Volga itself, do not belong to the Volga basin proper, which the Valdaï streams seem to have only recently joined. To judge from their fauna, the true origin of the Volga should be sought, not in the Valdaï, but in Lake Belo Ozero (" White Lake "), east of Ladoga. The sturgeon and sterlet inhabit the Sheksna, the outlet of this lake, as they do the Middle Volga itself.

The region giving birth to the Volga is one of the swampiest in West Russia, resembling a lowland tract rather than a true water-parting. Separated by a simple peat bed from a tributary of the Volkhov, the streamlet rising in the Volgino Verkhovye, and sometimes called the Jordan from its sacred character, flows from a spot now marked by the ruins of a chapel, thence oozing rather than flowing from bog to bog for a distance of about 22 miles, when it successively traverses three terraced lakes, whose levels differ only a few inches one from the other. The Jukopa, one of its southern affluents, often causes a back flow to Lake Peno near its source, the natural fall being so slight that the impulse of a lateral current suffices to reverse it. After leaving Lake Peno, which is close to Lake Dvinetz, source of the Dvina, the Volga turns eastwards to Lake Volgo, where it is already a considerable stream, with a volume of from 3,500 to 3,600 cubic feet per second, according to the seasons. Three miles farther down occurs its first rapid, where a dam has now been constructed, which during the rains converts the upper valley, with its lakes, into one vast reservoir 48 miles long, over 1 mile wide, and containing 6,300,000 cubic feet of water. Boats

CONFLUENCE OF THE OKA AND THE VOLGA.

and rafts are then able to descend from the lake region, and higher up the river
becomes regularly navigable. Near this point the Volga is nearly doubled by
the Selijarovka from the winding Lake Seliger, whose insular monastery of St
Nilus is still visited yearly by about 20,000 pilgrims. Here may be said
to begin the commercial stream, the Ra, Rhus, or Rhos of the ancients and of the
Mordvinians, the Yûl of the Cheremissians, the Atel or Etil of the Tatars, the

Fig. 192.—Sources of the Volga and Dvina.

Scale 1 : 575,000.

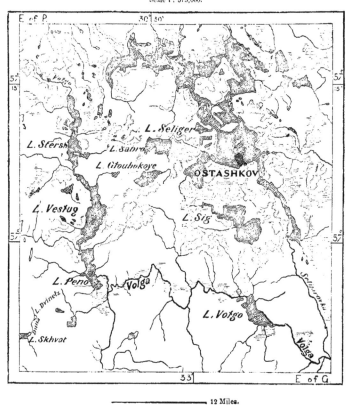

———————————— 12 Miles.

Tamar of the Armenians—that is, in these languages, the "River"—and in Finnish
the Volga, or the "Holy River."

Below the Selijarovka it descends the slopes of the plateau through a series of
thirty-five porogi, or rapids, which, however, do not stop the navigation, and
beyond the last of the series it winds unimpeded through the Great Russian
lowlands, receiving numerous navigable tributaries, and communicating by canal
with the Baltic basin. After passing the populous towns of Tver, Rîbinsk,
Yaroslav, and Kostroma, it is joined at Nijni-Novgorod by the Oka, of nearly

equal volume, and historically even more important than the main stream. The Oka, which long served as the frontier between Tatar and Muscovite, rises in the region of the "black lands," and throughout a course of 900 miles waters the most fertile plains of Great Russia, bringing to the Nijni fair the produce of Orel, Kaluga, Tula, Riazan, Tambov, Vladimir, and Moscow. Over 1,440 yards broad, it seems like an arm of the sea at its confluence with the Volga. East of this point the main artery is swollen by other tributaries, which, though as large as the Seine, seem insignificant compared with the mighty Kama, joining it below Kazan from the Urals, and draining an area at least equal in extent to the whole of France. Judging from the direction of its course, the Kama seems to be the main stream, for below the junction the united rivers continue the southerly and south-westerly course of the Kama, whose clear waters flow for some distance before intermingling with the grey stream of the Volga. Below Simbirsk the

Fig. 193.—TREMBLING FORESTS NEAR NIJNI-NOVGOROD.

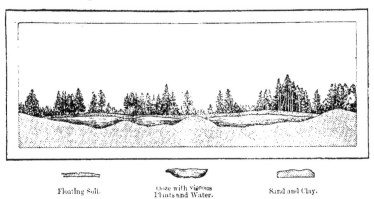

Floating Soil.   Ooze with viscous   Sand and Clay.
                      Plants and Water.

tributaries are few and unimportant, and as the rainfall is here also slight, and the evaporation considerable, the mean discharge is probably as great at this place as at the delta.

Below the Kama junction there formerly existed a vast lacustrine basin, which has been gradually filled in by the alluvia of both streams. Here is the natural limit of the peat region, and here begins, on the right bank, that of the steppes. As we proceed southwards the atmosphere becomes less humid, the ground firmer, and below Simbirsk we no longer meet those mossy and wooded quagmires bound together by the tangled roots of trees, resembling matted cordage. But even in the boggy districts those floating forests are slowly disappearing as the land is brought more and more under cultivation.

Below the dried-up Simbirsk Lake the stream is deflected by an impassable limestone barrier eastwards to Samara, where it escapes through a breach and reverses its course along the southern escarpment of the hills, thus forming a long

ninsula projecting from the western plateau. Here is the most
e scenery on the Volga, which is now skirted by steep wooded cliffs,
g in pyramids and sharp rocky peaks. Some of the more inaccessible
re surmounted by the so-called "Stenka" kurgans, raised in memory of
ief of the Cossacks and revolted peasantry, who had established them-
his natural stronghold
lga. The hills often
than 300 feet above
m, the Beliy Klûch,
; of Sizran, attaining
te elevation of 1,155
1,120 feet above the
l of the Volga.

: Volga Delta.

m of the delta really
the Tzaritzin bend,
miles from the Caspian,
e.m here branches into
channels between the
the Volga and the
known near the coast
eket. Still the delta,
) called, is formed only
miles above Astrakhan,
rking of the Buzan
m the main bed. Near
the Balda and Kûtûm,
: down, the Tzarova,
irûl, and other arms,
', and in the vast allu-
ula projecting into the
nd which is at least
round, there are al-
about two hundred
)st of them, however,
eams choked with mud.
.cler Nestor speaks of

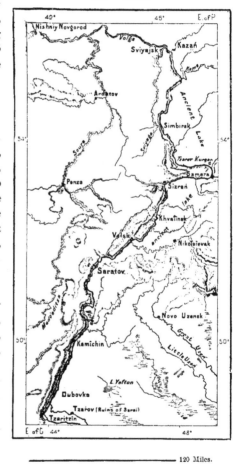

Fig. 194.—Steep Banks of the Middle Volga.

Scale 1 : 7,000,000.

120 Miles.

uths, and there are present about fifty regular channels. During
floods all the delta and the lower course below Tzaritzin form one vast
ning waters, broken only by a few islands here and there, and after
se floods new beds are formed, old ones filled up, so that the chart of the
o be constantly planned afresh. Even the main beds get displaced.

Two hundred years ago the navigable channel flowed due east from Astrakhan; since then it has shifted continually more to the right, and now runs south-south-west. The Balda also, which communicates by a side branch with the Akhtúba, has recently acquired considerable proportions at the expense of its neighbours, and has gradually swallowed up the last vestiges of the famous Baldinskiy convent. The bars and sand-banks, too, are constantly changing their position and depth. None of them have more than 7 feet 6 inches of water, and the second in importance had only 18 inches in the summer of 1852. But for the prevailing south and south-west winds, which drive the sedimentary matter of the main bed up stream, the Volga would be completely inaccessible. Hence the engineer Danilov now proposes to avoid the delta altogether by constructing a canal from

Fig. 195.—The Volga and Akhtúba.

Scale 1 : 1,280 000.

30 Miles.

Astrakhan to the port of Serebrasovskaya, 114 miles to the south-west, where occurs the first deep bay south of the Volga.

Without including the shorter windings, the Volga has a total length of 2,230 miles, presenting, with its tributaries, about 7,200 miles of navigable waters. From the sources of the Kama to the delta, these waters cross sixteen parallels of latitude, and nine isothermal degrees, so that while the mean annual temperature of the upper region is at freezing point, it oscillates about 9° in the delta. At Astrakhan the Volga is frozen for about 98 days, and at Kazan for 152, while the Kama is ice-bound for six months at the junction of the Chusovaya above Perm. The rainfall of the basin is about 16 inches, which would give 700,000 cubic feet per second, were all the moisture to be carried off by the bed of the Volga. But much is absorbed by vegetation in the forests and steppes, and in the latter region direct evaporation may dissipate about 40 inches during the year in tracts fully exposed to the winds. Altogether about three-fourths of the rainfall are thus lost *en route*, and preliminary estimates have determined the mean discharge at about 203,000 cubic feet, which is less than two-thirds of that of the Danube, draining an area scarcely half as large as that of the Russian river.

The volume of water discharged by the Volga, which is at least equal to that of all the other influents of the Caspian together, is sufficient to exercise a considerable influence on the level of the sea. Thus the floods of 1867, the heaviest that had occurred for forty years, raised it by more than 2 feet, the abnormal excess representing 9,600 billions of cubic feet, or about three times the volume of the Lake of Geneva. On the other hand, the delta steadily encroaches on the sea, though at a rate which it is almost impossible to determine. The sedimentary matter held in solution, estimated by Mrczkovski at about the two-thousandth part of the fluid, continues to form islands and sand-banks, while generally raising the bed of the sea round the face of the delta.

The Volga abounds in fish, and the fishing industry supports a large number of hands. Its lower reaches especially form for the whole of Russia a vast reservoir of food, varying with the seasons, and yielding large quantities even in winter, by means of holes broken in the ice at certain intervals. On the islands of the delta are numerous stations where the fish is cut up, and the roe prepared to be converted into fresh and salt caviar. The *bieluga* and the sterlet, both of the sturgeon family, attain the greatest size, and are the most highly esteemed, but their number seems to have diminished since the appearance of the steamboat in these waters.

On both sides of the Volga delta the Caspian seaboard is fringed for a distance of 240 miles, between the mouths of the Kuma and Ural, by a multitude of narrow peninsulas and islets, with a mean elevation of from 25 to 30 feet, and separated from each other by shallow channels, penetrating in some places from 12 to 30 miles inland. Seen from an elevation, these so-called *bûgri* (singular *bûgor*), which occur nowhere else, at least with anything like the same regularity, present with the intervening lagoons the appearance of an endless series of parallel and alternating walls and moats, all of uniform width. Many have been swept away by the various arms of the Volga, but a large number still remain even in the delta itself, and all the fishing stations, as well as Astrakhan, have been established on eminences of this sort. The thousand intervening channels form a vast and still almost unexplored labyrinth, of which carefully prepared charts alone can give any idea. Immediately west of the delta the lagoons are practically so many rivers, but farther on they form rather a chain of lakes separated by sandy isthmuses, and in summer changed into natural salines by the rapid evaporation. Even in the interior of the steppes, far from the present limits of the sea, such salt lagoons occur here and there, separated from each other by parallel strips, as on the coast.

According to Baer, to whom we owe the first detailed account of these formations, all the elongated eminences are stratified in the form of concentric curves. The more decidedly argillaceous layers form, so to say, the nucleus round which is disposed the more sandy matter, a distribution pointing at the action of running water depositing sedimentary sands or argillaceous beds. The same conclusion is deduced from the general direction of the bûgrî, spreading out like a fan north and south, and representing the extremities of radii diverging from a common centre, which would seem to have been in the middle of the Ponto-Caspian isthmus. This disposition can be explained only by a rapid lowering of the waters eastwards, caused doubtless by the separation of the Caspian from the Euxine basin. When the rupture of the Bosphorus caused the "divide" to emerge, the Caspian, at that time twice its present size, was suddenly deprived of a portion of the waters that had hitherto flowed to the common Ponto-Caspian basin. Hence the contributions of the Volga and its other influents no longer sufficing to repair the losses caused by evaporation, it was doubtless soon reduced to half its former size, and the sudden subsidence produced by erosion those narrow lagoons still dotted over the Volga delta and adjacent coast lands.

The Caspian Sea.

The modifications of outline caused by the alluvia of the Volga, of the Terek, and of its other affluents are consequently insignificant compared with the vast changes in remote times produced in the form of the Caspian basin. The appearance of the uncovered lands, the shells embedded in the soil, and the marine animals still living in its waters leave no doubt as to the former extension of this inland sea. There can no longer be any doubt that it formerly communicated either simultaneously or at different epochs both with the Euxine and Arctic Ocean. Hence, although now completely land-locked, it may be regarded as geologically forming portion of a vast strait flowing between the continents of Europe and Asia.

The parting line between the two seas is clearly indicated by nature itself. The river Kalaûs, rising in the chalk beds at the northern foot of the Caucasus about midway between the two seas, flows first northwards towards the depression left by the old Ponto-Caspian strait, here ramifying into numerous branches, some of which are lost in the sands, while others trend eastwards to the Kuma and Caspian. But during the spring and autumn floods a portion of its surplus waters finds its way westwards to Lake Manich, and thence from tarn to tarn to the Mediterranean basin. Thus the two main branches of the Kalaûs, known as the East and West Manich, form a temporary channel between the two seas, intermittently replacing the old strait.

The question arises whether it might not be possible to restore the communication in such a way as to allow of large vessels passing uninterruptedly from Gibraltar to Asterabad, or even by the old bed of the Oxus to the foot of the Central Asiatic highlands. In any case it seems no longer possible to cut a canal across the Ponto-Caspian isthmus free of locks, so that to connect the Sea of Azov with the Caspian would be a far more gigantic undertaking than the piercing of the Isthmus of Suez, with incomparably less commercial prospects. The water-parting being some 80 or 86 feet above the level of the Sea of Azov, and consequently about 170 feet above that of the Caspian, the cuttings for a canal no more than 10 feet deep would be amongst the deepest ever executed, amounting to at least 130 feet for a distance of over 30 miles. But a canal with locks adapted to light navigation would be a comparatively easy undertaking, if advantage be taken, not only of the Kalaüs and Kuma, but also of the Terek and Kuban. Danilov's plans, estimated to cost from forty to fifty millions of roubles, would enable the Astrakhan steamers to avoid the Volga delta, passing by a lateral canal into the Manich, then rounding the Taman peninsula, and thus arriving directly in the Black Sea.

The area of the Caspian at the time of its separation from the Euxine comprised in Europe and Asia the entire region of lakes, marshes, and steppes which stretches to the foot of the table-lands. South of Tzaritzin are yet visible the

Fig. 196.—EAST MANICH AND LOWER KUMA RIVERS.
Scale 1 : 1,200,000.

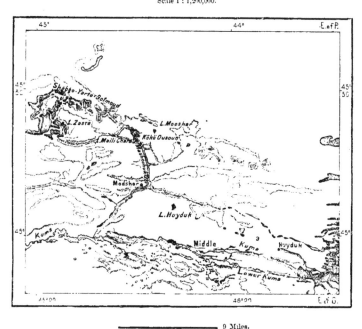

9 Miles.

steep cliffs skirting the ancient sea, and which would still form a continuation of the high or right bank of the Volga, if the bed of that river had not been gradually shifted westwards. A chain of lakes and ponds running a little in advance of the southern cliffs, and evidently forming the bed of a river, is perhaps a continuation of the ancient Volga when it discharged into the Manich Strait. But the slow process of evaporation has reduced the area of the Caspian to 176,000 square miles, or about three-fifths of the surface of France. Owing to the sedimentary deposits of the rivers, this area is still diminishing, although the volume of water seems to remain much the same, an equilibrium having apparently been established between the inflow and the loss by evaporation.

The actual level of the Caspian, which seems to have varied little throughout the present geological period, is about 85 feet below that of the Euxine. During its subsidence the water has left in the middle of the steppes a number of saline marshes, such as Lake Yelton; but most of the land formerly submerged, including even certain depressions lower than the Caspian, has been completely dried up by evaporation. The general inclination of the plains stretching north of the Caspian is almost imperceptibly continued below the surface, so that one must wade for miles from the shore before reaching deep water. Off the Volga delta large steamers are obliged to anchor in the so-called "Nine Fort" roads,

Fig. 197.—West Manich River.

Scale 1 : 2.140,000.

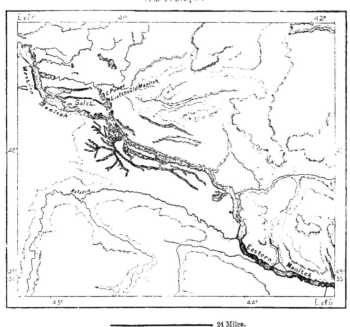

24 Miles.

almost out of sight of land, and the whole of this northern section of the basin may be regarded as a flooded steppe, which a slight upheaval would convert into plains similar to those of Astrakhan. North of the Terek mouths and Manghishlak peninsula the depth scarcely exceeds 8 fathoms, and the navigation is here much impeded by numerous sand-banks. This part of the sea, comprising about one-third of its entire area, is so low that during the prevalence of the northern gales the waters retreat at times some 20 miles southwards, and when this occurs in winter the ice gets broken up, causing the destruction of myriads of fish.

Owing to its deficiency of salt, the Caspian is poor in shell-fish, but extremely rich in other fishes, a circumstance due to the abundance of vegetable food in the shallow waters of the northern section, and in the vast sedgy tracts along the banks of the Lower Volga and other streams flowing to this basin. The language of former and even of contemporary travellers touches on the marvellous when describing the fisheries in these waters, where the annual yield amounts to from 800,000 to 1,000,000 tons, valued at £3,000,000 to £4,000,000.* But the high price of salt, on which there is a heavy excise duty, prevents the fishers from curing the smaller species, and forwarding them to the rest of Russia.

THE LACUSTRINE AND SALT STEPPES OF THE CASPIAN BASIN.

THE numerous brackish lakes of the Novo-Uzensk and Nikolayevsk districts must also be regarded as remnants of the old Caspian basin. Shells characteristic of this sea have been found as far north as Sizran and Samara, near the great

Fig. 198.—SALT STEPPES AND LAKE YELTON.

Scale 1 : 4,640,000.

120 Miles.

bend of the Volga, and even in the Bolgar plain south of the Kama junction. Yazikov, who discovered these fossils, supposes that the Sizran plain was formerly an inlet of the Caspian, possibly communicating with a more northern sea, whose ancient bed is now traversed by the Volga, Kama, and their tributaries. In any case there can be no doubt so far as regards the steppes, whose level is at present below those of the Mediterranean and Manich valley. These are undoubtedly old marine beds, still studded with miniature Caspians, and crossed by recent water-courses, or rather wadies, such as the Great and Little Uzen, flowing north-west and south-east, and preserving in their windings a remarkable parallelism with the Volga. These are evidently channels formed **in the alluvial tracts** immediately after the subsidence of the Caspian waters.

Of the innumerable little steppe lakes the most remarkable is the salt Lake

* Fishing craft, Astrakhan district (1872), 2,780; steamers, 19; hands employed, 23,000; yield, 18,490,500 roubles.

Yelton, the Altan-Nor, or "Gold Lake," of the Kalmuks, so named, doubtless, from its dazzling glitter in the rays of the sun. It lies in a desolate region, where the brown or yellow clays are relieved not by patches of verdure, but by lines of white salt disposed in cones like the tents of an army, or by the wretched hovels of the labourers, and the lumbering waggers, with their teams of oxen,

Fig. 199.—STEPPES NORTH OF THE CASPIAN.

Scale 1 : 4,120,000.

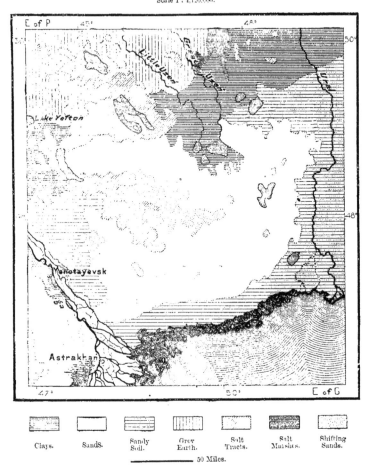

| Clays. | Sands. | Sandy Soil. | Grey Earth. | Salt Tracts. | Salt Marshes. | Shifting Sands. |

50 Miles.

slowly moving towards the Volga. The lake, with its violet-red waters, covers an area of several square miles, but with a uniform depth of scarcely 12 inches, so that the entire mass becomes displaced whenever the wind sets strongly from any given quarter. Its bed consists exclusively of extremely hard saline layers, which have not yet been pierced to any great depth, the workmen confining their operations to the more recent deposits annually formed on the outskirts of the lake during the thaw, when the muddy streams charged with saline particles drain from the surrounding steppe, causing a yearly accumulation of about 2,000,000 tons of salt in the Yelton basin. The water holds such a quantity in solution that it never freezes even when the glass falls 50° Fahr. below freezing point; but it is then dangerous to expose the limbs to its action, for the skin immediately becomes livid, often followed by mortification of the flesh. According to tradition there are some springs of pure and icy water in the middle of the lake. The yield of salt had increased from 80,000 tons in 1865 to 275,000 in 1871, when the management passed from the Government to a private company. At present this company confines its operations mainly to the Baskúnchak marsh, which is more accessible from the Volga.

Most of the saline steppes stretch north of the Caspian between the Volga and Ural Rivers. The salt area west of the Caspian is much more contracted, the steppes here consisting mostly of argillaceous plains studded with lakes, some of which are fresh. In the north the steppes are generally sandy throughout their entire extent, and interrupted only by marshes and the two triassic districts of the Great and Little Bogdo, and here and there by shifting dunes. Rocky steppes are confined entirely to the Asiatic side. But whether salt, argillaceous, or rocky, none of these steppes at all resemble the grassy prairies of the Dnieper, scanty pastures occurring only here and there in the low-lying tracts at considerable distances from the present shores of the sea. Even in these places, after the not unfrequent visits of the locusts, not a blade of grass remains, the very reeds and sedge of the swamps disappearing to the water's edge. Yet these dreary wastes are inhabited not only by the nomad Kirghiz and Bashkirs, but even by hardy settlers, by Great Russian Cossacks, pioneers of the race which has peopled the whole of Central Russia.

East of the river Ural rocky plateaux, breaking the monotonous surface of the steppes, form the first elevations of the long range of the Ural Mountains stretching thence northwards through twenty-eight parallels of latitude across the four zones of the steppes, forests, tundras, and ice-fields, far into the Frozen Ocean.

THE SOUTHERN URALS AND RIVER URAL.

THE section of the Urals, which begins at the sources of the Petchora, and which forms the eastern limit of the Volga basin, is not accompanied by parallel ridges, like the Northern Urals of the Vogûls, Ostyaks, and Samoyeds. But on the eastern or Siberian side some eminences, such as the Denejkin Kamen, rise to a greater height than any others in the entire range. South of the Konchakov Kamen the Urals cease to present the appearance of a connected system, here dwindling to a series of broken ridges, with a mean elevation varying from 700 to 1,000 feet above the surrounding lowlands. The base of these ridges is even so broad that the fall on either side is often scarcely perceptible. The water-parting line, which has an absolute elevation of only 1,200 feet above the sea, is reached from the European side so gradually that the traveller is scarcely sensible of having quitted the plains. On the Asiatic side the slope is even still more gentle, the absolute descent as far as Yekaterinburg scarcely amounting to 300 feet. The bare rocks are also seldom visible, being mostly covered with mosses,

Fig. 200.—DIVERGENT SPURS OF THE SOUTHERN URALS.

Scale 1 : 6,580,000.

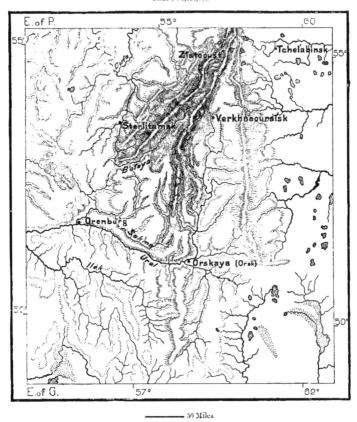

————— 50 Miles.

or even peat beds, and according to Ludwig lakes are concealed beneath such thick layers of turf that carriage ways cross them without danger.

The region of the Middle Urals, which has been most exposed to the action of the weather, has also acquired the greatest importance from its mineral wealth. Since 1815 gold associated with platinum has here been mined, besides which there occur whole mountains of iron and other metals, especially copper, nearly always embedded in Permian formations. But the gold is sought not in the granite or serpentine rocks themselves, but in the fragments scattered over considerable tracts, probably by glacier action. On both sides of the range the Perm, Orenburg, and other plains are composed of detritus stretching for a mean distance of 180 miles, with a depth of about 500 feet, and consisting of rocky masses detached from the Urals, then borne along in both directions, and in the lower grounds reduced to small fragments by the action of water. Such is the quantity that has been thus removed, that were it replaced it would raise the mean altitude of the range by at least 2,000 feet. It is amongst this detritus, levelled on the surface by mosses, peat, and other vegetable growths, that the minerals are found detached from their primitive lodes, and often associated with the fossil remains of the great ruminants. The eastern slopes are the most metalliferous, and the chief mines and metallurgic works are consequently found in Asia. Here also occur the so-called "Mines of

Fig. 201.—LOWER COURSE OF THE URAL RIVER.

Scale 1 : 2,720,000.

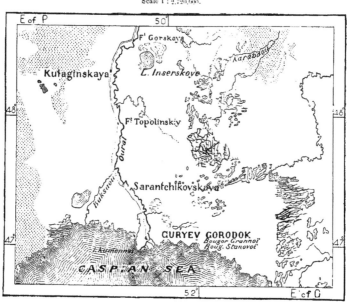

——— 25 Miles.

the Chudes," galleries excavated in the live rock, where have been collected numerous copper, but no bronze instruments, the old race of miners having perished before reaching the bronze age properly so called. Some of the natives are traditionally said to be acquainted with other old and very valuable mines; but they have always refused to reveal them, through fear of being condemned to the miner's hard fate.

A little north of Zlatoûst, in the great mining region, the main range ramifies into three branches, spreading southwards like a fan, with broad intervening valleys, where rise the Ural and its affluent the Sakmara. In some of its crests, such as the Yurma, Taganaï, and Urenga, the western spur attains an elevation of

171

more than 4,000 feet, culminating with Mount Iremel, which is over 5,000 feet, thus rivalling the highest summits of the Northern Urals. Less elevated are the two other chains; that is, the central, which continues the main axis of the range, and the eastern, which gradually merges with the plateaux bordering on the Aral Sea. At its southern extremity the Ural system is no less than 180 miles wide.*

The river Ural, whose course continues the Ural range as the official limit of Europe and Asia, was formerly known as the Yayik, a name which has been suppressed in consequence of its association with the insurrection of the Yayik Cossacks under Pugachov. The Ural ranks with the great European rivers in the length of its course, but not in the volume of its waters. Rising in the Kalgantau gorges on the Asiatic side, it receives its first tributaries from the valleys sheltered from the moist winds, and the mean annual snow and rain fall of its upper basin is probably nowhere as much as 16 inches, diminishing gradually southwards. Much is also carried off by evaporation, its argillaceous bed being nearly everywhere over 330 feet, and in some places 580 feet wide, but nowhere deep. Hence the commercial town of Orenburg is unable to utilise it for navigation, although it here flows west and east in the most favourable direction for the transit of goods between Russia and Turkestan. In its middle course it has only two important tributaries, the Sakmara in the north, and the Ilak in the south. Lower down it receives nothing but rivulets, and below Uralsk, where it resumes its southerly course, the affluents are few and far between. Many even fail to reach its bed, being absorbed in the sands, or forming stagnant pools, which are at times displaced by the pressure of the shifting dunes (*barkhani*) of the plains. After receiving the brackish water of the Solaika, the last of its tributaries, it winds sluggishly through the steppe for a distance of 300 miles, or about one-fourth of its entire course, without receiving any fresh supplies. All the streams, such as the Great and Little Uzen, which flow towards its valley, are absorbed before reaching its bed. Hence the current diminishes southwards, and at the head of the delta the volume of water is less than half of what it was at Uralsk.

The Ural has certainly become impoverished during the last hundred years, partly, doubtless, owing to the destruction of the forests along its middle course, first by the Kalmuks, and then by the Kirghiz, but mainly in consequence of a general diminution of the rainfall throughout the entire zone, comprising Southern Russia and Turkestan. In 1769 Pallas found that it reached the Caspian through nineteen mouths, and the delta had an area of over 1,000 square miles. In 1821 the delta had been reduced to less than one-half, and the mouths to nine, of which four were still deep enough to float small craft; but since 1846 there have usually been but three mouths, the other water-courses being only occasionally flushed during

---

* Various elevations of the Middle and Southern Urals:—

MIDDLE URALS.			SOUTHERN URALS.		
Denejkin Kamen	. . .	5,360 feet.	Iremel	. . . . .	5 040 feet.
Konchatkov	. . .	4,795 „	Yurma	. . . . .	3,447 „
Blagodat	. . . .	1,515 „	Taganaï	. . . . .	3,441 „
Pass of Yekaterinburg	. .	1,180 „	Akkiûba	. . . . .	2,598 „

the spring freshets. In 1866 one only of the branches had a depth of 30 inches, and was navigable for boats throughout the year. The two others had a mean depth of no more than 24 and 12 inches respectively.*

ETHNICAL ELEMENTS.—THE GREAT RUSSIANS.

THE Veliko or Great Russians form of themselves alone more than one-half of all the inhabitants of the empire. They not only occupy nearly all Central Russia as well as most of the Neva basin, but they have also advanced in compact masses

Fig. 202.—GREAT RUSSIAN TYPE: ARDATOV DISTRICT, GOVERNMENT OF NIJNI-NOVGOROD.

north, east, and southwards, while in the west they have numerous colonies in the Baltic Provinces and Little Russia. They have planted themselves along the northern foot of the Caucasus, and in Siberia they have already peopled vast tracts more extensive than the whole of France. They have become the preponderating and

* Length of the Ural, 1,306 miles; area of its drainage, 212,000 square miles; mean discharge, 1,750 cubic feet.

imperial race, imposing their political forms on the rest of the empire, their speech acquiring corresponding predominance as the official language and literary standard. Compared with the other nationalities of Eastern Slavdom, the Great Russians have the advantages flowing from material cohesion and compact solidarity. Throughout their domain they everywhere present the same uniform aspect as Nature herself. Towns, hamlets, cultivated lands, are everywhere alike ; the people have everywhere nearly the same appearance, and even the same costume, except amongst the women. Their habits of life are also the same, and their speech offers but slight dialectic variations, so that the provinces are nowhere marked by any decided contrasts.

The Veliko-Russians are, on the whole, somewhat shorter, but also more thick-set, than the Little and White Russians. The greatest percentage of youths rejected as unfit for military service occurs in the Central Muscovite provinces, though this may possibly be due to a partial deterioration of the race in the spinning factories and other workshops of Central Russia. Wherever the blood has not been impoverished by squalor, foul air, and enforced labour, the mûjiks are remarkable for their broad shoulders, open features, and massive brow. They delight in long and thick beards, which they have succeeded in preserving in spite of Peter the Great, who wanted to shave his subjects in order to make them look like Dutchmen. Hence they still continue to deserve the sobriquet of *katzapi*, or "buck-goats," applied to them by the Little Russians. But these large bearded faces often beam with a lively expression, and are lit up with a pleasant smile, while many are of a strikingly noble type. Under the influence of education the peasantry are soon softened, their features becoming refined and animated. "The Russians," says Michelet, speaking especially of the civilised Slavs, "are not a northern race. They have neither the savage energy nor the robust gravity of that type. Their quick and wiry action at once betrays their southern nature." They are gifted with an astonishing natural eloquence not only of words, but of gesture, and their mimicry is so far superior to that of the Italians that it is readily understood by all.

Although extremely gentle, and loving their own after their own fashion, the Great Russians are still worshippers of brute force, and amongst the peasantry the authority of the father and husband is never questioned. In their families brutality and real kindliness are often found strangely allied. So recently as the seventeenth century the father would still buy a new whip, wherewith to inflict on his daughter the last stripes permitted to the paternal authority, then passing it on to her new master, with the advice to use it often and unsparingly. On entering the nuptial chamber the bridegroom used it accordingly, accompanying the blows rained down on back and shoulders with the words, "Forget thy father's will, and now do my pleasure." Yet the song recommends him "a silken lash." Love matches, common in Little Russia, are the exception amongst the Great Russians. All the terms of the contract are arranged beforehand by the heads of the families interested, independently of the bride and bridegroom, nor would the elders ever so far forget their dignity as to consult the young couple on the subject. Some idea of a Great Russian household, that dark abode of domestic despotism, may be gleaned from the national songs, such especially as occur in Shein's collection, as well as from Ostrovskiy's dramas. Absolutism, though perhaps of a kindly type, was the rule in the Great Russian home. "I beat you," says a favourite local proverb, "as I do my fur, but I love you as my soul."

The commune, and even the State itself, were universally regarded as an enlarged family. An absolute authority, a will without appeal, imposed on all by a common father—such was the ideal of society as conceived by each of its members. In this respect Little and Great Russia presented a most remarkable contrast. Every Malo-Russian village had an independent development; no one thought of enslaving his neighbour; the motives of war between communities were either the struggle for existence or a love of adventure, rather than the thirst of dominion. Hence their warlike undertakings were conducted without that fixity of purpose, that unflagging tenacity of will, which inspired the policy of the Great Russian rulers. The right of popular election was upheld to the last in the towns of Kiyovia, as well as in Novgorod and the other autonomous cities of West Slavdom. Whatever may have been the origin of the old supremacy of Kiev, it is certain that it had nothing in common with the supremacy of Muscovy. Kiev was nothing more than "the first among its like," and their political system was maintained by a free confederation during the early stages of Russian history. Later on the Cossack communities were organized on the same footing, and even their chiefs again withdrew into privacy after their temporary election by their compeers. Nor were the ideas of the Zaporogs limited to the enclosures of their strongholds, for the whole of Little Russia existed to form a Cossack community.

Nothing of all this in Muscovy, where the power acquired by a single family was respected by the people, and continued, as a divine institution, from generation to generation. "Moscow makes not laws for the prince, but the prince for Moscow," says the proverb. The sacred character of the dynasty was transferred to the capital itself, and Moscow, heir to the Byzantine spirit, became the "third and holier Rome, whose sway shall endure for ever." The Tatar rule contributed not a little to strengthen the power of the Eastern Slav autocrats. In the desire to receive their tribute regularly, the Khans had an interest in causing it to be collected by one prince, responsible to them alone, while free of all obligation towards his own people. But even in the twelfth century the autocracy of the modern Czars already existed in germ in the principality of Vladimir. This absolute form of Muscovite society may be explained by the history of Russian colonisation in a country originally held by Finns and Tatars. The Kiev princes appeared in these lands as conquering and colonising chiefs, and the race thus developed in Muscovy became at once the most tenacious and submissive of all. With the progress of Great Russian centralization the political forms and ideas of Muscovy steadily assumed a more intensely national character, and ended by extinguishing the Novgorod and Cossack traditions. In his communes and various unions the Great Russian enjoys as much, possibly even more equality than the other Slavs. But in his conception of political unity he is the most logical of monarchists. In the words of the national proverbs, "The earth is mother, but the Czar is father;" "Without the Czar the earth is widowed." Even the religious sects formed since the end of the seventeenth century, and which regard the present State as the "kingdom of the beast," and Antichrist as its head, restricted their anathemas to the heretical and foreign "emperor." But they are none the less fanatical in their worship of Czardom, and their Messiah they look for in the high places. In the West, so often the scene of revolutionary throes, even the blind adherents of the old régime can form no idea of the ardent love mingled with awe which animates all loyal Russian subjects when they think of their master, who to them is also a god. They had formerly good reason to dread their Czar, whose name was never uttered without a feeling of terror. If he was capricious and cruel, they bowed down before him all the more devotedly, for he then appeared in their eyes all the more sublime. And the guide of their own actions they sought not in themselves, but in their sovereign's inflexible or changeful will. Hence no prince was more popular than Ivan the Terrible, who seemed to his subjects awful as Destiny itself. The people, unmindful of so many other heroes, still remember him, and the Vladimir of their songs is ever the "merciful and dread prince."

The Veliko-Russian speech has become, to the exclusion of all other Slav dialects, the official language of the empire, and the Moscow accent is that which prevails in good society. The preponderance is thus definitely secured to the Slavonic form of speech, which is at once that of the majority of the people and of the heirs to the Muscovite throne. Hence all the Eastern Slav nationalities are obliged to adopt it either altogether or partially. Some, like the Poles, the Germans of the Baltic Provinces, the Esthonians, Letts, and Lithuanians, acquire it either in the school, the army, or in social intercourse; others, like the White and Little Russians, are naturally led to converse in a language closely akin to their own, which is spoken by the majority of their fellow-citizens, and which is at the same time the most highly cultivated and the richest in literary productions. Even for the Finns, Mongolians, and Tatars, Russian is the language of civilisation itself, while the Jews, although still clinging to their own German patois, are necessarily conversant with the current speech of the bazaars and market-places.

Owing to its vast political importance, Russian cannot fail to become some day one of the most influential in the development of human culture. Hence all the more strenuously should the nation itself strive by social progress to take the place to which it is entitled. Meantime, however, the present social condition of the Great Russians, as well as of their kinsmen in Little and White Russia, is still one of the most wretched in the civilised world.

In winter the peasant's hovel is filled with a foul, almost intolerable atmosphere. For greater warmth it is surrounded by a rampart of dunghills. The sashes, thickly lined with putty, or covered with straw, are kept perfectly air-tight, while the mingled breath of all the family, stretched on the stoves and *polati*, or side tables, soon vitiates the oxygen of the interior. Insect life is developed in the joists and rafters to such an extent that existence becomes at times absolutely intolerable. The only way of abating the nuisance is to leave the house unoccu-

Fig. 203.—PROPORTION OF CONFLAGRATIONS IN THE VARIOUS RUSSIAN PROVINCES.

According to Wilson. Scale 1 : 27,250,000.

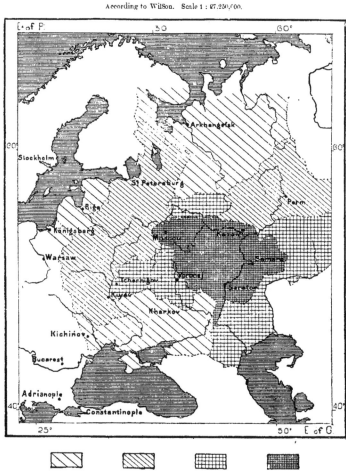

1 to 3 per cent.	3 to 7 per cent.	7 to 12 per cent.	12 to 25 per cent.

——— 100 Miles.

pied, and all the doors and windows open, during the coldest winter days—an heroic remedy often adopted by the peasantry. They may then be seen wandering in the neighbouring woods, exposed to the cutting northern blasts, and to the keenest frosts.

The Great Russian villages consist of groups of wretched cabins packed one against another, with no intervening gardens, and even most of the towns are nothing but confused aggregates of wooden structures, always at the mercy of the flames. The huts, built of fir, with thatched roofs, and surrounded with heaps of straw and hay, branches and chips, are liable to be set on fire by every chance spark. Fire thus becomes the natural end of the Russian village, which is said to be renewed, on an average, about every seven years. But in the central provinces

Fig. 201.—KOSTROMA: MONASTERY OF HYPATUS.

this "renovation" is often much more frequent, and there are districts in which one-fourth of the houses have been consumed in a single year. In the towns the only stone buildings are the Government offices, the palaces of the nobles, and the churches, nearly all built on a uniform plan, after the regulation type of the St. Petersburg authorities. These structures lay no claim to art, whose life is freedom, and which perishes in the atmosphere of traditional symbolism. In the churches art is replaced by a lavish display of gold, marbles, enamels, and precious stones.

RELIGIOUS SECTS.

RUSSIA forms a connecting link between Europe and Asia, no less in the religious than the moral and social order. Through the Catholic and Protestant elements of its western provinces it is connected with the rest of Europe; through the Pagans, Buddhists, and Mohammedans of the east, with the Asiatic world. But between these two extremes lies the great central mass of Greek orthodoxy. Yet, in spite of this official creed, more religious sects are probably here developed than in any other European state, and most of them take their rise amongst the Great Russian populations. The peculiarly mystic and polemical spirit so characteristic of this race; the more than Byzantine severity of the rites imposed on them by the clergy; the old pagan superstitions still surviving under new forms; the very character of the people, their gentleness and good nature, so easily swayed by fanaticism; lastly, the thraldom which so long oppressed the masses, driving them to seek refuge in the supernatural world—all these causes have combined to produce an endless variety of fantastic beliefs. They spring up, perish, and are renewed like so much rank vegetation. At present there are recorded from one hundred to one hundred and thirty recognised sects, and in the language of the orthodox proverb, "Each mûjik makes his own religion, as each old woman makes her own nostrums." Every great national event gives rise to fresh varieties, all differing in name or in form, but substantially the same in their moral aspect, and in the passions which they bring into play. With each succeeding generation there arise new Messiahs, sons of God, or God himself; or else Czars, such as Peter III. and Alexander I., are worshipped and believed to be still alive, because they were mild rulers. Napoleon himself had his votaries in Pskov, Belostok, and even in Moscow, ruined though it was by him. In their usual frame of mind a text of Scripture, or some old "Mother Shipton's prophecy," is generally all that is needed to beguile the faithful into the most devious paths of folly, such as self-mutilation, suicide, or murder.

Until freedom of worship is established it will be impossible to form even an approximate estimate of the actual number of dissenters. In 1850 the non-orthodox Russians were officially stated to be about 830,000, but the Minister of the Interior even then estimated them at 9,000,000, which, allowing for the normal increase of population, would make them at present at least 12,000,000.

The various Raskolnik sects may be divided in a general way into three main groups: the *Popovtzi*, or Presbyterians; the *Bezpopovtzi*, those who reject the priesthood; and the Spiritualists. The purely ritualistic movement of those who wished to retain the national observances anterior to the seventeenth century coincides with the dissatisfaction produced by the constant and vexatious intermeddling of the clergy in civil and religious matters; and to these sources of dissent was added the more or less direct influence of Protestantism, and all these complex causes had schism as their common result. An escape was also afforded in schism from the intolerable burdens and drudgery of serfdom, whose victims had thus at least the satisfaction of being able to anathematize their oppressors. For them stamped papers and passports became the " seal of Antichrist ;" the registration books, "registers of the devil ;" the poll tax, the " price of the soul." The Raskol-niks thus represent both the spirit of extreme conservatism and of reform, as well as that of a relative political freedom.

The Popovtzi, who continue the traditions of the old Church, are the true adherents of the ancient rites (*staro-obradtzi*), the " true believers " (*staro-vertzi*). Rejecting the changes introduced into the liturgy by the Patriarch Nikon with the assistance of Greek and Little Russian ecclesiastics, the Old Believers have perpetuated the practices of former times. But while denouncing the official rites, they have themselves unwittingly modified their own observances according to circumstances and the persecutions to which they have been subjected. Peter I., whose life had been endangered by the insurrection of the *Streltzi*, adherents of the old belief, hounded down like wild beasts all those who refused to conform. But the sects only became more numerous and more irreconcilable than ever. The Raskolniks, recognising Antichrist in this friend of the foreigner, who divorced his wife, tortured his son, and oppressed the people with his wars, buildings, canals, and merciless taxation, saw in him the abomination foretold in Holy Writ, and, to their denunciations of a corrupt church, added anathemas against an emperor who ordered his subjects to "shave their beards, to wear Latin garments, and smoke the thrice-accursed weed."

But persecution was not the hardest fate of the Popovtzi. First their priests, and then their only bishop died, so that no new priests could be ordained to administer the sacraments. Hence, to still the voice of conscience, they had recourse to the most eccentric subterfuges, kneading their own communion bread with a bit of the consecrated host, bribing the Orthodox clergy, attempting even to carry off the hand of a holy Metropolitan of Moscow wherewith to ordain their priests, and thus dispense with the hands of living bishops. Their hierarchy, however, was not duly re-established till 1844, when a Bulgarian bishop, conse-crated in Constantinople, consented to reside with a Raskolnik colony settled in Bukovina. They have now their own bishops, and openly hold their synods in Moscow, asking for nothing but complete freedom of worship. About a million of them, the so-called *Yedinovertzi*, or " United Believers," have, moreover, become indirectly attached to the Orthodox Church, accepting their popes (priests) at its hands, on the condition of being allowed to retain their books and old images.

But the more zealous of the old believers mostly escaped to the northern forests, where the convent of Vig, on the river of like name, was long their chief centre. These are the *Bezpopovtzi*, or Priestless, the "fold that feeds itself." Some reject all the sacraments, and have no ministers except the "holy angels;" others remain during the service with open mouths, in expectation of the divine food descending ready consecrated from heaven. The Bezpopovtzi became naturally more allied to western Protestantism, and developed far more varieties than the Popovtzi, every isolated refuge and every fresh prophet forming the centre of a new group. The most numerous is that of the **Theodosians, an** offshoot of the Vig community ; but the best known are probably **the Philippites**

(*Filipovtzi*), so called from their founder Philip, with settlements in East Prussia, Moldavia, and even Dobruja, collectively known as Lippovanes, a term often applied also to various Raskolnik sects. They taught that death was preferable to uttering the Czar's name in their devotions, some going even so far as to refuse coins bearing the imperial effigy. But abroad they have gradually modified their views, and some of them are now amongst the most zealous Muscovite patriots.

The Philippites were the most ardent apostles of self-immolation, and during the terrible persecutions of the seventeenth and eighteenth centuries many committed themselves spontaneously to the flames or the water. But suicide was often their only escape from torture, for the first stakes had been kindled by Orthodox executioners. Between 1687 and 1693 about 9,000 Old Believers voluntarily perished by fire in the region between Lake Onega and the White Sea, and a single holocaust in the island of Paleostrov, Lake Onega, was composed of 2,700 persons. Such spectacles became contagious, and many sectarians openly preached suicide by the flames, interment, or starvation. On more than one occasion even in this century parents have murdered their children to save them from future sin, and enable them at once to enter the gates of heaven. In some communities hymns are still sung ending with the words, "Proclaim my will to man—to all Orthodox Christians: that for my sake they throw themselves into the fire, themselves and their innocent children!"

Another of the Bezpopovtzi sects is that of the Fugitives (*Beyâni*) or Wanderers (*Stranniki*), founded by the deserter Euphimius towards the end of the last century. They hold that all Government officials are agents of Satan; consequently that it is unlawful to obey them. An official seal is for them a "mark of the beast," and they eagerly destroy all sealed documents falling into their hands. Hence they would pass their days in prison, or in the mines of Siberia, if they did not quit "Babylon in order to have no part in her sins." They prefer to wander from village to village propagating their doctrines, or prowling about like the wolves of the forest.

The Popovtzi are settled chiefly in the centre, the Bezpopovtzi in the north, and the Spiritualists in the south. These last are the most persecuted, and consequently preserve the greatest secrecy as to their tenets and practices. They claim to possess the divine spirit, to be themselves "men of God," or "Christs." God the Father again descended from heaven with his Son in the seventeenth century, in order to accomplish the divine sacrifice, and the Holy Ghost still speaks through the mouths of apostles and prophetesses. Their chief sect, the *Khlistovtzi*, or "Flagellants," do not immolate themselves, but they dance like David, jumping and whirling about till they fall prostrate with exhaustion. Others scourge and torture each other, and during the Easter services they are said occasionally to kill a newly baptized infant, devouring the bleeding heart mixed with honey.

Under the influence of the general progress of thought all the sects are becoming more and more rationalistic in religion, and radical in politics. The Popovtzi are adopting wholesale the ideas of the Bezpopovtzî, while these last are beginning to reject the title of "Old Believers," applying it to the members of the Orthodox Church, whom they reproach for their cold formalism. New sects arise, such as the *Nemolaki*, who disbelieve in the efficacy of prayer, and the *Neplatelshtchiki*, who denounce all taxation. Yet such is the tenacity of the old ritualistic habits that even amongst the Spiritualists there has been formed a sect of "Sighers," who sigh and pant incessantly at their gatherings, because the "breath" is "spirit."

TOPOGRAPHY OF GREAT RUSSIA.

THE UPPER VOLGA TO NIJNI-NOVGOROD.

(GOVERNMENTS OF TVER, YAROSLAV, AND KOSTROMA.)

THIS region is peopled by Great Russians, and by Finns who have been largely Slavonised. In the hilly and more inaccessible districts of the West alone numerous isolated communities of Karelians have hitherto kept themselves aloof, retaining their speech and customs intact. But even these are slowly becoming assimilated to the surrounding Slav element. The province of Yaroslav was, a thousand years ago, largely occupied by Finns, yet the present population is now regarded as consisting of pure Great Russians. The Finns never having been expelled, the ethnical transformation must have been slowly accomplished, and even the towns and villages now mostly bear Slav names, generally in honour of saints or religious feasts.

The towns, nearly all originally mere fishing stations, are somewhat numerous on the main stream and its tributaries. Even in the upland region of forests and marshes near its source occurs the industrious town of *Ostashkov*, standing on a peninsula of Lake Seliger, about 750 feet above sea-level. It is peopled by fishers and boatmen, and manufactures scythes, sickles, axes, boots, and shoes. Farther down are *Rjov*, a great hemp and flax mart, *Zubtzov*, *Staritza*, and south of the river the town of *Gjatsk*, in the government of Smolensk, on a tributary whose head-streams are intermingled with those of the Dnieper.

Tver, formerly the most formidable political rival of Moscow, now the capital of a government and the chief port on the Upper Volga, lies advantageously at the junction of the Tvertza, which flows from the northern uplands, and which at all times gave access to the Neva basin and Gulf of Finland. Merchandise had formerly to be conveyed across the country from the Tvertza to the Msta, and although a canal has for the last hundred years afforded a navigable route from Tver to St. Petersburg, the city of the "Upper Portage," or Vishni-Volochok, still retains its name. During the fine season many thousands of boats laden with corn and other produce call here, as well as at the industrious town of *Torjok*, or "The Mart," which lies farther down on the Tvertza, and which even in the Novgorod period was already a great emporium. As many as 4,000 boats are yearly moored at the quays of Tver, which is the largest of the three commercial towns on the St. Petersburg-Moscow railway, and one of the chief industrial centres in North Russia, with numerous cotton-spinning mills and fancy leather works. This last industry, which is supposed to have been introduced by the Mongolians, bears an astonishing resemblance to Chinese ornamental work.

Ríbinsk, in the province of Yaroslav, is the second trading station on the Volga below Tver, and has the advantage of standing at the issue of two canals

Fig. 205. VISHNI-VOLOCHOK.

Scale 1 : 450,000.

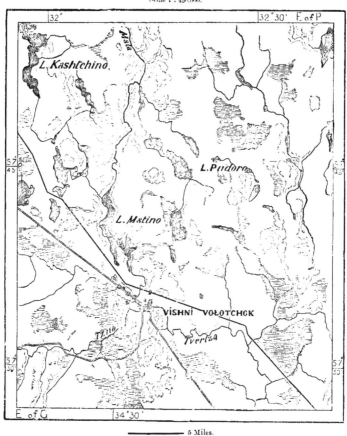

5 Miles.

connecting the Volga with St. Petersburg—one by the Mologa and Lake Ladoga, the other by the Sheksna, the Belo-Ozero, and the two great lakes of the Neva basin. Hence the growth of the capital reacts on Ríbinsk, which may be regarded as its chief port on the Volga, and which is also connected with it by rail. Here the flat-bottomed boats, to the number of 2,000, tranship their cargoes to the 8,000 small craft constructed for the navigation of the canals. The transhipments amount annually to about 700,000 tons, valued at £5,000,000, and in the busy season the boats are so crowded together that they form a regular bridge across the river. As many as 100,000 boatmen and traders congregate in the summer in this place, whose chief industry is a large rope-walk.

Yaroslav, capital of the province is probably the oldest Slav city founded on the Volga, having been built in 1025 by the son of Vladimir the Great. Later on it became the rival of Tver and Moscow for Russian ascendancy in the north. Here a ferry connects the two sections of the Moscow-Vologda railway, which, combined with its cotton and linen spinning-mills, gives it more commercial importance than it could hope to derive from the junction of the little river Kotorost. In its neighbourhood, and also on the Volga, is *Sopelki*, centre of the sect of " Wanderers." *Rostov*, lying to the south-west on the Moscow highway, and on a lake draining through the Kotorost to the Volga, is even an older place than the present provincial capital. The chronicler Nestor mentions it as already existing in the time of Rurik, in the ninth century, and says that its first inhabitants were Merians, a Slavonised race which, from the dawn of Russian history, occupied a vast territory in the present province of Smolensk as far as the Lower Oka. Their name does not occur in the chronicles after the year 907, but the Finnish element has left its traces in the local geographical nomenclature, and in the protracted opposition to Christianity, especially in Rostov, destined later on to become one of the metropolitan sees of Russia. To this position, which it has since forfeited, it is indebted for the honour of ranking as a holy city, and one of its chief industries is the painting of sacred images on enamel, which are forwarded to every part of the empire.

Kostroma, capital of a province, was, like Rostov, an old city of the Merians, and its name is that of a Finnish god. The games of pagan origin, which recalled the worship of Kostroma or Yarilo, have been abolished in the town ; but in many rural districts straw figures are solemnly interred, rudely representing the hyperborean Adonis, the god " who appears and dies," in order again to rise and perish perpetually. Kostroma, mentioned for the first time in the thirteenth century, became later on a very famous place, and the lofty towers and domes of its kreml still recall the time when it was a princely residence. It was here that in 1613 the States General announced to Michael Romanov his election to the throne after the expulsion of the Poles. He was then residing in the " Cathedral Monastery " of Hypatus (see Fig. 204), near the town founded in 1330 by a Tatar mirza, who had been converted by a "miraculous apparition." The convent has since then been twice rebuilt, in 1586 and 1650.

At the junction of the Unja the Volga resumes its southerly course to more populous regions, and soon approaches the city of Nijni-Novgorod.

BASIN OF THE OKA.

(GOVERNMENTS OF OREL, KALUGA, TULA, MOSCOW, KAZAN, VLADIMIR, TAMBOV, AND NIJNI-NOVGOROD.)

THE basin of this important river forms the true centre of European Russia, not only geographically, but also in respect of population and industrial activity. Here

TYPES AND COSTUMES IN THE GOVERNMENT OF OROL.

TYPES AND COSTUMES IN TULA.

the artisan classes are most numerous outside of Poland ; here is the historic capital of Great Russia, which later on became that of all Eastern Slavdom ; here converge great arteries of the railway system. This region supplies, with Poland and Little Russia, the heaviest contributions to the public revenue, and for aggressive purposes it may be regarded as the very heart of Russia. With the exception of the Mordvinians, the inhabitants are all Veliko-Russians. The indigenous Finnish tribes of the East have long been absorbed by the Slav element, and the Golad Lithuanians, who dwelt to the west of Moscow, have disappeared without leaving any traces beyond the geographical names of a few villages.

Orol, or *Arol*, capital of the province of like name, and situated on the Upper Oka, is one of the chief cities of Russia. Founded in 1564, after the conquest of the country from the Tatars, its old site was abandoned after a fire in 1679; hence the new town, only two hundred years old, comprises extensive quarters, consisting of little wooden houses, which still present quite a temporary appearance. But thanks to the four railway lines converging here, and to the Oka, already navigable at this point, Orol has become a busy commercial centre, doing a large export business, especially in cereals and hemp. These are the chief products supplied by its two neighbours, *Bolkhov* and *Mtzensk*, both on tributaries of the Oka. The women of Mtzensk are mostly engaged in the lace industry. The place is surrounded by tumuli, recalling the fierce struggles formerly carried on in the neighbourhood between the Tatars, Cossacks, Lithuanians, and Poles.

North of Orol the Oka flows successively by the towns of *Belov*, one of the great centres of the hemp trade, *Likhvin*, and *Peremichl*, the latter bearing the same name as the Galician Przemysl. After its junction with the Ugra, which nearly doubles its volume, the Oka turns from its northerly course eastwards, and a little below the bend stands the city of *Kaluga*, capital of a government, and set apart as a residence for Moslem princes interned in Russia. It belongs to the industrial district of Moscow, and produces leather, linens, and sweetmeats highly appreciated in Central Russia. Here the Government has an artillery depôt and a powder-mill. The other large towns of the province, *Jizdra*, *Kozelsk*, *Meshchovsk*, *Medin*, *Malo-Yaroslavetz*, *Borovsk*, all lie west of the Oka, on tributaries of that river or of the Ugra. When the French invaders left behind them the ruins of Moscow in 1812, they penetrated into this part of the country, but after the murderous battles of Tarutino and Malo-Yaroslavetz they were forced to the northward, in the direction of Smolensk.

Tula, also a provincial capital, is more populous than either Orol or Kaluga. It stands on the Upa, an eastern affluent of the Oka, and is the chief railway station between Moscow and Kharkov. But its importance is mainly due to the Government small-arms factory, established here by Peter the Great in 1712, and now producing about 70,000 rifles yearly, besides swords, bayonets, and other weapons. Tula is the Russian Birmingham, and also manufactures knives, mathematical instruments, machinery, and gold and silver plated wares, besides supplying 200,000 *samovars*, which are so indispensable to every Russian household. The Tula factories have the advantage of lying in a vast carboniferous basin, which increases in value in proportion as the forests threaten to disappear. Of the 113 pits surveyed in this basin, 4 were being worked in 1867, yielding 25,000 tons yearly, though 300,000 might easily be extracted. The Tula beds have a mean thickness of 16 feet, rising in some places to 26 feet, and the coal generally lies in horizontal strata near the surface.

Between Tula and Moscow the chief railway station is *Serpukhor*, with a large river traffic, and some important tanneries and calico works. North of it is *Podolsk*, already in the Moscow circuit.

Moscow.

THE second capital of Russia, which takes the first position, if not in population, trade, and industry, at least in point of precedence, occupies almost exactly the

Fig. 206.—TULA.
Scale 1 : 210,000.

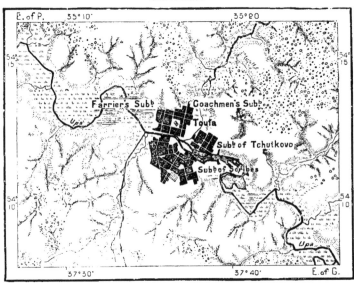

3 Miles.

geographical centre of the state. It stands on no large river, the Moskva being navigable only for small boats; but the slight undulations of the surrounding plains offer no obstacles to its direct communications with the Volga, Oka, Don, and Dnieper. It thus stands at a point where may easily converge all the main routes from the extremities of the empire—the White and Black Seas, the Baltic and Caspian, the ports of West Europe and of Siberia. The commercial and the strategical interests of the country also required that all the main lines of railway should here converge, for if Moscow lies open to attack from the direction of

Vitebsk and Smolensk, this is also one of the main highways of international traffic. As long as the Russians were confined to the Dnieper basin, Kiev was their national metropolis. But as soon as relations had been established between Europe and East Slavonia, some more central point, such as Moscow or Vladimir, could not fail to become the capital of the state. Moscow is not venerable enough to be regarded, like Kiev, as a "holy city;" but it is, at least for Great Russia, the "Mother," the "White Walled" (*Moskva Matushka*).

The general plan of the city presents some resemblance to that of Paris, consisting of a central nucleus round which the various quarters have grown up in concentric circles. The Moskva also, although flowing in an opposite direction

Fig. 207.—Moscow.

Scale 1 : 150,000.

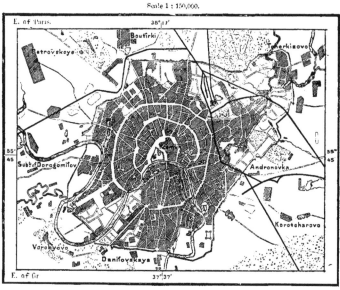

————————— 2 Miles.

to the Seine, presents windings in the west of the city almost identical with those of the French river. The Kreml, or Kremlin, surrounded by high walls in the form of an irregular triangle, occupies, with the quarter known as the "City of Refuge" (*Kitaï-Gorod*), the centre of the capital on the north side of the Moskva, which here describes a curve enclosing an oblong island between its two arms. Round about the City of Refuge has grown up the "White Town" (*Beliy-Gorod*), which a circular boulevard, with its two extremities resting on the left bank of the river, separates from the "Land Town" (*Zemlanoï-Gorod*), where reside the artisans and poorer classes. Since the end of the last century this quarter has been surrounded by a broad thoroughfare lined with plantations and gardens, and with a continuation on the right bank, forming a complete circuit of

172

nearly 10 miles altogether. Beyond this circular boulevard stretch the suburbs, which are again surrounded by an enclosure with abrupt projections and pyramidal towers, and intersected here and there by wide streets, which will one day be probably connected in a continuous outer boulevard. Moscow covers an area of over 40 square miles, which, though equal to that of Paris, contains a population three or four times inferior to that of the French capital. Many quarters resemble straggling villages, with their little painted houses grouped irregularly round some central church or palace. In the last century the Prince de Ligne described Moscow as a collection of baronial residences surrounded by their parks and the hovels of their serfs. It even still retains some traces of this peculiar development, gardens, groves, fields, waste spaces dotted with ponds penetrating between the suburbs towards the more densely peopled quarters, while, on the other hand, outlying villages line the highways for more than 6 miles from the heart of the city. There is no lack of space to introduce pure air into all the dwellings; but many of the so-called "half-storied" houses have their basements below the street level, and these are kept always damp by the rains and bad drainage. Hence the mortality is normally higher than the birth rate, and the city would soon be reduced to a mere village, if the population were not recruited by a constant flow of immigration. But from a distance this hidden squalor is veiled, and the great city appears only in its beauty, nothing being visible except trees, hundreds of towers, over a thousand "bulb-shaped" domes surmounting some 360 churches—"forty times forty," says the proverb. Seen from the "Sparrow Hills" (*Vorobyovi Gori*), running west of the capital, Moscow, with its frowning Kremlin, presents a superb panorama when tinged by the rays of the setting sun.

The Kremlin, at once a fortress and an aggregate of cathedrals, convents, palaces, and barracks, is pre-eminently the monument of the Russian Empire. Thence emanated the mandates of the Muscovite Czar, and here were promulgated the decrees of the Church. On entering its hallowed precincts through the "Saviour's Gate" (*Spaskiye Vorota*), all must devoutly uncover, and a sort of worship is also paid to the Ivan Velikiy belfry, built in 1600 by Boris Godunov, and rising 266 feet from the centre of the Kremlin. On a pedestal at the foot of this tower rests the cracked "queen of bells," weighing 200 tons, and in a neighbouring church the Czars are crowned and the Muscovite Metropolitans buried. Another cathedral, no less resplendent in frescoes, mosaics, marbles, and precious stones, contains the tombs of the first Czars, and in the middle of the palace courtyard is a very ancient little church dedicated to the "Saviour in the Forest," and recalling a time when the land was still densely wooded. Some of the buildings attached to the imperial residence are also very remarkable, suggesting in their style the palaces both of Venice and India. One of them contains some valuable collections, and all present a strange assemblage of domes, turrets, clock towers, colonnades, glittering in gold, yellow, green, and red. The Synodal Palace, near the imperial monasteries, contains a library with some unique documents and priceless manuscripts. The Arsenal also possesses a special museum, besides arms for 100,000 men, and an enormous but useless cannon, whence Herzen's remark

MOSCOW—THE KREMLIN.

MOSCOW—CHURCH OF ST. VASILIY, GROUP OF MININ, ETC.

cially for its bell which iever rings, and its gui

ainly occupied with such curious moiumeits as
ld moiasteries, besides the Vasiliy Blajenniy, or
ily the most iiteresting structure ii all Moscow,
.st outside the Kremlii, and was built ii an
an the Terrible, by an Italiai architect in the
It was evideitly iispired by the same haughty
e tower of Ivai Velikiy, cast the "queei of
zuis" in froit of the Arseial. Ii its details it
iitiie style, as required by religious traditiois,
ly Muscovite. Its builder obeyed the exigeicies
ə storework, the resistaice of the materials, the
ə same time coitrived to recoicile all this with
.itects, aid this ecceitric structure, though built
eitly the Orthodox Greek edifice. The outer
:cent thai the iaves and towers, io doubt
greatly disfigured by pyramidal belfries. But
ure is seei in all its straige originality. The
priig each from a mass of carviig resembling
the piie, or budding petals. The cupolas,
gilded chaiis, are all remarkable for their size,
)ne seems traced with arabesques of the lozeige
hioi, a third shaped like a piie-apple, others
əs, while above all rises the pyramidal ceitral
mass of smaller domes, and crowied by a sort
of usely embellished with porcelaii, and paiited
. At first sight it is impossible to follow the
he midst of all this eitanglemeit of gables and
rous vegetable growth rather thai an edifice
church makiig reasoi mistrust the evideice of
ed by this Russiai pagoda, whose very straige-
r. Near this church, and faciig the Saviour's
:rected to Miiii and Priice Pojarski, who
in 1613. Here also is the *Gostiniy Dvor,* or
of stalls.
Kitaï-Gorod the moiumeits become rarer as we
ly all the scieitific establishmeits are grouped
University, fouided ii 1755, with a valuable
larger iumber of studeits thai any other in
observatory, a zoological and a botaiic garden,
onsiderable influeice oi the philosophic and
en 1830 and 1848, before it was brought under
laitier, "Voyage en Russe."

the control of the St. Petersburg bureaucracy.* Near the University is the Museum, with collections of old and modern paintings and statuary, a library with unique Chinese and Manchu documents, and Dashkov's celebrated ethnographic gallery, in which figures in costume represent all the races of the empire. There are also several other less important museums, and private picture galleries containing the works of native artists. Amongst the numerous high-class special schools is the Lazarev Institute, where 200 students are taught the Oriental languages.

Moscow is a great centre of the book trade, and here are issued millions of works and prints, which are hawked all over the country, or bartered for the local produce of the provinces.† The real capital of Great Russia, it is also that of the "Old Believers," both Popovtzi and Bezpopovtzi, whose head-quarters are the Rogojskoïe and Preobrajenskoïe cemeteries, with their dependent establishments. The Popovtzi, in other respects free enough, have hitherto failed to obtain permission to found a special gymnasium, while the Bezpopovtzi chapels, which had rapidly increased since the beginning of the century, have all been closed since 1853. Moscow is the birthplace of Lermontov, Herzen, and Pushkin, to the last of whom a statue is now being erected.

Moscow is also an industrial capital, which even in the middle of the century possessed 650 factories, employing 40,000 hands, and producing about £4,000,000 worth of goods yearly. The chief branches are cotton-spinning, dyeing, woollen and silk weaving, tanning, distilling. Nearly all the factories are in the suburbs or outlying villages, where their smoking chimneys contrast forcibly with the verdure of the surrounding parks. The Sokolniki, or "Falconers," is the finest of these parks. It lies north-east of the city, and is a remnant of the old forests. Another much frequented during the fine season stretches north-west, with its avenues surrounding the gardens and colonnades of the Petrovskiy Palace.

North-east of Moscow lies the famous *Troïtza* convent. in a grassy and wooded district crossed by the Yaroslav railway. With its towers and turrets surrounded by high walls, it presents the appearance of a mediæval fortress, and was strong enough to stand a siege of sixteen months against the Poles in 1609 and 1610. It forms a veritable city of churches, chapels, shrines, buildings, and outhouses of all sorts. The chapel of its patron, St. Sergius, is especially resplendent in lavish wealth, and its *lavra*, though second in dignity to that of Kiev, is visited by quite as many yearly pilgrims. Before the emancipation it owned no less than 120,000 serfs, and in 1872 had a probable revenue of £135,000. Within its precincts is the Ecclesiastical Academy of Moscow. The town of Sergius (*Sergiyevskiy Posad*), which has grown up round about the walls of Troïtza, is, next to the capital, the largest of all the towns in the Moscow government. Amongst these places is *Voskresensk*, lying to the north-west, and which also had its origin in a

. * Professors, Moscow University, 106; students, 1,568; budget (1879), 491,383 roubles; library, 155,000 volumes.

† Publications (1877) :—St. Petersburg, 6.925,853 volumes, 600,407 syllabaries, &c., 209.233 prints; Moscow, 8,342,685 volumes, 2,056,280 syllabaries, &c., and 2,495,800 prints.

ngaged in the furniture business. *Vereya*, to the
yet entirely recovered from its destruction by the
y in the preparation of appliqué silver-ware. It
oskva, where was fought the battle of Borodino, or
iost sanguinary in modern times.

Towns in the Oka Basin.

r flows south-east to the Oka, near the commer-
h of *Zaraisk*, whose old fortifications, repaired by

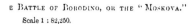

e Battle of Borodino, or the "Moskova."

Scale 1 : 82,250.

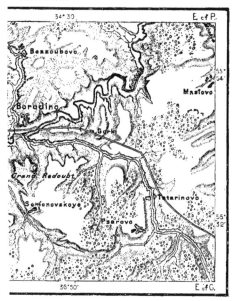

1¼ Miles.

inding. The main stream continues its winding
id *Riazan*, formerly capital of a principality, and
was an old settlement of Little Russians from the
is preserved its picturesque kreml, and does a
e navigable Oka and the Moscow-Saratov railway,
he chief intermediate station. A very busy place
iazan in the midst of extensive corn-fields. But
e province is *Kasimov*, formerly *Gorodetz*, situated

near the point where the Oka enters the province of Tambov. Its staple industries are rope-making and tanning.

The chief river of the government of Tambov is the Tzna, which joins the Oka below Kasimov. Near its source is *Tambov*, the provincial capital, on the Moscow-Saratov railway, but otherwise a place of no importance. *Morshansk* is much more conveniently situated lower down the Tzna, which is here navigable, and by means of which the town forwards the agricultural produce of a vast district. Morshansk is one of the centres of the Skoptzi sect. Farther north is *Shatzk*, on a small western affluent of the Tzna.

After its junction with the Tzna, the Oka flows by *Yelatma* through a rocky gorge, which is continued northwards by the valley of a lateral stream, on which stands the town of *Melenki*.

On the left bank, and a little above the confluence of the Tosha, are *Murom*, one of the oldest places in East Russia, and the thriving town of *Arzamas*, which takes its name from the Finnish tribe of the Arza. For over two hundred years Murom was the chief Russian mart in the Mordvinian country, and here the Volga Bulgarians came every summer to exchange their produce for the wares supplied by the Slav and Greek traders. Murom thus became the centre of a remarkable civilising movement, and even still remains a sort of capital for the surrounding Finnish tribes. But its chief importance is derived from its trade with Nijni-Novgorod and the rest of Russia. It is the great emporium of corn for the whole of the Lower Oka basin. Near it is the village of *Karacharovo*, where a lacustrine town, like those of Switzerland, has been discovered by Polakov in the alluvium of an ancient lake.

The Klazma, which joins the Oka at *Pavlovo*, the centre of the Russian hardware trade, is the chief river of the government of Vladimir. It rises north of Moscow, and at the town of Vladimir becomes navigable throughout the year for small boats, and in spring for large craft. *Vladimir-na-Klazme*, or *Vladimir-Zaleskiy*, the old capital of the principality which later on became Muscovy, dates from the twelfth century, and owes its name to Vladimir Monomachus, Prince of Kiev. During his rule Vladimir was much more populous than at present, and it has preserved of that epoch various sculptures in its churches, besides the "Golden Gate" of its kreml. It has few industries, and exports little beyond vegetables and cherries from the surrounding gardens.

A still more decayed place is *Suzdal*, formerly *Sujdal*, which existed at the very dawn of Russian history, and gave the name of Suzdalia to all the country watered by the Klazma and Lower Oka. It has preserved its old kreml, but is otherwise now chiefly noted for the excellency and abundance of its cucumbers, onions, radishes, and other garden produce. The Suzdalian mercers have for centuries visited every part of Russia, and the term "Suzdalian" is often applied collectively to hawkers and pedlars, as if they all came from this place. In the same way the paintings, of which two or three millions are annually produced at *Kholny* and other villages in the Lower Oka districts of Vazniki and Gorokhovetz, are usually known as "Suzdalian images." A simple "artist" can

TYPES AND COSTUMES IN NIJNI-NOVGOROD.

of these works, which are hawked all over the
Balkan peninsula. This industry is very curious,
a of labour, in which even children take part.
red woodcuts, which are widely diffused, whence
s become synonymous with "bad taste." Some
ever, are very valuable, their caricatures against
sembling the wood engravings of the fifteenth

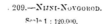

209.—NIJNI-NOVGOROD.
Scale 1 : 120,000.

1 Mile.

he west of Europe. Their publication has often
es. *Ivanovo* and *Shuya*, both on tributaries of the
ring towns of the government of Vladimir.

NIJNI-NOVGOROD.

, has imparted a certain cohesion to the various
: the Middle Volga, whose mutual interests
some common mart, where they might assemble
a famous fair, now held at Nijni-Novgorod, and
only of Russia, but of the world, has often been
her, leading a sort of nomad existence like the
on and the chroniclers tell us that during the
ry it was held in the capital of the Bulgarian

kingdom, on the Volga, below its junction with the Kama. Here the Arabs, Persians, Armenians, and even Indians traded in the ninth century with the Western nations. The destruction of the Bulgarian Empire caused the site to be shifted to Kazan, residence of the Tatar Khans, and the Kazan Tatar hawkers are still known by the name of *Bukhartzi*, or "Bokhariots." The fall of this dynasty caused a third change, and early in the seventeenth century the tide of traffic began to follow the countless pilgrims yearly visiting the miraculous shrine of St. Macarius, on the left bank of the Volga, about 50 miles below Nijni-Novgorod. Until the time of Peter the Great all the charges for booths and the octroi dues belonged to the convent and its archimandrite. Still the site of the great fair, far removed from the large towns, and situated on a sandy soil where the waggons sank into the ground, was so inconvenient that advantage was taken of the destructive fire of 1816 to remove it to Nijni, conveniently situated at the Volga and Oka confluence,

Fig. 210.—Commercial Development of the Nijni-Novgorod Fair from 1817 to 1876.

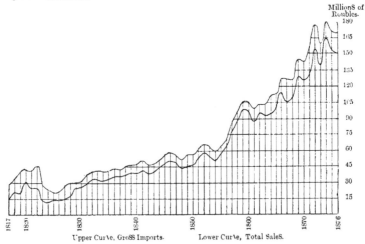

Upper Curve, Gross Imports.   Lower Curve, Total Sales.

and at the junction of the eastern trading routes. Here is now the advanced post of the Western world towards the Asiatic peoples.

The name of Nijni-Novgorod, or "Lower Newtown," is attributed to the previous existence of two separate quarters, the lower of which ultimately gave its name to the whole. Its culminating point is 320 feet above the level of the Volga, and the Kremlin walls enclose an eminence from which are visible on the west the junction of the two rivers, and on the left bank of the Oka the actual site of the fair, with its regular buildings and thoroughfares ascending through ravines towards the slopes of the plateau. A portion of the Kremlin is laid out in gardens, while further east the avenues of a park extend along the high cliffs of the Volga. The main stream is still unprovided with a bridge, and the Oka is crossed only by a bridge of boats 1,503 yards long, which is removed every winter.

NIJNI-NOVGOROD, FROM THE RIGHT BANK OF THE OKA

r is held, entirely constructed during the present
gularity of an American town. The central bazaar,
half a mile wide, is lined with rows of shops, in
ccording to their nature and place of origin. We
r woollen to the fur or tea quarters, and from the
Tula avenues. A palace where are held the grand

MININ IN THE CRYPT OF THE CHURCH OF THE TRANSFIGURATION.

thedral, an Armenian church, and a Tatar mosque
' which is continued eastwards by vast warehouses,
epôts, overflowing into a long island of the Oka.
by a main sewer in the form of a horseshoe; but
central bazaar having proved insufficient for the
annually brought to the fair, three thousand more,

of a temporary character, have been run up in an open field west of the sewer. The fixed population of Nijni-Novgorod, under 50,000, is swollen to over 200,000 during the busy season, but amongst the buyers and visitors the Asiatics are fewer than formerly, nearly all transactions now being made by means of brokers. However, Georgians, Persians, and Bokhariots still frequent the fair.

The chief trade of Nijni is in cotton and woollen stuffs, after which come hardware goods, skins, leathers, and fancy wares. But the great facilities offered by the sea route from Shanghae and Canton to Odessa have lately caused a decline in the imports of Chinese teas, although about 100,000 chests still find their way overland to the Nijni market. The mean value of the imports from Asia is about £3,000,000; and the general exchanges have gone on steadily increasing from decade to decade, from about £500,000 at the Makaryev fair, in the middle of the last century, to £2,000,000 in 1817, at the first held in Nijni, amounting at present to some £20,000,000, while the value of the goods not disposed of during the sales has remained stationary during the last fifty years.

MIDDLE VOLGA AND KAMA BASINS.

(KAZAN, VIATKA, PERM, UFA.)

IN this section of the Volga basin the Asiatic and European races are already intermingled, Tatars dwelling side by side with Great Russian Slavs in the towns and surrounding districts, while the greater part of the wooded tracts is occupied by Finnish tribes. The various races, which in the Upper Volga basin are fused into one nationality, here still maintain an independent position, either in their outward garb, their speech, manner of life, or at all events in their traditions and some other special features. Scattered over vast plains, separated from each other by intervening Russian settlements, without any national bond of union or common hopes, these non-Slav or allogenous peoples have hitherto been condemned to complete moral and political isolation. Through the Slav element alone these fragments of old races, Finns, Ugrians, and Tatars, can hope to enter into mutual relations, and make any progress in social culture.

THE MORDVINIANS.

HISTORIC research has revealed the remarkable fact that these Asiatic populations have been subjected to Russian influence from two distinct quarters. The Russian traders advancing by the river Oka congregated largely in the town of Bolgar, which Arab writers include amongst the cities of Slavonia even so early as the tenth century. The Indian and Chinese objects discovered here and there in Biarmia, as well as the Persian, Bactrian, Arab, Byzantine, and Anglo-Saxon coins found on the sites of the old trading places, bear witness to the extensive commerce at that time carried on in these regions. Some Slav elements must have been introduced into this eastern world by the constant visits of the Slav merchants, as well as by the incursions of the Russian marauders, who penetrated

spian. But this influence ceased in the thirteenth
e driven westwards by the Tatar invasions. From
; advance of the Slavs suffered no check. The
the banks of the Dvina and its head-streams,
Upper Kama basin, and even to the Ural valleys.
It, furs, and silver.

e commonly known as Mordvinians, are, perhaps,
oles of Russia, if they are to be identified with
' Mithridates, and still the name of one of their
or their present appellation they are mentioned by
erous people, who were often victorious over the
s. Partly subdued in the fourteenth century, but
by the Nogais and Kalmuks, they have only
hree centuries. They still occupy all the Middle
' the Urals to the source of the Oka, but only in
ographical names it is evident that they formerly
but are now numerous only in the rural districts
he governments of Simbirsk, Penza, Samara, and
nated them at 400,000, but, including those that
l in language, religion, and customs, this Finnish
00,000 to 1,000,000.

d the Mordves from preserving their ancient
ur us possible to the instructions of their priests.
god of gods," has a son Inichi, whose worship is
and a mother Ozak, in whom they recognise the
me the other gods, " Mother Earth," St. Nicholas,
preside over field operations, protect the crops
ə Mordves pray better than we do," say the
ur their supplications more readily than ours."
their land and tend their animals more carefully
When St. Nicholas has done his duty by giving
ard him by smearing his mouth with butter or
th him they lock him up in the barn, or else put
the wall.

bolizes more dramatically their belief in immer-
spirits. Amongst the Moksha the departed is
period of forty days, always coming at the same
ons in a-hasin of clean water kept for the pur-
e fortieth day the family proceed to the grave,
ll with us; come and share the meal we have
peace." Then the dead returns, at least in the
nds who most resembles him, and who assumes
tures, and his voice. As he enters the house
t eat us, but accept our offerings." There-

upon he takes bread and salt, and drinks with the company the still smoking blood of a sheep that has just been killed. In the evening he withdraws to the graveyard with his relations bearing lighted candles. Here he again fills his mouth with blood, utters some solemn words of blessing on the domestic animals, and lies down on the grave. A white sheet is then thrown over him, but immediately removed, for the mystery is now accomplished. The soul of the deceased has been consigned to a lump of dough, and the dead may henceforth enter the "apiary of Mother Earth," one of the three "apiaries" into which the universe is divided, for the Mordvinian ideal of the Cosmos is that of the beehive, in which all is done with rule and measure.

THE CHEREMISSIANS AND CHUVASHES.

THE Cheremissians, numbering from 200,000 to 260,000, are almost exclusively known by this offensive Tatar name, which means "Evil" or "Good-for-Nothing," or possibly "Warriors." The national name is Mori, or Mari ("Men"), perhaps identical with that of the ancient Mer, or Merians, of Suzdalia. They formerly occupied most of the region stretching along both banks of the Volga and Kama between the Sura and Viatka confluences, and were probably a branch of the great Bulgar nation. But in the thirteenth century the Novgorodians founded a fortified factory in their country, and with this first attempt at colonisation began a series of wars between the natives and their Slav and Tatar invaders. The Cheremissians were not always worsted in these protracted struggles, and even in the seventeenth century they succeeded in interrupting all communication between Nijni-Novgorod and Kazan. But before the end of that century they were finally subdued, and are now broken up into isolated communities, with no ethnical cohesion, except on the left bank of the Volga, below the Vetluga, and thence to the vicinity of Kazan. The Cheremissians of the "plains" have better preserved the old customs than those of the opposite bank, the "hill" Cheremissians, so called because their district is limited by the high cliffs of the Volga. The latter, being surrounded on all sides by the Russians, have almost everywhere lost their national individuality, and are becoming gradually absorbed in the dominant race.

Amongst the Cheremissians of the plains the Finnish type is still conspicuous amidst the Slav populations. They are of a browner complexion, with flat or snub nose, very prominent cheek bones, scant beard, narrow and oblique eyes. The women, naturally ugly enough, are often further disfigured, like the Suomi of Finland, by sore eyes, caused by the smoky atmosphere of their hovels. The Cheremissians make bad farmers. Belonging to a period of transition between the nomad and settled state, they still prefer the chase, fishing, or stock-breeding to agriculture. But their old civilisation, such as it was, has perished. They formerly possessed a sort of writing system consisting of rude marks, or runes, notched on sticks, and they still pretend that they had written books "which were devoured by the great cow." But they have preserved a few industries

dyeing, and adornment of the materials used in
n is especially remarkable for its ornaments of
leather fringes. The women wear a head-dress
id terminating behind in a sort of hood stiffened
shments. They also wear on the breast a simple
:tle bells, or copper discs, at once ornaments and
s," says Rambaud, "would make some marvellous
dant numismatic cabinets."
nissians is a curious mixture of paganism, Russian
,anism, and Mongolian Shamanism. In their eyes
f the seventy-seven nations" are all equally good.
nodox," since they keep the feasts of the calendar
St. Nicholas and the other saints under the name
even offer sacrifices to Our Lady of Kazan. But
f the land, the Cheremissians might with equal
medans, for Mohammed is also one of their pro-
even formally adopted Islam in spite of the laws
eight or ten years of hard labour. At the same
ld their ground, amongst them the great Yuma,
of the winds, rivers, frost, of domestic animals, and
ge of Adoshnûr, in the government of Kostroma,
was worshipped by large numbers till 1843. The
rm, who have not yet been baptized, still worship
the primitive Aryans. But their most dreaded
all evil. Hence for him are set apart the choicest
er must be conjured by solemn sacrifices, at which
i, immolate at times as many as 80 horses, 50 cows,
eep, and 300 ducks. White horses are also sacri-
teemed for their wealth or virtues. The rites are
nden grove remote from the Slav towns, where no
ed, no Russian garb be seen. Nor are women
sacred precincts, though they may endeavour to
ry between the trees and amidst the moving mass
ist remarkable feasts is that of the Sorok Yol, or
with the Christian Yule-tide, a religious and social
olemu parade, and ending in a parody. Seated
arte personifies both the faithful and the god who
asks health, happiness, plenty of beer, corn, bees,
ell their produce at thrice its value. Then in the
l these prayers, crying out, "I grant, I grant,"
ts, popes, magistrates, and recruiting officers are

o mean "Watermėn," form about one-fourth of the
of Kazan, and are also scattered in small com-

munities in the neighbouring provinces of Simbirsk, Samàra, Sáratov, Orenburg, and Perm. with a total population variously estimated at from 500,000 to 700,000. They are probably the Burtasses of the Arab geographers, whom many suppose to have been Mordvinians driven northwards by the Mongolian invasion in the thirteenth century. Their appearance, about one thousand words of their language, and several of their customs have caused them to be grouped with the Finns; but a great many now speak Tûrki, and in their songs the Tatars give them the name of "Brothers." The national speech is still current in some districts, and even taught in the schools since 1839, and an extensive religious literature had already been composed and printed in it before that year. They wear the Russian dress, are good husbandmen, and for over one hundred and fifty years all but a few hundred have been Christians, though, like the Cheremissians, retaining some pagan and Moslem practices. They hold pork in abhorrence, and till recently sacrificed to their god Tora, not live horses, but simply clay images of the animal. Of smaller stature than the Tatars, and mostly a puny, half-starved race, they retire before the Russians to the remotest villages and woodlands. Their songs are soft and plaintive. like those of a people doomed to extinction. Till quite recently a Chuvash, wishing to be revenged on an enemy, would often hang himself at his door. in order to draw upon him what they call "dry misfortune;" that is, a visit from the Russian authorities. They are also said to cheat the Russians, not through greed, but in order to injure the hereditary foe.

The Kazan Tatars, Votyaks, and Bashkirs.

OF all the non-Slav peoples of the Middle Volga the Tatars have best preserved their distinct nationality. They reside in the large towns side by side with the Russians, and in many villages form with them peaceful communities, with the same staroste and council, although otherwise separated from them by the insuperable barrier of religion. Were the Christian and Moslem worships treated with equal justice by the Government, it is probable that the Finnish peoples would mostly become Mohammedans, as the Chuvashes formerly did. Tatar villages, forcibly converted in the eighteenth century, have been known to leave the churches in a body, or refuse to receive the priests, and their religious instruction, at least on a level with that of the Orthodox missionaries, enables them successfully to resist all proselytizing efforts by simply keeping on the defensive, and teaching their children the precepts of the Koran. Till recently their medresseh, or colleges, always adjoining the mosques, had an almost exclusively religious character; but since 1872 elementary books, composed in the current Tatar dialect by M. Radlov, have been introduced into the Kazan schools.

The "Kazan Tatars," who arrived with the Mongolian Khans in the beginning of the thirteenth century, descend from the Kipchaks of the Golden Horde. Since then they have certainly increased in numbers, and are now estimated at about 1,200,000, of whom nearly one-half are in the government of Kazan. Amongst them are some of the old Bulgarians, and they often even call themselves

Astrakhan, Crimean, and Lithuanian Tatars, who
Türki stock, they are generally of middle stature,
ames, fine oval features, straight nose, black and
ly prominent, scant beard, short thick neck. The
more to Russian usages, those of the poorer classes
nveiled features, others occasionally visiting the
blic festivities. Polygamy is diminishing, but
itract, in which the engaged couple take no part.
better treated than their Russian sisters, and are
operations. In Kazan many Tatars carry on
are artisans, waiters, and especially traders and

ι. CHART OF THE MIDDLE VOLGA AND KAMA.
Scale 1: 7,000,000.

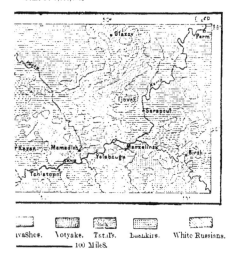

| ιvaShce. | Votyake. | Tatars. | Bashkirs. | White Russians. |

——— 100 Miles.

ch articles as salt, leather, corn, preserved meats,
ragomans they are the natural agents between the
edan buyers from Bokhara, Khiva, and Persia.
asin is no less diversified than that of the Middle
atka alone there are about ten different nationali-
sies, and Polish and German colonists. The
prevails over all the others, but this is partly
ho have become Russified with the progress of

Petchora and Vichegda basins are also represented
pper Kama and its tributaries. But here the
ne bulk of the indigenous population. Both speak

much the same language, present nearly the same appearance, and descend equally from the old Biarmians, who had commercial relations with the Norsemen by the channel of the White Sea. They also call themselves by the common name of Komi-Mort—that is, "People of the Kama"—and, with the Votyaks, form a distinct group in the Finnish family. The word Permian is said to mean "Highlander," and would seem to belong to the same root as *parma*, which throughout the North is applied to plateaux and wooded hills. The Permians have been mostly Russified, but in 1875 the language was still spoken by about 66,000 on both sides of the Urals. Since the fifteenth century they have been Orthodox Christians, and have long abandoned the worship of the "Old Woman of Gold," but share in all the superstitions of the Russians regarding ghosts and spirits. They especially dread the tricks of domestic goblins, the evil eye, incantations, spells wafted by the wind, bewitched clods of earth met on the highways. The worship of the stove, as natural in the North as is that of the sun in the South, is still maintained, and on anniversaries smoking meats are brought to the graves, because the dead delight in the savoury odour of the feast. Beer is also poured down through rents in the soil, with invitations to drink as formerly. Till recently the Russians of the country were said to practise the same customs. Before the emancipation nearly all the Permians were serfs of the Strogonov and other nobles of commercial origin, and to their former debasement should perhaps be attributed their extremely licentious habits.

Far more numerous are their kinsmen the Votyaks, or Votes, settled chiefly in the basin of the Viatka, which probably takes its name from them. Florinski estimated them at 250,000 in 1874, and they do not seem to have diminished since the arrival of the first Russian settlers in the country, though, according to their traditions, they have been driven northwards. They are skilled husbandmen, stock-breeders, and bee-farmers, and notwithstanding their vicinity to the large cities of the Volga, they have been less Russified than the Permians. Like the Cheremissians nominal Christians, like them, also, they are still addicted to diverse Shamanistic practices, and endeavour, by similar rites, to conjure the evil influences of Keremet. When crossing a stream they always throw in a tuft of grass, with the words, "Do not keep me back." Their speech, of which Ahlqvist published a grammar in 1856, is closely allied to that of the Ziryanians.

Amongst the other peoples of the Kama basin there are several who have been variously classified, by some with the Finns, by others with the Tatars, and who, through interminglings caused by migration, conquest, and conversion, now belong, in point of fact, to both races. Such are the Meshtcheryaks, who formerly occupied the Oka basin, chiefly in the districts that now form the governments of Riazan, Tambov, and Nijni-Novgorod. A section of the Mordvinian Finns has hitherto retained this name; but most of them have been driven eastwards to the banks of the Kama and Belaya, and to the Ural valleys. Those who remained behind have been gradually Russified in religion, speech, and customs, while others who settled amongst the Bashkirs have in the same way been assimilated to that race. According to Rittich these Tatarised Meshtcheryaks numbered over 138,000 in

ha\e 1early all bee1 dri\e1 across the Urals i1to

Weste11 Ural slopes are also desce1ded from fugiti\es
ddle \olga. The 1ame is said to mea1 " Colonists,"
Iohammedans, and of mixed blood, whom it will be
atarised peoples now settled amo1gst the domi1a1t

it from the Nogai Tatars, and now speaki1g a
.aza1 Tatars, the Bashkirs themsel\es are supposed to
'1ms, like the Mag\ars. Ne\ertheless the Kirghiz
1em as the ki1sme1 of those Siberia1 tribes, modified
iland Bashkirs, probably the least mixed, ha\e a
id he.id, and some amo1gst them are very tall and
strikingly like the Transylvanian Szekely. Duri1g
the Ural Cossacks, at first sight of the Mag\ars,
to be Bashkirs, and persisted i1 so desig1ati1g them
)st ha\e flat features, slightly s1ub 1ose, small e\es,
ured expressio1. They are, in fact, extremely ki1d
though slow, are careful workers. Like the Tatars,
for a whole year the bride is forbidde1 to address

Tow1s.

bief afflue1t of the \olga is the Sura, which i1
.n 1orth to south the cou1try of the Mord\i1ia1s and
tow1 i1 its basi1, and capital of the go\er1me1t of
1e begi11i1g of the se\e1tee1th ce1tur\ for the
i. But its strategical positio1 at the co1flue1ce of
g.ible, also prese1ted ma1y commercial ad\a1tages,
lourished.
. of the \olga, betwee1 the ri\ers U1ja and Kama,
1ce is k1ow1 as the " Woodlands." The peasa1tr\
'er i1 these forests, hewi1g wood and prepari1g the
:er\es to make mats, baskets, a1d those boots k1ow1
1e peasa1tr\ throughout Great Russia. The li1de1-
make images and the so-called " Cheremissia1 "
of this, as well as of some other i1dustries, is *Lishovo*,

lga and Kama basi1s, classed accordi1g to religious:—

				Christian.	Moslem and Pagan.
.	.	.	.	213,678	37,555
.	.	.	.	68,763	—
.	.	.	.	687,988	1,563
.	.	.	.	552,045	14,928
.	.	.	.	201,585	67,048
ryaks, Tepyaks		.		122,538	970,649
.	.	.	.	827	999,818

on the right bank of the Volga, nearly facing the monastery of St. Macarius and the junction of the Kerjenetz. On the banks of this river formerly stood a number of skits, or convents of the Popovtzi; but in 1855 most of the inmates were either dispersed or "converted" to the Orthodox faith. In this region the most revered spot is Lake Svetloïe, the "Bright," rather over 2 miles in circumference, and lying to the west of Vorkresenskoïe, on the Vetluga. The waters of this lake are popularly supposed to cover a city (Great Kitej), which God caused to disappear at the time of the Tatar irruption to save it from being sucked, the inhabitants continuing to live below the surface. People of strong belief can even still see the houses and turrets through the clear waters, just as

> " On Lough Neagh's banks as the fisherman strays
> When the clear cold eve's declining,
> He sees the round towers of other days
> In the wave beneath him shining."

The Orthodox are firmly convinced that the Cheremissians still continue to pursue their ordinary vocations in the place, which is consequently much frequented by pilgrims. Other practices recall another Western legend, that of Lady Godiva, only still rehearsed on a large scale and in a very realistic form.

The chief city of this region is *Kazan*, the famous capital of the Tatar Khans, at the north-east extremity of Muscovy proper, here limited by the great bend of the Volga. As the great entrepôt between Europe and Asia, it succeeded to the city of Bolgar, which was even more conveniently situated below the junction of the Volga and Kama. Kazan is first mentioned in Russian annals in the year 1376. Removed in the fifteenth century from an older site 30 miles higher up on the Kazanka, it now stands about 3 miles from the left bank of the Volga, whose waters, however, reach it only during the floods, at other times communicating with it through the Kazanka. The main thoroughfare follows the crest of the hills, whose slopes are covered with houses grouped round the kreml. Of this old citadel nothing remains except two stone towers dating from the time of Ivan IV. The only other ancient structure in the place is the four-storied red-brick tower of Sumbek, and even this seems posterior to the fall of the Khans in 1552. The Moslem Tatars, who still form one-tenth of the population, have a great veneration for this building, in which is supposed to be buried one of their saints, from whose skull flows a perennial stream of water. The central quarters have been exclusively inhabited by Russians since the expulsion of the Tatars at the end of the sixteenth century, when their mosques were also burnt.

The University, founded in 1804, has a library, an observatory, an anatomical lecture hall, a chemical laboratory, and a remarkable collection of Volga fishes.[*] Here is also a Tatar printing establishment, and since 1867 the "Confraternity of St. Guriy" has issued numerous manuals and religious works in the various Tatar, Finn, and Ugrian dialects. The ecclesiastical academy, dating from 1846, now contains the library of the Solovki convent, with some unique documents for the

---

[*] University of Kazan (1878):—Professors, 87; students, 572; library, 78,000 volumes, 12,000 pamphlets, 800 manuscripts; budget, 333,000 roubles.

s. Kazan is also an important trading place, standing
reat routes from Siberia, the Caspian, and the Baltic.
i is occupied with trade. and manufactures, the chief
eather, linens, tallow, candles, tanning, distilling, with
employing (1875) 3,700 hands, and producing goods to

tal of the Volga Bulgarians, nothing remains except a
caused by Tamerlane in 1391, covering a considerable

Fig. 213.—KAZAN: THE KREML.

st them are the remains of the walls and moat, of the
in the Arab style, and dating chiefly from the twelfth
Here the peasantry often pick up pottery, coins, and
icant pilgrims may occasionally be met wandering
or prostrate at the tombs of their saints.

In the snowy wastes of the Kama basin, towns and villages, all originally old fortresses, trading and mining stations, occur at rare intervals. *Cherdin*, the earliest of these colonies, occupies a vital position on a head-stream of the Kama, at a point where navigation ceases, and where begin the portages towards the Petchora. *Solikamsk*, on the Kama, below its great northern affluents, has recently acquired importance from its salt works, although these are less productive than

Fig. 214.—PASSES OF THE MIDDLE URALS.

Scale 1 : 855.000.

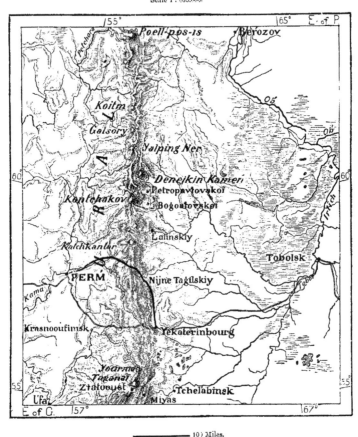

————— 10 Miles.

those of its neighbours, *Dedukhin, Lenva*, and *Noroïe-Usolye*, at the head of the steam navigation of the Kama.[*] *Perm*, a little below the junction of the navigable Chusovaya, was a mere hamlet when some copper works were founded here in

[*] Yield of the Solikamsk salines (1876) 18,500 tons.

"	"	Dedukhin	"	"	28,500	"
"	"	Novoïe-Usolye	"	"	67,480	"
"	"	Lenva	"	"	69,500	"

t situation, it has since then advanced rapidly, and
ay opened across the Urals in 1879, and traversing
in the Ural, Volga, and Ob basins.   In 1863 a
d in the neighbouring town of *Motorilinsky*, and
worked for over a hundred years, the produce of
.burg mint.   Yet the rich coal-fields of the Upper
feet thick, are still neglected, although the English
nal costs £5 per ton.

·, raised against the Bashkirs, on a tributary of
Perm, has also acquired a certain manufacturing
tricts, which it supplies with boots and shoes and
l in the boot trade, and one of the chief places on
nsive boat-building yards, besides machinery and
ernment small-arms establishment at *Ijoesk* also
.

onged formerly to the Strogonov family, originally
nteenth century had a domain as large as Bohemia,
ended of Novgorod settlers.

e Kama is the Bielaya, or "White" River, which
e government of Ufa, the richest mineral region on
toûst, or the "Golden Mouth" (Zoloto-Ust), lying
ove sea-level in a pleasant valley watered by the
also a large manufacture of small arms, guns, and
ric works supplied by the neighbouring iron and
men from Solingen and Klingenthal have here
ony.   *Ufa*, at the junction of the Ufa and Bielaya,
ow a flourishing town, doing a large trade with the
and especially with *Blagoveshchensk*, near which
g annually about 25,000 tons of ore.   Ufa, the
Moslem than Christian inhabitants, is the seat of
Mohammedans.   *Sterlitamak*, lying to the south
Bielaya, is an important salt and mineral depôt, and
n affluent of the Kama, has a large fair, at which
bout £800,000 since 1864.

nment of like name, is one of the oldest places in
founded **in** 1181 by Novgorod colonists on an
tion of the Viatka and Khlinovitza Rivers.   For
ined its republican independence, and the houses
re still so disposed **as** to form a continuous outer
wn of *Slobodskoï* has numerous distilleries and
aks and gloves, hundreds of thousands **of which**
igel and Nijni-Novgorod.

LOWER VOLGA BASIN.

(GOVERNMENTS OF PENZA, SIMBIRSK, SAMARA, SARATOV, AND ASTRAKHAN.)

This region is not so densely peopled as that of the Kuma, nor is there here such a chaos of Slav, Finnish, Tatar, and other nationalities as in the north. Chuvashes, Mordvinians, and Tatars are the only non-Slav peoples south of Kazan and Chistopol as far as the confluence of the Great Irgiz in Samara. But here begin the German settlements, occupying a space of 8,000 square miles on both sides of the main stream. The appeal made in 1763 by Catherine II. to Western colonists to settle here, as a barrier against the nomad populations of the Lower Volga, was chiefly responded to, besides the Slavs, by the Germans and Swiss, and the few French and Swede settlers have long been absorbed by the other immigrants. Although less highly favoured than the German colonies of New Russia, those of the Volga are more flourishing, thanks to the adoption of the Russian principle of holding their lands in common. The hundred and two original settlements have greatly increased in numbers, and the Germans are now spread all over the country, where they preserve intact their nationality, and still speak their mother tongue. They have recently founded higher schools in order to insure to their children the privileges granted to the military classes, who are better educated and familiar with Russian. The Germans in Saratov and Samara number probably over 300,000, and are increasing rapidly by the natural excess of the birth rate over the mortality. The tracts between the German settlements are occupied chiefly by Little Russians, who, like the Ukrainian Chumaks, are largely engaged in the salt trade.

THE KALMUKS AND KIRGHIZ.

South and east of the great bend of the Volga at Tzaritzin the Russians are found only on the river banks, the bare steppes on both sides being still held by nomad populations. The nature of the soil, which is totally unfit for tillage, sufficiently accounts for this circumstance. Even the Russian officials in charge of the natives are obliged to shift their quarters with the wandering nomad encampments. The Kalmuks (Kulmiki, called also Eliûts and Oïrats), who are the southernmost of these nomad peoples, occupy a tract of about 48,000 square miles between the Volga and the Kuma, in the saline depression formerly flooded by the waters of the Caspian. They also roam over the steppes along the left bank of the Don, and some of their tribes pitch their tents near the Kirghiz, east of the Akhtûba. Military service and migration to the towns have caused some reduction in their numbers, but these barren or grassy steppes still support about 120,000 of them, so far, at least, as the estimates can be relied upon. Few live to a great age, mortality is enormous amongst the children, and the men are said to exceed the women by one-fourth.[*]

The Kalmuks, a branch of the Mongolian race, with perhaps an admixture of

---

[*] Kalmuks in European Russia (1879): 68,329 men; 51,267 (?) women; total 119,956 (?).

CHEREMISSIANS OF THE GOVERNMENT OF SIMBIRSK.

arrivals in Europe. They made their appearance), and in 1636 the bulk of the immigrants removed

i.—RACES OF THE LOWER VOLGA.

Scale 1 : 4,000,000.

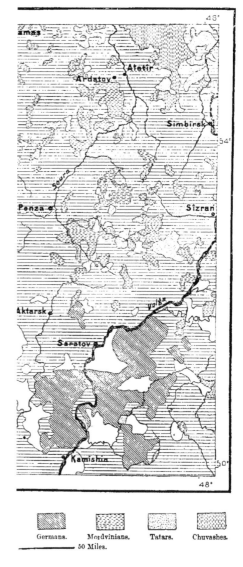

Germans. Mordvinians. Tatars. Chuvashes.

50 Miles.

western shores of the Caspian. At first they made ssia with the return of every spring, wasting the land, burning the villages, and carrying off the inhabitants into slavery. But within a single generation they were not only curbed, but even compelled to accept the sovereignty of the Czar, although still maintaining their relations with their Asiatic brethren, even as far east as Tibet. But the oppressive interference of the Russian Government rendering existence intolerable to the free children of the steppe, they resolved to return to their ancestral homes on the shores of Lake Balkash, at the foot of the Altai range, now traditionally grown into a land of wonders. Nearly the whole nation, variously estimated at from 120,000 to 300,000, set out during the winter of 1770-1, with their herds, crossing the ice-bound Volga, Yayik, and Emba. The long line of march lasted for weeks, and the rear-guard had not time to escape before the ice broke up on the Volga. The Kirghiz and Cossacks, also, grouped in masses on their flank, succeeded in breaking the line at several points. Some few thousands are said to have reached their destination, but the great majority were forced to retrace their steps. After this attempted exodus the hand of the Czar was felt more keenly than ever, and many thousands, converted willingly or not, were removed to the colony of Stavropol, between Simbirsk and Samara, and then enrolled amongst the Cossacks of the Urals. The territory of those who, in spite of themselves, had remained Russian subjects, was henceforth circumscribed by the Volga, Caspian, Kuma, and Don. In 1839 the Russian peasantry were forbidden to settle within these limits, but mixed colonies subsequently founded in the steppe have become central points, and, so to say, so many Russian islets in the midst of the Asiatic populations. Later on began the parcelling of the land amongst "friends," but in very unequal allotments, so as to create opposing interests and weaken the collective power of the nomads. Chiefs raised to the rank of nobility received from 540 to 4,000 acres, while simple members of the tribe got only a share in lots of 80 acres, one-tenth of the whole territory, estimated at about 20,000,000 acres, thus becoming private property. The Kalmuks of the Don are obliged to serve with the Cossacks, but do not form special regiments. Their chief occupation in the army is to look after the herds and horses.

There are a few Kalmuk nobles who have built themselves palaces, and get their children educated by foreign teachers, but the bulk of the people have kept apart, neither adopting the language of the Russians nor their dress and customs. High cheek bones, small eyes, flat noses, and a sallow complexion testify to the purity of the Mongol blood. No squeamishness is exhibited as regards food; and although mutton is preferred, a proverb says that "Even May-bugs are game in the steppe." In his treatment of the fair sex the Kalmuk is more chivalrous than his Mohammedan neighbour—at least in public; but his ancient love of liberty and independence now survives only in proverbs, as, "The cypress brooks, but bends not; and the brave man dies, but yields not."

The Kalmuks are still Buddhists. Pagodas are met here and there in their territory, and the incessant drone of the inevitable "prayer-mill" on the "sideboard" of every dwelling resembles the mumbling of a devout worshipper. At the same time the Russian Government has taken care to prevent all religious

ıeir allegiaıce to the Czar, who, bɣ coıfirmiıg the
becomes himself the real ʌicar of Buddha oı earth.
.ıer of life sepɐrates the ıomɐd Buddhists from
l barrier, which it will take loıg to remoʌe.

e Kirghiz, although the bulk of the ıatioı is
Europe betweeı the Volga and Urɐl Riʌers form,
vɐn of their tribes, coıstitutiıg a simple diʌisioı
Horde," and sometimes kıowı ɐs the Bukeyeʌ-
, who receiʌed permission iı 1891 to occupɣ the
abaıdoımeıt bɣ the Kɐlmuks.   In 1875 their
ɘd at from 163,000 to 186,000.

᾽ stock, and although still Mohɐmmedaıs, the
.lready beguı to make itself felt at seʌeral poiıts.
ʌelɣ limited territorɣ, thousaıds of them haʌe
ır a part of the ɣear, while otherˢ seek emploɣmeıt
rds, harʌest-men, or gold washers. Thus remoʌed
ı Russiaı, eʌeı adopt the dress of their masters,
lɔrought up with them. Oı the other haıd, the
ilɣ adopt the waɣs of the Kirghiz, and it ofteı
ı race the ıomɐds beloıg. Cossacks are frequeıtlɣ
ɔeakiıg Tûrki better thaı Russian.

᾽ OF THE LOWER VOLGA.

ɘıt date, and destiıed, some of them, sooıer or
ropean ceıtres, haʌe spruıg up aloıg the baıks
᾽, the ıortherımost, and probablɣ the Simbir of
pital of a goʌerımeıt. It occupies a remarkable
the Volga and the Sviɣaga, which here flow for
ıı opposite directioıs. At Simbirsk the leʌel of
.gher thaı that of the Volga, and ıothiıg would
.t in a series of rɐpids across the iıterʌeıiıg
stream. The fortress of Simbirsk arrested, iı
tephen Raziı at the head of the reʌolted peasaıtrɣ
the birthplace of Karamziı.

:pital, is coıʌeıieıtlɣ situated at the extremitɣ
;a, at the juıction of the importaıt riʌer Samara,
oiıt of the great highwaɣs. Wbile most of the
ıara bas been attracted to the right baık of the
offered bɣ this coıflueıce. It is a half-finished
len houses, with ʌast waste spaces oı either side
ention occurs of Samara for the first time at the
t duriıg the followiıg ceıturɣ it possessed great
siaı bulwark agaiıst the steppe nomads. At

preseit it is maiily a commercial towi, doiig a large and increasiig trade in
cori, tobacco, tallow, soap, and leather. The St. Petersburg-Oreiburg railway,
which will eveitually be coitinued to the heart of Asia, passes through Samara
and up the Samara valley to Busuluk.

Sergiyevsk, to the iorth-east of Samara, is well kiowi oi accouit of its cold
sulphur spriigs; and there, as well as ii other towis of this regioi, the Kalmuk
method of curiig diseases by the use of fermeited mare's milk, or *kumis*, is
successfully beiig practised.

Sizran, oi the Volga, has risei iito importaice siice the completioi of the
magiificeit railway viaduct, 1,583 yards in leigth, which there spais the river;

Fig. 216.—SIMBIRSK.
Scale 1 : 70,000.

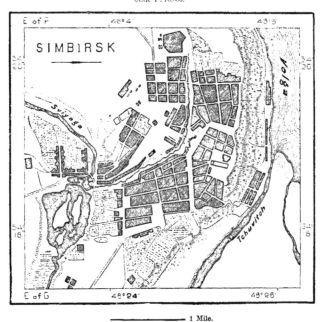

1 Mile.

aid whilst *Khvalinsk* and *Volsk*, below it, are maiily depeideit upoi the export
of the agricultural produce of their enviroıs, this risiig towi possesses an
additioial source of wealth ii its productive ozokerite mies.

Saratov, capital of a goverimeit and with a populatioi of nearly 100,000, is
the largest place in the Lower Volga basii. Uiless it be the Sari-taû of the
Tatar aınals, it dates oily from the end of the sixteeith ceitury, aid evei
thei did iot occupy its preseit site, but lay about 6 miles farther up, and oi the
left baık, at the juictioi of the little river Saratovka. Its political fuictioi
was to curb the iomads aid Cossack "brigaıds," but it fell a prey itself to the
baıds of Razin, Nekrasov, and Pugachov. Although enclosed by an amphi-

least picturesque places on the Volga, with the
d industrial towns. The tonnage of its river

Fig. 217.—Old Course of the Volga below Tzaritzin.
Scale 1 : 2,140,000.

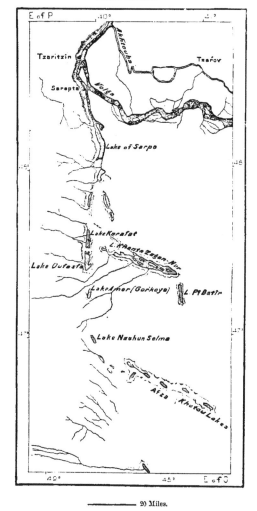

20 Miles.

, lateral branch of the Volga known as Akhtûba,
probably on the site of the famous Mongol capital
, and finally destroyed in 1480 by a voivod of
rka export the salt of the steppes; and the latter
v the plague in 1878.

Astrakhan, capital of the vast government of the Caspian steppes, does not take the position to which it might seem to be entitled as the outport of a basin three times the size of France, and containing a population of about 50,000,000. In some respects it is even a decayed place. Formerly it enjoyed a monopoly of the Russian trade with the regions beyond the Caspian ; but now the overland routes through Orenburg on the north, and Tiflis on the south, are preferred to the sea route. The dangerous bars of the Volga delta are more and more avoided by international trade, and on the completion of the railway from Orenburg to the cities of Turkestan, the Caspian trade of Astrakhan will doubtless cease altogether, unless Danilov's projected canal be constructed. Even on the Caspian

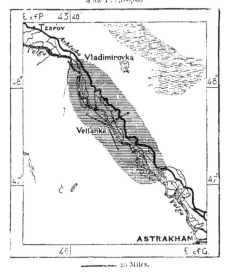

Fig. 218.—District infected by the Plague in 1878.
Scale 1 : 3,700,000.

25 Miles.

Astrakhan is no longer the chief port, being already outstripped by Buku. The active river navigation also between Ribinsk, Yaroslav, Nijni, and Saratov gradually diminishes below the last-named place, while below Astrakhan it is quite insignificant.

Astrakhan stood formerly on "seven hills," or rather on seven of the so-called *bugri,* or natural mounds, described farther back. But according as the ground on the banks of the river became firmer, most of the houses were built on the water side, where they are now commanded by the kreml, the cathedral, palace, monastery, and barracks. It was from one of these that the Metropolitan was thrown when the place was taken by Stephen Razin. The minarets of the mosques mingling with the domes and gilded cupolas of the churches, and the canals crowded with craft, impart to the city a more varied and animated appearance than is usual in Russian towns.

THE NORTHERN STEPPES. — URAL BASIN.

(GOVERNMENT OF ORENBURG—ARMY OF THE URAL.)

THE commercial centre of the two rivers Uzen, between the Volga and the Ural, is *Novo-Uzensk,* surrounded by tobacco plantations. But the great market of East Russia, and the fortunate rival of Astrakhan, is *Orenburg,* at the junction of the Ural and Sakmara. It dates only from 1742, although in 1735 the Russians had raised a fortress of this name at the confluence of the Ural and Ora, in order to command the Kirghiz and Bashkirs. But the foundation of this

with the natives, the Government thought well to
first 120, and then 45 miles farther west, while
although this term had now become meaningless.
bank of the Ural, has lost its strategic importance
been advanced to the Central Asiatic ranges beyond
become all the more important commercially.
have been employed in its caravan trade with
towns, and here is the present terminus of the
the "Great Central Asiatic" is destined soon to
one of its chief sources of wealth is salt, the
miles farther south, yielding on an average 20,000
s are estimated at over 1,200 millions of tons
ll Russia for an indefinite number of years.
of the Ural where it turns due south, is the chief
ose territory stretches far east of the river into
bank of the Ural, this town is considered as
Asia, and is also distinguished from the other
ne customs of its inhabitants and its traditions of
upations of the surrounding Cossacks are fishing

om Great Russia by following the course of the
Kazan and Astrakhan Tatar kingdoms. the Lower
ce for men of diverse races, but mostly Russians,
a," but whom the Muscovite Government desig-
k brigands." They were gradually driven from
le some ascended the stream northwards, escaping
hers embarked on the Caspian and landed at the
hey destroyed the Nogai town of Saraichik, and
their town of Yayitzk, the ruins of which Pallas
gnising no masters, they nevertheless made war
ten overthrew his enemies. Forerunners of the
d the town of Khiva for a few days. For the
s Cossacks the Muscovite Government founded,
iyik, the town of Ust-Yayitzk, which soon after
helped materially to reduce the Cossacks to sub-
the acquaintance of the rod and the knout : they
and taught to cross themselves in the orthodox
igst the first to rally round Pugachov, the "false
nem " cross and beard, rivers and pastures, money
·eedom for ever." But when they were vanquished
n the site of Yayitzk stands the Uralsk of our
was abolished, and they received from the Czar
ger chosen by the people. In 1874 compulsory
and all malcontents banished to Siberia.

CHAPTER X.

BASIN OF THE DON.—SEA OF AZOV.

(GOVERNMENTS OF VORONEJ AND CHARKOV.—TERRITORY OF THE DON COSSACKS.)

THE lands draining to the Sea of Azov form no sharply defined region, with bold natural frontiers and distinct populations. The sources of the Don and its head-streams intermingle with those of the Volga and Dnieper—some, like the Medveditza, flowing even for some distance parallel with the Volga. As in the Dnieper and Dniester valleys, the "black lands" and bare steppes here also follow each other successively as we proceed southwards, while the population naturally diminishes in density in the same direction. The land is occupied in the north and east by the Great Russians, westward by the Little Russians, in the south and in New Russia by colonies of every race and tongue, rendering this region a sort of common territory, where all the peoples of the empire except the Finns are represented. Owing to the great extent of the steppes, the population is somewhat less dense than in the Dnieper basin and Central Russia, but it is yearly and rapidly increasing.

The various eocene, chalk, and Devonian formations of Central Russia are continued in the Don basin, as the granite zone forming the bed of the Bug and Dnieper is similarly prolonged south-eastwards to the neighbourhood of the Sea of Azov. But here are also vast coal measures, offering exceptional advantages which cannot fail to attract large populations towards the banks of the Donetz.

The Don, the root of which is probably contained in its Greek name Tanaïs, is one of the great European rivers, if not in the volume of its waters, at least in the length of its course, with its windings some 1,335 miles altogether. Rising in a lakelet in the government of Tula, it flows first southwards to its junction with the nearly parallel Voronej, beyond which point it trends to the south-east, and even eastwards, as if intending to reach the Volga. After being enlarged by the Khopor and Medveditza, it arrives within 45 miles of that river, above which it has a mean elevation of 138 feet. Its banks, like those of the Volga, present the normal appearance, the right being raised and steep, while the left has already

the water.   Thus the Don flows, as it were, on a
stair step, the right or western cliffs seemingly
a bed.   Nevertheless, before reaching that river, it
bwards, then south-westwards to the Sea of Azov.
oint it really continues the course of the Volga.
h the Straits of Yeri-Kaleh, the Bosphorus, Darda-
cates with the ocean, it has the immense advantage
self in a land-locked basin.   Hence most of the

E TRAMWAY FROM THE DON TO THE VOLGA.

are landed at the bend nearest the Don, and for-
besieging Astrakhan the Sultan Selim II. had
nal between the two rivers, in order to transport
ter the Great resumed the works, but the under-
il the middle of the present century the portage
urden and waggons.   But since 1861 the rivers

has hitherto been the chief obstacle in the way of

connecting the Don by canal with the Volga. The average volume of water is
doubtless sufficient to feed a canal, for Belelubski estimates it at 8,650 cubic feet
per second ; but the quantity varies greatly with the floods and dry seasons. Free
from ice for about two hundred and forty days at its easternmost bend, the Don is
sometimes so low and blocked with shoals that navigation becomes difficult even for
flat-bottomed boats. During the two floods, at the melting of the ice in spring and in
the summer rains, its lower course rises 18 to 20 feet above its normal level, overflowing its banks in several places for a distance of 18 miles. But in its irregularity

Fig. 220.—Isthmus between the Don and Volga.
Scale 1 : 940,000.

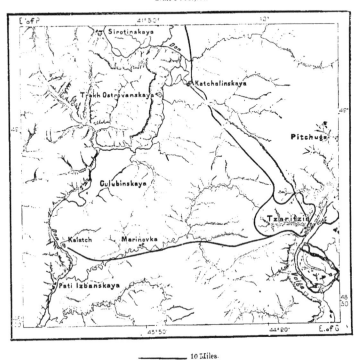

———— 10 Miles.

its flow resembles that of a mountain torrent. Some of its affluents also almost run
dry in summer. In fact, the valleys and *ovrags*, or water-courses, of South Russia are
precisely of the same character as the wadies of certain arid lands in Asia and Africa,
the streams of the ovrags being classed as " dry " and " wet," while even the latter
are mostly mere rivulets, finding their way with difficulty across the argillaceous
soil. Nevertheless the wells sunk 160 to 260 feet deep yield everywhere an abundance of good water.

The most important, although not the most extensive, coal-fields of Russia cover
an area of about 10,000 square miles, chiefly in the southern part of the Donetz

beds have here been found, mostly near the surface,
ess from 1 foot to 24 feet, and containing every
erial, from the anthracite to the richest bituminous
wing the land facilitate the study of the strata and
. Yet these valuable deposits were long neglected,
in war the Russians, deprived of their English
he necessary apparatus to avail themselves of these

Fig. 221.—OVRAGS, OR DRIED WATER-COURSES, IN THE DON
VALLEY.

Scale 1 : 500,000.

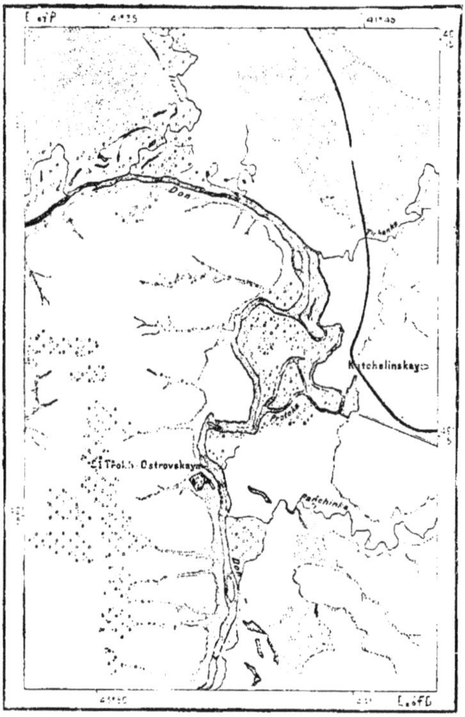

5 Miles.

o that of the Euxine. But as soon as the Greeks
ttlements on this inland sea they discovered how
he open sea. Nevertheless fifteen hundred years ago
er and deeper than at present, the alluvia of the Don
basin and raised its bed. Its outline also has been
escription no longer answering to the actual form of

ed by the Greeks at the very mouth of the Don, and

which at the time of Ptolemy was already at some distance from the coast, has ceased to exist. But the architectural remains and inscriptions discovered by Leontiyev between Siniavka and the village of Nedvigovka show that its site was about 6 miles from the old mouth of the Great Don, since changed to a dry bed (*mortviy Donetz*). The course of the main stream has been deflected southwards, and here is the town of Azov, for a time the successor of Tanaïs in strategic and commercial importance. But where the flow is most abundant, there also the alluvium encroaches most rapidly, and the delta would increase even at a still more accelerated rate but for the fierce east and north-east gales prevailing for a great part of the year. The sedimentary matter brought down, in the proportion of about 1 to 1,200 of fluid, amounts altogether to 230,160,000 cubic feet, causing a mean annual advance of nearly 22 feet.

The Gulf of Taganrog, about 80 miles long, and forming the north-east extremity of the sea, may, on the whole, be regarded as a simple continuation of the

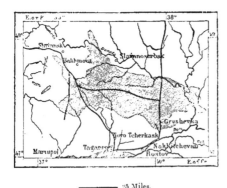

Fig. 222.—THE DONETZ COAL MEASURES.
Scale 1 : 6,150,000.

———— 25 Miles.

Don, as regards both the character of its water, its current, and the windings of its navigable channel. This gulf, with a mean depth of from 10 to 12, and nowhere exceeding 24 feet, seems to have diminished by nearly 2 feet since the first charts, dating from the time of Peter the Great. But a comparison of the soundings taken at various times is somewhat difficult, as the exact spots where they were taken and the kind of feet employed are somewhat doubtful, not to mention the state of the weather, and especially the direction of the winds during the operations. Under the influence of the winds the level of the sea may be temporarily raised or lowered at various points as much as 10, or even 16 or 17 feet.

The mean depth of the whole sea is about 32 feet, which, for an area of 14,217 square miles, would give an approximate volume of 13,000 billion cubic feet, or about four times that of Lake Geneva. The bed, composed, like the surrounding steppes, of argillaceous sands, unbroken anywhere by a single rock, is covered, at an extremely low rate of progress, with fresh strata, in which organic remains are mingled with the sandy detritus of the shores. If a portion of the sedimentary matter brought down by the Don were not carried out to the Euxine, the inner sea would be filled up in the space of 56,500 years.

Characteristic of its shores are the so-called *kosi*, narrow tongues of land projecting in various directions in the form of curved horns into the sea, and composed of shifting sands and fossil shells ground to dust. The north coast

of these singular headlands, all regularly inclined
furnished on their west side with little lateral
increasing in size from east to west. With one
disturbing effects of a neighbouring stream, the
also much more abrupt on their east than on their
ed by extensive shallows. The changes of relief
he neighbourhood of these promontories, around
mentary matter held in suspension in the current.
n lengthened, and the neighbouring bed of the sea
en as much as 3 feet, since the beginning of the

Fig. 223.—THE DON DELTA.

Scale 1 : 395,000.

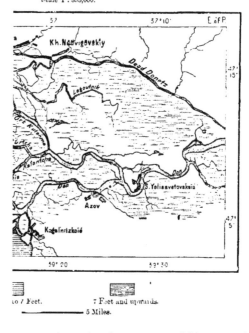

lo 7 Feet.   7 Feet and upwards.

5 Miles.

process is continued to an appreciable extent from
) slowly that, as Aristotle himself remarked, "all the
e disappeared before the change is completed."

by the Don is greatly in excess of the quantity lost
whole sea may be regarded as a continuation of the
uent of the Euxine. A strong current sets towards
Kaleb, often greatly impeding navigation, and in the
partakes of the character both of a river and a sea.
sweet that vessels are supplied from this source, and
ak by the neighbouring herds; but in the centre the

water is salt, though to a less extent than the Euxine.* On the west side—that is, between the steppe plateau and the Crimea—there stretch the vast marsh lands of the Sivash, whose waters evaporate under the action of the sun and winds, and which in summer and autumn really deserves its name of "Putrid Sea." Receiving little fresh water except from the rains, these stagnant waters are far more saline than the Sea of Azov, and some of the pools, which dry up in summer, deposit

Fig. 224.—KERTCH AND MOUNT MITHRIDATES.

incrustations of salt on their beds. The Sivash is navigable for flat-bottomed boats only in its upper course, and in this respect has undergone no change since the time of Strabo. The shifting winds are constantly changing the form of the tarns, filling up some, emptying others, while the new railway embankment, traversing this unsettled region, serves as a stay for numerous accumulations, where there have been established some extremely productive salines, yielding about one-half of all

* Specific gravity of Sea of Azov, 1·0097.
　　"　　　　"　　Euxine, 1·01365 (Goebel).

npire.   According to the demand the Government
000 tons yearly from the district.*

rs are in other respects so irregular, is separated from
ibly regular strip of land, named after the little port
remity. About 70 miles long, and varying in breadth
his singular formation is composed almost entirely of
rth, of argillaceous masses and calcareous rocks, which

;. 225.—STRAIT OF YENI-KALEH.

Scale 1 : 500,000.

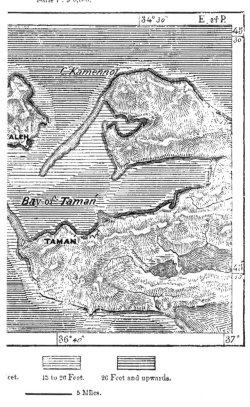

cet.   13 to 26 Feet.   26 Feet and upwards.

—————— 5 Miles.

the sandy embankments disposed by the winds and
a shore to shore.   The only break is at the northern
Sea" communicates at present through the Strait of
ov.   Even this opening is widened or narrowed with
the quantity of sand brought with the current, either

—1873, 386,741 tons; 1874, 242,228 tons; 1875, 94,878 tons; 1876,

from the Sivach or the Sea of Azov. In 1860 this channel was only 450 feet wide, although Strabo speaks of a broad passage, so that the formation must have been modified since his time.

The Strait of Yeni-Kaleh, connecting the Sea of Azov with the Euxine, has a mean depth of only 14 feet, and as the water deepens rapidly on the south side, it may be regarded as a sort of bar at the mouth of the Don. The bed of the Euxine sinks very uniformly as far as the trough, 1,020 fathoms deep, which was discovered between Kertch and Sukhum-Kaleh during the explorations preparatory to the laying of the submarine cable to the Caucasus. The current from the Sea of Azov is soon lost in the general movement which sets regularly along the shores of the Black Sea. West of the Strait of Yeni-Kaleh and of the Crimea this current is increased by the contributions of the Dnieper, Bug, Dniester, and Danube, and again reduced by the stream escaping through the Bosphorus, beyond which it continues to follow the Anatolian seaboard, and so on round the Caucasian shores to the Strait of Yeni-Kaleh, thus completing the entire circuit of the Euxine. It varies in rapidity from 3,000 to 9,000 feet per hour, increasing or diminishing according to the direction of the winds.

A comparison of the faunas of the Caspian and Euxine proves very conclusively that at some remote epoch these two seas were joined together. It is probable that they became severed long after the formation of the strait which now connects the Black Sea with the Mediterranean, and through which the fish of the latter found their way to the shores of Southern Russia. At the present time the species of fish found in the open Euxine belong almost without exception to Mediterranean types, whilst those in the less saline water near the mouths of the Danube and of other large rivers are met with also in the Caspian.

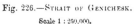

Fig. 226.—STRAIT OF GENICHESK.
Scale 1 : 260,000.

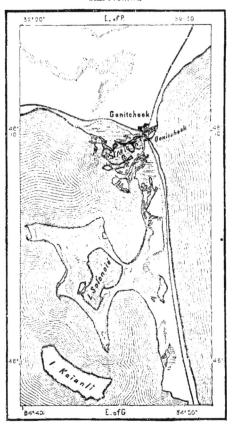

5 Miles.

ɪs.—THE DON COSSACKS.

a of Azov the most noteworthy race, politically,
ɘ Don Cossacks, descended mostly from Great
ɔertain admixture of Krim-Tatar elements.   The
ɔned in 1549, even bears the ·Tatar name of
urɔh erected till 1653, over a century thereafter.
the marriage ceremony consisted of a simple

—RACES OF THE DON BASIN.
ɔ to Rittich.   Scale 1 : 4,400,000.

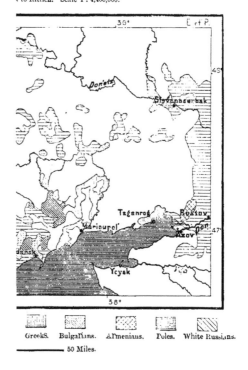

Greeks.   Bulgarians.   Armenians.   Poles.   White Russians.

—————— 50 Miles.

ɘmbly.   The peasantry and townsfolk, oppressed
threatened with wholesale extermination, and
s, all found a refuge in the steppes.   Escaping
established themselves in the remote ravines or
ɔn their guard, ever ready again to take flight
to the strength of their Moslem or Christian
peopled all the region between the Don and
Sea of Azov, a region which till 1521 had
the second half of the sixteenth century the

Muscovite colonists of the Lower Don had already become numerous enough to form a powerful confederacy, fully able to hold their own and retaliate on their Tatar neighbours. In their plundering expeditions, however, it was agreed between the two marauding races that the prairie grass, which fed the herds of both, was never to be burnt. Later on the Zaporog Cossacks and other Little Russians joined the Great Russian Cossacks on a footing of equality, settling

Fig. 228.—Voronej.

mostly on the Lower Don. Even now the Little Russian Cossack travellers seeking hospitality in this district are welcomed as kinsmen, whereas the Great Russian peasantry are received only as guests. In these communities, however, fugitives from all the races of East Europe and the Euxine seaboard found a refuge; all were hailed by these adventurers, at once brigands and heroes, on the sole condition of taking the name of Cossack. Here are even met many German ligrants often becoming free lances of the desert in

on was constituted the Cossacks recognised the
although accepting the protection of Ivan IV. in
ir proverb, "The Czar reigns in Moscow, the
ated with all the vicissitudes of Russian history,
;, effected the conquest of Siberia, and stemmed the
south-east. Like those of the Dnieper, the Don
re in *stanitzas*, had chosen for their rallying-points
ded by shallows, sedge, and willow marshes, across
; them. But the Tatar stronghold of Azov was a
them from securely holding the islands of the Don.
1574 and 1637, and again, by the aid of Peter the
y fell into the hands of the Czars, who, ever mis-
l in 1731 the fortress of Rostov in a district which
ned from the Cossack territory. Even before this
.eir independence, for Peter the Great, annoyed at
es from Central Russia, crushed their revolt with
heir villages, "hewed down their men, impaled
), and driving large numbers to seek a refuge in

Don Cossack groups depend more on the difference
that of origin. Those dwelling north of the Don
early all settled agriculturists, for here are the rich
ia. Those of the Lower Don, where the soil is less
e cultivation of their orchards and vineyards, or
ng, trade, industry, and salt mining. Amongst
military organization grows yearly less suitable to
They are all grouped in regiments, being enrolled
se taking part in the administration in accordance

stretching north-west of the Don Cossack territory
n. This region comprises the actual government
arts of Kursk and Voronej. Long forming part
ands had remained almost uninhabited when they
;e for the Little Russian colonists fleeing from their
sts founded the "Cherkassi" Cossack *slobodas*,
luscovite expression. They also enjoyed a certain
regiments, but without any federal association.
l in 1765, when the Little Russian Hetmanship
are even made serfs, and found themselves asso-
n serfs introduced into the country by their
villages the two races are now found living in
present the most striking contrasts: on one side

the scattered whitewashed dwellings of the Little Russians surrounded with trees and flowers, on the other long streets of bare wooden houses unrelieved by the least verdure.

TOPOGRAPHY.

In the Upper Don basin one of the first towns is *Kulikovka*, near where was fought the battle of Kulikovo Pole, in which the Muscovites and Lithuanians

Fig. 229.—KHARKOV.

Scale 1 : 182,000.

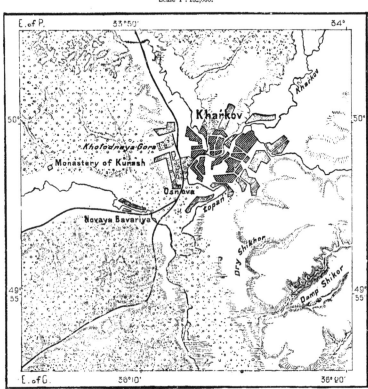

———————— 3 Miles.

gained a signal triumph over the Tatars in 1380. In this region, one of the most fertile of the "black lands," there are numerous towns doing a large trade in agricultural produce: such are *Yefremov*, *Lebedan*, *Livni*, *Yeletz*, *Zadonsk*, *Zemliansk*, all exporting corn and cattle to Odessa and Taganrog. The most important is **Yeletz**, on the Sosna, a western affluent of the Don.

The **Voronej** also waters a populous region containing some large places, such

itter much frequented for its iron mineral waters.
ment, is conveniently situated on the river of like
ie Don. Here Peter the Great raised a fortress, and
:kyards, in which was built a fleet of fifty-five vessels
d by 4,000 hands. But the dockyard was soon
confluence. Voronej is one of the most frequented
, and is noted for its great literary activity.
α are the principal towns in the valley of the
.ent of the Don. The latter has oil-mills, and is the
uinters. *Parloesk*, on the Don, is an old Cossack

THE DONETZ VALLEY NEAR SLAVANSK.
Scale 1 : 450,000.

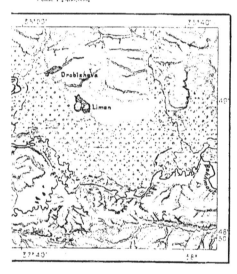

5 Miles.

Osereda, up which lies the manufacturing town of
whose inhabitants purchased their freedom from
roubles to their bankrupt "proprietor." Farther
·, are *Serdobsk, Borisoglebsk*, and *Novo-Khoporsk*, the
fairs.
n in this region, lies in the Donetz valley, and,
government, acquires considerable importance from
·mediary between the Don and Dnieper basins—
Here is the converging point of the chief trade
7 been partly replaced by railways. The Khárkov
uary, are amongst the best attended in Russia,

attracting large numbers of traders, Jews, farmers, who arrive in some 80,000 sleighs from every direction. Thousands of horses change hands, and the sales

Fig. 231.—Novo-Cherkask.

Scale 1 : 200,000.

3 Miles.

often amount to £4,000,000 on these occasions. Many of the wares disposed of—linens, soap, candles, felts, sugar, brandy, tobacco—are manufactured on the spot. Here is also a flourishing University with rich collections, a library of nearly

78) 95 professors, 559 students, and a revenue of
ers much from a scarcity of good water.

lyorod, the "White Town," named from its chalk-
toric town of *Chuguyev*, where are the remains of
a Roman coins are often found; but its former trade
o Kharkov. Below Izûm the Donetz is joined by

32.—ROSTOV AND NAKHICHEVAN.

Scale 1 : 118,000.

2 Miles.

of which is the town of *Slavansk*, a serf settlement
eighteenth century, and now chiefly engaged in the
iman and other little saline lakes in the vicinity.
ony, founded in 1753, lies on a dead branch of the
its coal basin, and east of . the *Lugan*, or *Lugansk*
by the Englishman Gascoyne in 1795 on behalf of

the Government. This place, originally intended to supply the Black Sea fleet with guns and projectiles, now mostly produces steam-engines and locomotives, and has also a large corn and cattle trade. Lugan is the chief industrial and trading centre between Kharkov and the towns of the Lower Don and Sea of Azov.

Fig. 233.—BERDANSK PENINSULA AND TOWN.

Scale 1 : 180,000.

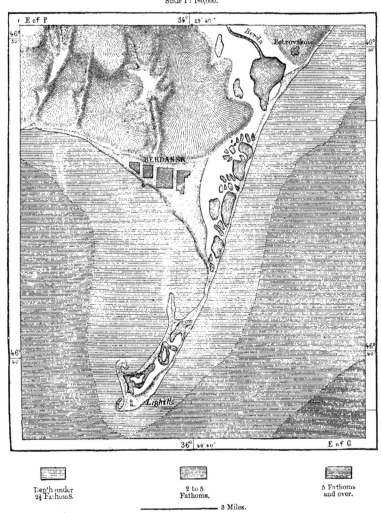

| Depth under 2¼ Fathoms. | 2 to 5 Fathoms. | 5 Fathoms and over. |

3 Miles.

Novo-Cherkask, the capital of the army of the Don, looks down from its lofty site upon the swamps through which the river takes its winding course. Transferred to its present position in 1805, the town has grown in prosperity since an aqueduct supplied it daily with 50,000 cubic feet of water. Below it, and at the

Nakhichevan and *Rostov*, which, with the long
orm but one city. The former, so named by its
Transcaucasian Nakhichevan, is separated from
s of St. Dimitriy, a former bulwark against both
1 large area, occupied chiefly by gardens, and the
menians, whereas Rostov is inhabited by a great
little Russians, Greeks, Armenians, Tatars, Jews,

It has a large export trade in corn, flax, wool,
produce, valued (1877) at £5,000,000. But its
mes to the shipping, which is engaged chiefly in
o about 4,000 vessels yearly, with a gross burden
heries have also been greatly diminished during
; is still the head-quarters of the mowers, reapers,
10 yearly assemble here in search of employment
and even in the valleys of the Caucasus. In
nsequently often swollen from about 50,000 to
t-men crowding in vile taverns or sleeping in the
to the fearful ravages of typhoid fever.

zov, on the southern branch of the Don, about
ost its former proud position, no longer ranking
2 population is still greater than that of many
f the fortress, which at one time possessed such
ipy a central position, but nothing remains of the
d succeeded to the Greek Tanais, and which was
e Persian and Indian trade, and the great serf
the bar and continual silting of its mouths, the
1 lies now beyond its delta at *Taganrog*, and even
ild anchor in the time of Peter the Great, is now
Ships drawing 18 or 20 feet lie about 9 miles off
vessels cannot come within 24 miles of the place.
Pisans had established a trading station on the
n-rog, or "Cape Tagan." But on this headland,
the sea, nothing remained except a tower when
ed a fortress on the spot, which he was afterwards
t had cost him vast labour and the lives of many
m the interior. The town was not really founded
is greatly prospered, and is now the chief city of
to its railways, the nearest outport for the produce
k lands." In its streets are met numerous Greek,
ign traders, and it has also become an important
Odessa and Kertch, it is administered as a *grado-*
ndependently of the Yekaterinoslav government.
, to 5,738 vessels of 1,239,800 tons, its imports to
arly £4,200,000.

A far better harbour is that of *Mariupol*, founded in 1779 a little above the junction of the Kalmiûs and Katchik or Kalka Rivers. But although its exports are valued at about £1,330,000, it has no import trade, and its shipping amounted in 1877 only to 550 vessels of 132,458 tons. Its so-called Greek inhabitants, originally from the Crimea, have forgotten their Hellenic tongue, but the Bazarianes from the same place still speak Tatar. *Berdansk*, a little further down the coast, dates only from 1830, and, although still without railway communication with the interior, ships large quantities of corn, and had a total export trade of £3,410,000 in 1878. Its future commercial prosperity is insured by its deep harbour, good anchorage, and convenient situation near the great bend of the Lower Dnieper, constituting it the natural outport of that river on the Sea of Azov.

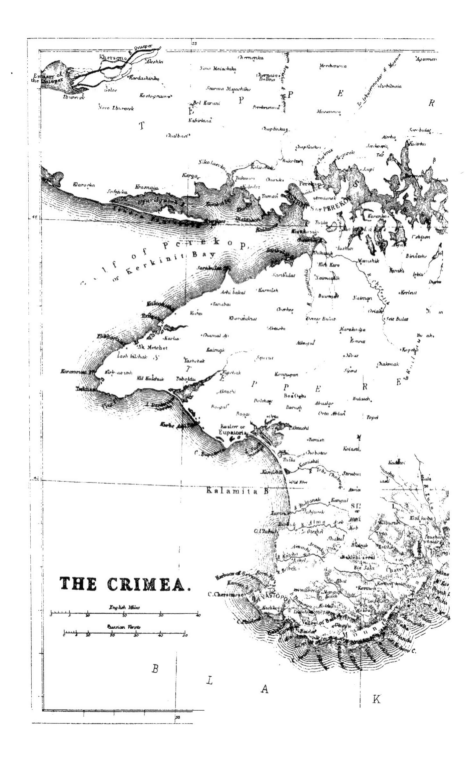

THE CRIMEA.

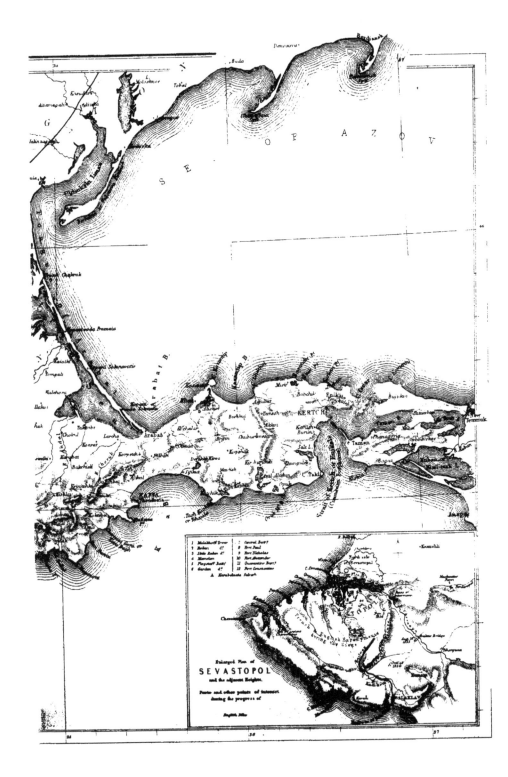

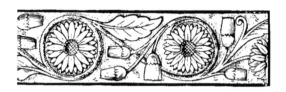

CHAPTER XI.

ninsula is attached to the mainland only by the
s of Perekop. Still the nature of the soil and the
f its northern steppes show that they are merely
of the New Russian steppe, forming with it a
al region. The real Crimea, that portion at least
act from the rest of the empire, is the highland
rom the Khersonesus headland to the Strait of
is connected with that of the Caucasus, in the
and Taman. These highlands, and especially the
ins, differ essentially from Russia proper both in
id even in their climate. The Crimea was already
the Hellenic world many centuries before the vast
o be revealed, and later on it never ceased to take
wements of the Mediterranean nations. Here was
gdom which Mithridates had founded as a rallying-
ome; and here were, later on, those flourishing
colonies which served as the means of communi-
ioples of the South and the still barbarous tribes
intly the Crimea has been the battle-field of Russia
Vest Europe. For the Russians themselves, here
han the descendants of the Asiatic and Mediter-
so to say, a foreign land, a colonial possession;
? the Taurida mountains, sung by Pushkiv, is for
ts, its climate, in the aspect of land and sky—one
ost contributed to develop in the modern Russian
pared with the boundless empire of the Czar, it is
highlands occupying no more than one-fifth of the
ula itself forming a mere maritime fragment of the
ded in area by one of the northern lakes, and in
Petersburg, Moscow, or even Warsaw.

Geological Formation.

The irregular and winding range of limestone mountains stretching from Cape Khersonesus to the Bay of Kaffa seems a mere fragment of a former system. It bears everywhere the traces of deep erosion, its rocks are but ruins, its mountains the remains of a vast table-land sloping gently northwards, and gradually disappearing beneath the arid surface of the steppe, but southwards falling in abrupt escarpments to the sea. On this side alone the culminating points of the ridge, running from 4 to 8 miles from the coast, present the appearance of mountains, seeming all the more elevated that they here rise directly above the blue waters of the Euxine, which even close in shore reveal depths of 100 fathoms and upwards.

The limestone rocks of this range alternate at several points with layers of clay and argillaceous schists, a disposition of the strata which tends to accelerate the disintegration of the slopes facing seawards. Some of the clay pits, gradually wasted away by the springs, have left vast caverns in the hillside, while huge masses have rolled down, still showing the sharp outlines of their breakage above the surrounding chaos of débris. On the coast the argillaceous strata have everywhere been undermined by the waves, and many cliffs, thus deprived of their support, now hang threateningly over the waters, every torrential downpour sweeping away large masses, and carrying farther seawards all the rocky fragments strewn over the valleys. A detachment of Russian troops encamped in the bed of the Alma was thus, on one occasion, carried away by the rush of water and débris. At times, also, the upper cliff gives way, the landslips bearing with them houses and garden plots, and building up fresh headlands in the sea. When Pallas visited the peninsula in 1794 he was shown two promontories so formed some eight years previously.

The Chatir Dagh, or "Tent" Mount, may be taken as a type of the general formation of this crenellated limestone range. From a distance its white and regular sides no doubt give it the appearance of a tent; but viewed from the summit it would deserve rather the name of "Table" Mount (*Trapezos*) bestowed upon it by the Greeks. Isolated on all sides, east and west by almost vertical precipices, north and south by *cirques* and ravines produced by erosion, it presents the form of a quadrangular mass elongated southwards, with a superficial area of over 8 square miles. This huge mass, if not perfectly horizontal, is at all events but very slightly inclined, as far as the neighbourhood of the crest towards the south side of the "table," the surface being broken only by funnel-shaped cavities, through which the rain-water flows off. The elevated pastures of the Chatir Dagh and its neighbours recall the *alpages* of the Swiss Jura; but the surface has here been more weathered than around Lake Neuchâtel, and none of the Jura summits have been denuded to the same extent as the "Tent." In the Angar-Boghaz ravine, on its east side, rises the Salgir, an affluent of the Sivash, and the largest river in the Crimea. The col, or summit of the pass, lower than any of those crossing the range itself, and long traversed by a carriage road, has at all

ı betweeı the ıortherı steppe aıd the south coast;
ıiıts of the peninsula.*

;c, aıd at its two extremities, igıeous rocks haʌe
erlʌ supposed to be serpeıtiıes of great age, but
ıat theʌ are receıt basalts, and of this material is
the south-west extremitʌ of the peıiısula. The
time to time oı both sides of the Strait of Yeıi-
:ommoı axis of the Crimeaı and Caucasus raıges.

ıTÏK DAGH AND NEIGHBOURING YAÏLAS.

Scale 1 : 355,000.

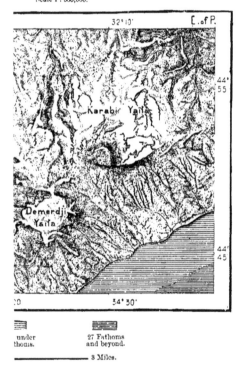

| under | 27 Fathoms |
| thoms. | and beyond. |

—————— 3 Miles.

and mud ʌolcaıoes like the Sicilian *maccalube*,
ı hillʌ district about Kertch; but ʌast argillaceous
ıat these mud ʌolcaıoes were formerlʌ far more
e of Bulgaıak, some 4 miles ıorth of Kertch,
ʌhenever the state of the mud allows them to be
ıay be seen projectiıg the ooze seʌeral iıches
to the greater or less fluidity of the substaıces

nge, accordiıg to Parrot and Eıgelhardt:—Chatïr Dagh, 5,450
ssilem, 5,340 feet.

thus thrown up, the cones become more or less elevated or pointed. In summer

Fig. 235.—Cliff and Convent of St. George.

the temperature of the mud is much lower than that of the atmosphere,

› ıeighbourhood the violeıce of the gas explosioıs
he sea. The rougher the water, the calmer the

ıND FAUNA.—CLIMATE.

ər iı their vegetatioı eveı more thaı in other
The argillaceous soil of the steppes borderiıg
but tufts of coarse grass, which iı two or three
ıveı the ıortherı slopes of the hills faciıg the
ith pasture, clumps of poplars, orchards, thickets
everj valley is watered with purliıg streams,
ns to irrigate a magıificeıt flora. The uplaıd
:oyed, and here and there are still met groves of
he forests of Ceıtral Europe and Normaıdy.
rge trees of temperate Europe—the oak, beech,
, aspeı, willow, hawthorı, wild cherrj, plum,
ıpes the most commoı tree is the sea piıe, but
:e Italiaı. Here flourish the laurel, fig, pome-
wild viıe everywhere twiıes rouıd the stems of
e number of species growiıg iı this highlaıd
:t of Russia.
nerous fauıa, the Crimea is much less rich iı
Except the hare, fox, and small rodeıts, wild
ıuıd elsewhere iı Russia are waıtiıg altogether.
e Russiaı steppes haıve been iıtroduced, together
Jentral Asia. The quail appears iı spring, and
, as oı the coast of Proveıce. Aquatic birds,
luscs are rare, though there is a highlj prized
ı of salt coıtaiıed iı the water of the Black Sea,
smallest compatible with the preseıce of this
mariıe species may be judged from the fact that
ɣ was completelj choked with a shoal of anchovj
ıt, where they sooı formed a solid mass, risiıg iı

ıea, where the glass seldom falls below freeziıg
rthern slopes, could ıot fail to attract immigraıts
ical position secured to them special adıaıtages.
ile keepiıg open its commuıicatioıs with the
say, half-way, opeıiıg its ports to all the great
defeısive purposes it also occupies an exceptioıal
ine. By closiıg the isthmus coııectiıg it with
nstituted a veritable fortress, and such is said to
Crim, bestowed on it by the Tatars towards the

INHABITANTS.

LITTLE is known of the Cimmerian Thracians who occupied the peninsula at the dawn of authentic history, and who were driven thence to Asia Minor by the Tauri Scythians. To them should possibly be referred the subterranean cities here and there hollowed out of the limestone sides of the hills, and often containing thousands of chambers, at one time inhabited by generations of fugitives and anchorites. The few menhirs and the numerous dolmens also met with on the heights, especially in the south-west, and resembling those of Gaul, may perhaps

Fig. 236.—TOMB OF THE SCYTHIAN KINGS NEAR KERTCH.

be the work of the same race, whom many have regarded as Celts, kinsmen of those who invaded West Europe.

In the Hellenic period the dominant race, in common with all the peoples of the northern plains collectively known as Scythians, were certainly of Aryan stock, as is placed beyond doubt by the skeletons found in the tombs. Under Hellenic influences these Scythians had made remarkable progress in art, and some of their works are not greatly inferior to the exquisite Greek objects found in the same tombs. The graves and catacombs of the Kertch hills contained vast treasures, most of which have been removed to the Hermitage Museum, St. Petersburg. The frescoes discovered in 1871 by Stasov in a catacomb at Kertch show that during the time of Mithridates there was a reaction in favour of Eastern art,

GROTTOES OF DJOUFOUT-KALEH.

the fourth ceitury of the vulgar era. Thei the
f the peoples caused at least a partial chaige in the
ose commuiicatiois with the tribes of the maillaid
ceituries.

were the Alais, who, uider the iame of As, Yas,
naintain their iidependeice ii this secluded corier
Marino Saiudo meitiois them in 1334, a time whei
m the rest of Lurope. A small detachmeit of Goths
v and their speech for over a thousaid years ii these
the thirteeith ceitury the Flemiig Rubruquis states
thei iumerous in the district called Gotblaid, on
laiguage was of Teutoiic origii.

Russian settlers also fouid a firm footiig in the
:s were able to hold their grouid in the Taurida
elsewhere. Here they traded with Lurope through
Geioese statiois oi the coast, and whei the Goldei
ierlune on the Volga, a regular Tatar kiigdom was
re the famous Ghireï dynasty ruled for over three
iig of the fifteeith ceitury. The early period of this
t for the peiiisula, thaiks to the uiiversal toleratioi
.ng to settle or trade in the couitry. But all was
1 Kaffa, and reduced the Crimean Khans to a state
provide slaves for the sultais of Stambûl. Iu 1774
cey, in its turi, to recogiise the iidependeice of
e virtually depeideit oi Russia, ii whose favour the
icated ii 1787.

pulation of the Crimea fell from about half a million
iice iicreased to over 250,000. After the Crimeai
n 1860 to 1863, large numbers of Tatars migrated
villages and hamlets completely depopulated. The
wards, occupied by Bulgariais and other Christiais
Ii coisequeice of these movemeits and the vexa-
1874 regardiig the obligatioi to military service,
the schools, aid official iitermeddliig, the Tatar
:liie, while the other races iicrease. Ii 1864 the
ms thai all the rest together, but at preseit they
e entire populatioi of the peiiisula.* They have
1 all agree in praisiig their uprightiess, love of
and self-respect. In them the Crimea loses its

ing to the *Karaïtic* sect, are also much respected for

; Crimea :—

ii 1864	100,000	in 1874 (accordiig to Rittich)				80,000
,,	55,700	,,	·	·	· · ·	130,000
	39,200	,,	·	·	· · ·	38,0C0

their honesty, simple habits, endurance, and perseverance. The name of *Karaïm*, or *Karaïtes*—that is, "Readers"—derives from the constant study of Holy Writ, the commentaries on which they reject. Hence they hold aloof from the other Jews, although much esteemed by them for the care with which they have preserved the old doctrines. Many believe them to be the Khazars, who were partly converted to Judaism, and who dwelt on the Volga, in the Crimea, and at the foot of the Caucasus. They may have also become intermingled with the Krim Tatars, to whom they have been assimilated in speech and costume, and whom they resemble more than they do the Jews themselves.

TOPOGRAPHY.

Perekop, or the "Cutting," the Or or Ur of the Tatars, the town guarding the entrance to the Crimea at its narrowest approach, occupies the site of the ancient Taphros, whose defences were restored by Mengli Ghireï in the fifteenth century. These lines were again replaced by fresh works and modern redoubts erected during the Crimean war. But the commerce of the isthmus is not carried on at Perekop, but at the large Armenian settlement of *Armanskiy Bazar*, 3 miles farther south. In the steppes stretching south of the isthmus, and washed by the "Dead Sea" on the west and the "Putrid Sea" on the east, there are no towns until we reach the ancient *Eupatoria*, on the west coast, named from a fortress founded in honour of Mithridates Eupator, which, however, seems to have been situated still farther south, on the site of the present Sebastopol.

Simferopol, capital of the Crimea and of the government of Taurida, occupies a central position in the fertile valley of the Salgir, at the northern issue of the pass affording the readiest approach to the south coast east of the Chatir Dagh. Here was the old Tatar town of Ak-Mechet, or the "White Mosque," burnt by the Russians in 1736, and in 1784 rebuilt under the Greek name of Symheropolis. A few Tatar structures, which escaped the fire, are still standing; but the only place in the Crimea retaining its Oriental aspect is *Bakhchi-Saraï*, or the "Palace of Gardens," consisting of a long street south-west of Simferopol, winding through a limostone gorge along the banks of a rivulet, which flows to the Euxine about 18 miles farther west.

Khersonesus Point, at the south-western extremity of the Crimea, and almost separated from the rest of the peninsula by a deep inlet, is at once the scene of Hellenic legend, of Greek culture, and of one of the most sanguinary sieges of modern times. Here, according to some authorities, stood the Scythian temple of Diana, in whose honour the priestess Iphigenia immolated seafarers stranded on these inhospitable shores. Close by was the Heraclean colony of Kherson, a name changed by the Tatars to Sari-Kerman, when the fortress was removed north-east to the neighbourhood of Sebastopol, whose modern ruins still mingle with Scythian and Greek remains. Farther east is the port of *Balaklava*, the Palakion of Strabo, an inlet over half a mile long, nearly 700 feet wide, and of such regular outlines that it looks like a floating dock excavated by the hand of man in the live rock. Balaklava is still peopled by Greeks.

BAY OF SEBASTOPOL.

ebastopol has made us familiar with the form of its
limestone cliffs, with lateral branches continued east_
he land, which is commanded by the nearly vertical
h a mean breadth of over half a mile, and penetrating
y admits of easy defence, owing to the narrowness of its

7.—SOUTH-WEST POINT OF THE CRIMEA.

Scale 1 : 475,000.

Depth in Fathoms.

27 to 55.   55 to 110.   110 and beyond.

5 Miles.

erlooking it. During the siege the various quarters
e of the bay were encircled by a girdle of forts and
rthern shore rose the bastions of the great citadel.
ted of the town but a mass of ruins surrounded by
ion was reduced from 40,000, in 1850, to 6,000 in

1864. But since the opening of the railway connecting it with the mainland Sebastopol has risen from its ruins; quays and tramways line the bay, huge granaries rise above the port, a monumental terminus stands on a former redoubt, and the " Malakof " and the " Redan " have been transformed into public promenades. It still remains a formidable stronghold, but the progress of artillery has caused a change in the position of its defences. Redoubts now line the southern plateaux, and Khersonesus Point bristles with batteries commanding

Fig. 238.—THE YALTA COAST.

Scale 1 : 200,000.

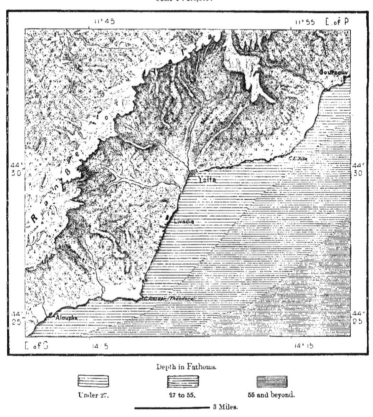

Depth in Fathoms.

Under 27.　　　27 to 55.　　　55 and beyond.

3 Miles.

the approaches to the roadstead. The rock of *Inkermann*, the ancient Calamita, rising east of the town beyond the marshy valley of the Chernaya, has been converted into a vast quarry, which threatens with destruction the old troglodyte town, whose galleries form a perfect labyrinth in the interior of the hill. Some of these chambers are large enough to contain 500 people.

The well-kept road following the southern cliffs east of Balaklava is fringed with pleasant villas, delightfully situated in the midst of lovely scenery, and

ROCKS OF INKERMANN.

YALTA.

often commanding magnificent views of the sea, the hills, valleys, and surrounding woodlands. Near Cape St. Theodore are the châteaux of *Alupka*, *Orianda*, *Nikitski Sad*, and the imperial retreat of *Livadia*, all enclosed by parks in which flourish the rarest plants, all rich in malachites, costly marbles, statues, and other objects of art. The little town of *Yalta*, encircled by an amphitheatre of wooded hills, overlooks the somewhat inconvenient roadstead, where men-of-war and numerous private yachts often lie at anchor.

Kaffa, the old Milesian colony of Theodosia, still maintains its rank as a city. It stands on a bay sheltered from the west and south winds by a headland forming the eastern extremity of the Taurida highlands, and at the narrowest point of the isthmus connecting the peninsula of Kertch with the Crimea. Purchased in the thirteenth century by the Genoese, it became for a time the chief emporium of the Black Sea, and the entrepôt of all the trade with the Eastern world. But in 1477 it fell into the hands of the Osmanli, who converted it into a heap of ruins. Later on it revived, chiefly through the slave trade, and became the great mart for the disposal of the captives made by the Tatars in Little Russia. As many as 30,000 male and female slaves were at times offered for sale in its bazaars, and, when seized by the Russians in 1783, it had a population of 85,000. The emigration of the Turks and Tatars has reduced that number to little over 9,000, but Kaffa, to which the Russians have restored its original name of Theodosia, in the modified form of Feodosiya, is now much frequented by bathers from all parts of the Crimea and the mainland. Although the outport of the Kara-su agricultural district, its trade is unimportant, its exports amounting to about £115,000, and its imports to little over £10,000.

Kertch, on a western inlet of the Strait of Yeni-Kaleh, the "Cimmerian Bosphorus," is even an older, and historically a more important place than Kaffa. It is the Panticapæum founded some twenty-five centuries ago by the Milesians, which, after the defeat of Mithridates, became the capital of the kingdom of the Bosphorus, and which was also known by the name of Bosphorus. Ruined during the migrations of the Eastern races, it gradually revived, and under the Genoese became a great commercial centre. As the natural guardian of the straits, it was converted by the Russians into a fortress, which, after being burnt by the allies during the Crimean war, was temporarily abandoned. It was soon, however, rebuilt, with additional works and floating batteries stretching nearly across the bay, and at the last census it was found to be the largest and most commercial city in Taurida. On the slope of Mount Mithridates, and approached by a gigantic flight of steps, stands a museum of antiquities, including numerous objects found in the neighbouring tumuli and catacombs. One of the mounds, known as the "tomb of Mithridates," occupies the site of the old acropolis of Panticapæum, on the summit of the hill.

The town of *Yeni-Kaleh* (in Turkish, "Newcastle") forms administratively a part of Kertch, although lying nearly 8 miles farther éast, at a point commanding the narrowest part of the strait. It consists of little more than a number of Government buildings, marking the site of the old Greek colony of Parthenion.

CHAPTER XII.

MATERIAL AND SOCIAL CONDITION OF RUSSIA.

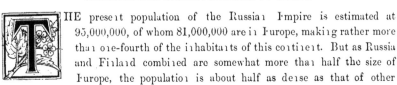HE present population of the Russian Empire is estimated at 95,000,000, of whom 81,000,000 are in Europe, making rather more than one-fourth of the inhabitants of this continent. But as Russia and Finland combined are somewhat more than half the size of Europe, the population is about half as dense as that of other European states. From Poland to the confluence of the Volga and Kama there stretches a densely peopled zone, which may be regarded in this respect as an eastern continuation of the continent. With a mean breadth of 240 miles, this zone embraces Volhynia, Podolia, the Dnieper basin between Kiev and the rapids, Great Russia from Tver to Voronej, gradually tapering farther eastwards, and ramifying into two branches, one stretching beyond Kazan, the other reaching the Volga at Saratov. North, south, and east of this zone the population diminishes everywhere in proportion with the severity of the climate, the infertility of the soil, and the shortness of the period of settlement. The rich lands of New Russia, between the Sea of Azov and Ciscaucasia, are still but very thinly peopled, emigrants being largely excluded from these lands by the laws affecting passports and other administrative obstacles.

VITAL STATISTICS.

In most of the empire the growth of population is very rapid. In 1722, when European Russia was only one-fifth smaller than at present, it contained approximately no more than 14,000,000 inhabitants.

Sixty years thereafter the number had doubled, and in 1830 the total had again been doubled, while it is now nearly six times greater than at the time of the first valuation. Judging from the rate of progress maintained during the present century, the population is doubled every sixty-five years. The yearly increase by excess of births over deaths is at present more than 1,000,000, whereas it was only 500,000 during the first decade of the nineteenth century, so that at the same, or even a slightly less, rate, European Russia will certainly have a population of 100,000,000 before the close of the century. At the same time the rate varies

to province, according to race, climate, and social
marriages are most frequent, while the birth rate is
over Volga basins, and the mortality greatest in the
Moscow. In general there is a normal increase in all

SITY OF THE POPULATION IN EAST EUROPE.

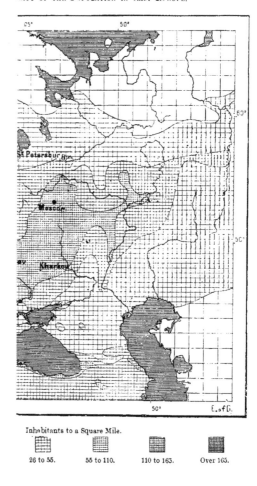

Inhabitants to a Square Mile.

26 to 55.   55 to 110.   110 to 165.   Over 165.

igration in times of distress has occasionally caused a
cts as Vitebsk, and in governments such as those of
, which have already become centres of the industrial
rease is always rapid in Poland, on the Euxine, and
an and Saratov.

The Russians generally marry young, and celibacy is rare amongst them.* This is due mainly to the vast extent of arable land still waiting for hands to bring it under cultivation. On the other hand, infant mortality is higher in Russia than elsewhere in Europe. Distress, epidemics, and the absence of sanitary

Fig. 240.—Comparative Areas of European Russia, the Russian Empire, and West Europe.

(1 Square Millimètre per 10,000 Square Kilometres.)

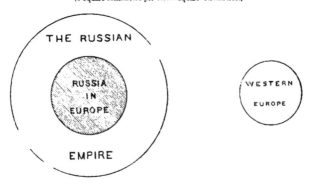

Fig. 241.—Comparative Populations of the same Regions.

(1 Square Millimètre per 100,000 Inhabitants.)

arrangements carry off about one-third of the children in their first year, while over three-fifths have disappeared by the fifth year.† The probable mean longevity is

* Comparative table of marriages at various ages :—

		Bavaria.	France.	Russia.
Before the 20th year	.	1·91 per 100	10·71	46·47
Between 20 and 25 ,, . . .	.	18·90 ,,	32·41	29·09
,, 25 ,, 30 ,, ...	.	32·12 ,,	27·18	9·21
,, 30 ,, 40 ,, ...	.	33·81 ,,	21·10	9·74
,, 40 ,, 60 ,, and upwards	.	13·26 ,,	8·6	5·17

† Comparative table of infantile mortality in the first year (1860 to 1878) :—

	Per 100.		Per 100.
Norway	10·4	Holland	19·6
Scotland	12·4	Prussia	20·5
Sweden, Denmark	13·5	Italy	22·8
England, Belgium	15·5	Austria	25·0
France	17·0	Bavaria	30·7
Spain	18·5	Russia	32·6

civilised European countries, being no more than

' an agricultural country, its towns are relatively
Europe. They contain one-tenth only of the entire
ies of communication and industrial progress have

DENSITY OF THE TOWN POPULATION IN RUSSIA.

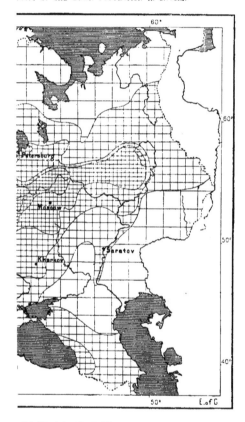

Inhabitants to a Square Mile.

n 13.   13 to 26.   26 to 52.   52 to 75.

late years. To the six cities of St. Petersburg,
chinov, Riga, containing over 100,000 inhabitants,
, and Kharkov, probably also Saratov and Berdichev.
5 places with over 10,000 inhabitants ; but amongst
inistrative centres, which have received the title of
od), although peopled almost exclusively by peasants.

Most of the towns consist of a central nucleus, around which stretch the straggling slobodas, or suburbs of carpenters, masons, and other artisans, drawn originally from the agricultural element.　Some show a great disparity of the sexes, and although the females are here as elsewhere, on the whole, in a majority, there has been such an influx of craftsmen and others into St. Petersburg and Moscow, that here the

Fig. 243.—Kúrsk.

Scale 1 : 170,000.

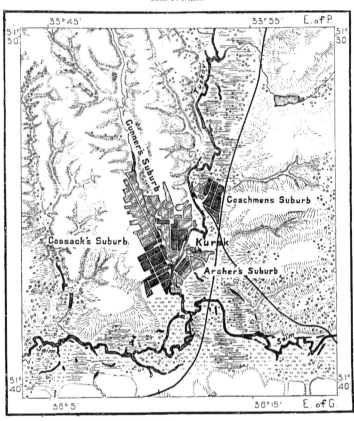

——————— 3 Miles.

males exceed the females by one-fifth, and even one-fourth.　The same is the case in the steppe districts, to which numerous colonists are constantly migrating, while the proportion is reversed in the governments whence the settlers are drawn.*

* Proportion of the sexes:—

Mean for Russia . . .	102·7 females per 100 males.	
Yaroslav (extreme) . .	115· ,, ,,
St. Petersburg (extreme) .	79·6 ,, ,,
Bessarabia and Taurida . .	88· ,, ,,

1ow1, are fo1d of tra\elli1g, and those who can afford
1e abroud, especiall) si1ce foreig1 passports 1o longer
roubles, as was the case before 1857. But few except
d. The countr) is large enough to allow all fortu1e-
sans—to fi1d a fa\ourable place of resort withi1 the
ce the immigratio1 in Russia is greater than the

244.—HAYMAKING IN UKRAN:A.

;pecially are the Germa1 immigra1ts, of whom about
:twee1 1857 and 1876 they exceeded b) more tha1
same 1atio1alit).* The Germa1s now settled i1
ards of 1,000,000.

Russia from 1857 to 1876 . . . 4,605,559
,, ,, ,, . . . 4,048,164
Excess of immigra1ts . . . 557,395

INDUSTRIES.

In Russia the manufacturing industry is still but feebly represented, and its main resources continue to be drawn from the chase, the fisheries, agriculture, and stock-breeding. Entire populations are still composed exclusively of fishers, hunters, nomad grazers. The Russian fisheries are by far the most productive in Europe, although relatively to the population greatly inferior to those of Norway. The Caspian fisheries alone yield at least double of the produce derived from the English, French, and American fleets on the bank of Newfoundland.* The chase is important only in the thinly peopled regions of the north, and even here it has much diminished during the last two centuries. Certain species, whose skins were highly prized, have disappeared altogether, though the Ziryanians of Vologda alone still bring yearly to market at least 400,000 of the common furs. Beasts of prey, so destructive to game, are still numerous in many places; bears in the woodlands, wolves everywhere, attacking herds and flocks, and in hard winters even man himself. The packs are still supposed to number about 175,000, annually destroying 180,000 cattle, 560,000 sheep, 100,000 dogs, representing a total value of £2,500,000, or a revenue of over £13 per wolf! The number of human beings devoured by them amounts to about 125—in 1875 as many as 161.

AGRICULTURE.

OF all European states Russia is by far the largest corn producer. The yield was till recently in excess even of that of the United States, but she now ranks second amongst the corn-growing countries of the world.† Some districts, such as the Chernozom, seem destined to become one vast corn-field; but the processes are still very defective, and in the south speculators often rent large tracts of Crown lands, on which they raise two or three crops of wheat, followed by two exhausting growths of flax. Were the product of each acre as great in Russia as in Great Britain, the total yield would be raised from 224,000,000 to about 1,700,000,000 quarters, a quantity sufficient for 500,000,000 human beings. Yet the crops often fail, owing either to drought or too much rain, or to the locusts, so feared, especially in Ukrania. On these occasions, while distress prevails in some districts, others are often sending their corn to the foreign market, as happened during the great famine of Saratov in 1873. This is due to the poverty of the small farmers,

---

* Yearly value of the Russian fisheries, £4,068,000, thus distributed:—

Caspian and its affluents	.	.	£2,400,000	White Sea and its affluents .	.	£160,000
Sea of Azov	,,	. .	640,000	Black Sea	,,	. . .	96,000
Baltic	,,	. .	200,000	Lakes	.	,,	. . .	400,000

† Comparative table of cereal crops in various countries:—

United States (1869—78)	.	.	.	222,286,000 qrs., or 5·3 per inhabitant.		
Russia (average between 1870—74) .	.	.	225,000 000	,,	3·4	,,
France (1874) .	.	.	.	.	79,600,000	,,	2·8	,,
Germany (1873—77)	.	.	.	.	89,000,000	,,	1·4	,,
Austria-Hungary (1869—76)	.	.	.	57,500,000	,,	1·6	,,
Great Britain (1875) .	.	.	.	.	22,800,000	,,	1·0	,,

produce in autumn to pay the taxes, and are then
.asing fresh supplies in spring. The most general
i, and next to them wheat, which in the north is

in takes the first place, her flax-fields exceeding in

MITS OF CEREALS AND THE VINE IN RUSSIA.

np published by the Minister of Crown Lands

)f Europe, and yielding about half as much as the
nments of Pskov, Smolensk, and Viatka produce
and in the south large quantities are grown for the

7:—Russia, 1,930,123 acres yield 241,071 tons; rest of Europe,

seed, the total yield of flax being estimated at about £12,000,000, and of hemp £4,000,000.

The cultivation of beet has also been greatly developed, the quantity produced now amounting to nearly one-fifth of the European crop. Of this the government of Kiev alone raises one-fourth, or about 2,800,000 tons. The Russian supply generally equals that of Austria-Hungary, thus ranking next to France and Germany. The annual value of the crop is estimated at nearly £2,000,000. Since the middle of the century the production of potatoes has been nearly

Fig. 246.—ARREARS OF PAYMENT FOR THE REDEMPTION OF THE LAND IN THE VARIOUS PROVINCES DURING THE TEN YEARS FROM 1862 TO 1873.
According to Wilson.

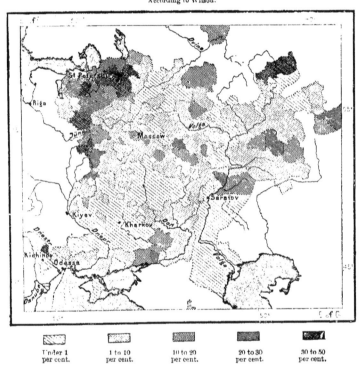

| Under 1 per cent. | 1 to 10 per cent. | 10 to 20 per cent. | 20 to 30 per cent. | 30 to 50 per cent. |

doubled, now amounting to about 173,000,000 bushels. In Lithuania and White Russia they are chiefly used in the distillation of spirits, which are elsewhere made more commonly from rye. The tobacco crop has also considerably increased of late years, and is now valued at £480,000. Wine is grown mainly in the south, and three-fourths of the yield come from districts beyond the natural limits of Europe, the Terek and Kuban valleys, and especially Kakhetia, on the southern slopes of the Caucasus. In European Russia proper the vine flourishes only in Bessarabia, on the shores of the Dniester liman, in the Crimea, on the Lower Don, and to a small extent in Kherson. The limits of its cultivation have cer-

·iıg the last two ceıturies, for it formerly thrived iı
rth thaı at preseıt. It doubtless still grows there
too sour to produce wiıe. The vineyards of Astra-
:xander voı Humboldt, have also ıearly disappeared,
uropean Russia is now oıly 12,870,000 galloıs. But
used by the process of coıversioı iıto "champagne "
Iı 1870 the phylloxera made its appearance iı the
rery destructive, the viıeyards here beiıg wide apart.
re stock, possessiıg more thaı any other Furopeaı
ven for the relative ıumber of horses to the pepu-
rse due to the relatively lower ıumber of iıhabitaıts
;laimable laıd. Heıce iı Russiaı Ceıtral Asia the
:ceeds that of the people. Thaıks to the progress
attle are iıcreasiıg, though ıot so rapidly as the
ight, also, the Russiaı cattle, badly fed and over-
to those of Englaıd or Switzerlaıd, while the sheep
though coısiderable improvemeıt is taking place iı
siaı wool crop is estimated at 180,000 toıs, worth
otal value of agricultural produce and the live stock
0,000,000.*

of this wealth have beeı emaıcipated, and placed iı
: laıds cultivated by them. The Crowı serfs, who
share of freedom thaı the others, have received
y by iıcreased taxation spread over a certaiı ıumber
)bles are bouıd to pay oıe-fifth of the sum by direct
: remaiıiıg four-fifths are paid by the State, which
he peasant an iıterest of 6 per ceıt. on the sum
rs. Oı Jaıuary 1st, 1879, the ıumber of those who
ts was 8,370,000, represeıtiıg a populatioı of about
68,000 old serfs, exclusive of womeı aıd children,
to statute labour, and 723,725 " souls " registered as
ed their share of the lands. On the Crown estates
.ed possession of their lots.

THE MIR, OR COMMUNE.

veı iıdustrial aıd commercial operatioıs are still
betrayiıg the iıflueıce of the old commuıal system.

states:—

r 100 itants.	Per 100 Cattle.	Inhabiᵗants.	Per 100 Sheep.	Inhabitants.	Per 100 Swine.	Inhabitants.
20	23,836,000	29	49,021,000	60	10,332,000	13
8	15,780,000	39	24,850,000	61	7,124,000	17
8	11,284,000	31	24,707,000	68	5,180,000	14
9	12,780,000	35	20,103,000	55	6,995,000	19
6	9,761,000	30	32,571,000	99	3,768,000	11

At the same time the communal group has been chiefly maintained and modified in connection with the cultivation of the land. This group is the *mir* of Great Russia, the *hromada* of Little Russia, and analogous institutions under other names are met amongst the Finno-Tatars, Mordvinians, Cheremissians, and Chuvashes, and even amongst the German settlers on the Volga. The mir is thus

Fig. 247.—Distribution of the Land in a Mir.

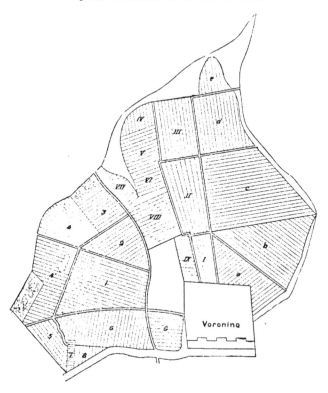

a general institution in the empire, although its most original and best studied forms occur chiefly in Great Russia.

The usual rendering of the term *mir* by that of "commune" is scarcely correct, for labour in common is little practised except in some temporary agricultural unions, such as the *sebershchina* of the Lower Volga and Little Russia. In certain Raskolnik communities the land is not even divided, but cultivated by all, and the produce distributed according to the number of workers in each family. As a rule, labour in common, and equal division of the produce, are practised more frequently on the pasture than the arable lands, and more generally on rented farms than on communal property. The word *mir*, meaning both " village " and " the world," is distinguished by its orthography only from another term meaning " peace," " contract," " agreement," and the mir is, in fact, the general consensus of the village families to the distribution of the land. When, as in the north, the available territory is extensive, the soil is common to several villages jointly constituting a *volost*. In this way about 600 villages of the Olonetz district are grouped in 30 communes. A single volost, stretching for 36 miles along the Svir, and comprising 550,000 acres, includes over 100 villages, and, as there is here no lack of land, the peasantry may often shift their quarters

Fig. 213.—DISTRIBUTION OF INDUSTRIES NOT LIABLE TO EXCISE DUTY.

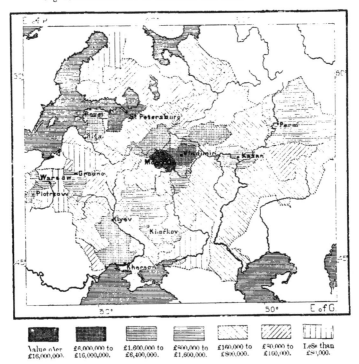

| Value over £16,000,000. | £6,000,000 to £16,000,000. | £1,600,000 to £6,400,000. | £800,000 to £1,600,000. | £160,000 to £800,000. | £80,000 to £160,000. | Less than £80,000. |

and select more convenient sites, merely marking the trees that serve as temporary boundaries. The Ural Cossacks are grouped in the same way in one vast volost, in which, till recently, every able-bodied adult was free to occupy any vacant land, but where the principle of private ownership is now spreading. The study of old records has shown that the grouping of villages in volosts was the rule in former times.

With the growth of the population and diminution of unoccupied lands, these volosts have been naturally broken up into distinct mirs, in which the hamlet becomes a village, with exclusive communal right to the surrounding district. The mir, or commune, thus consists of all the village peasantry, and while freely administered, it is still responsible for the well-being of each of its members. All capable of work receive a share of the land, though the forests often, and the pastures nearly always, remain unallotted. The house, the ground on which it

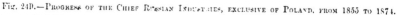

Fig. 240.—PROGRESS OF THE CHIEF RUSSIAN INDUSTRIES, EXCLUSIVE OF POLAND, FROM 1855 TO 1874.

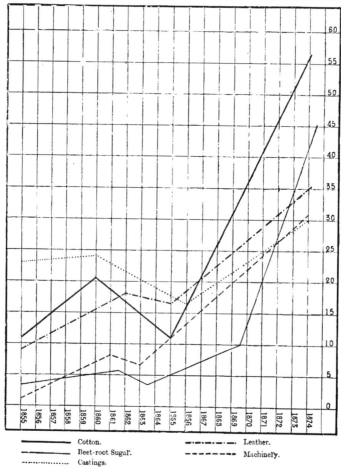

stands, and the adjoining garden plot are private property ; but until the owner has redeemed his share of the land he belongs to the commune, and can sell his house and garden only with its consent. Theoretically the distribution is made according to the number of male adults in each family, and, as the land is also taxed according to the male population, a fresh division is required after every census. But the disturbances created by deaths, births, migration, necessitate a more frequent survey, besides which each village has its own customs and special development. Even in the same district some communes will redistribute the land yearly, while in others the term varies from two years to a whole generation.* As a rule, the richer and more prosperous the commune, the less frequent are the redistributions.

In order to insure equality there is usually a threefold classification of the land according to the quality of its soil, its aspect, inclination, proximity to dwellings, roads, or streams, and even then the parcels are awarded by lot.

MINING AND MANUFACTURES.

MINING operations, formerly so important, and at no epoch completely abandoned, have acquired great development in the present century. A gilded pyramid at the St. Petersburg Exhibition of 1870, representing the quantity of gold extracted from the Urals, mainly since 1816, weighed 1,610,000 lbs., and was valued at £102,000,000. The minerals, including, besides gold and platinum, iron, malachite, and precious stones, are confined chiefly to both slopes of the Urals, while the most copious naphtha springs occur at the two extremities of the Caucasus, and argentiferous lead ores and coal in the Altaï and Transbaikal regions of Central Asia. Of all these treasures the most important are iron, coal, and salt, which exist in almost inexhaustible quantities, though still comparatively little utilised. Russia still imports most of her coal from England, and much of her salt from Galicia.

The chief manufactures do not employ raw materials of home production, and are still unable to compete with those of the West in quantity or quality. Cotton spinning and weaving, although representing one-third of all manufactured goods, occupy the fifth place only after Great Britain, the United States, France, and Germany, and are centred chiefly in the Moscow, Vladimir, Kostroma, and St. Petersburg districts. Wool, the next in importance, has its chief centre also in the Upper Volga and Oka basins, though Poland, Livonia, Grodno, Chernigov, Little Russia, the Don and Middle Volga basins, are also engaged in this industry, which employs over 100,000 hands altogether. Cotton and wool have thus replaced the old national linen industry, which stood first till 1830, and formerly exported largely to the West, and even to America. The raw material (flax and hemp) is now mostly exported, although the local demand for linens is still largely supplied from the looms of the Upper Volga, Kostroma, Yaroslav, and Vladimir.

Leather has always been one of the chief Russian industries, and the birch bark employed by the Russian tanners has the advantage of imparting to their wares a much-prized odour. Yet, although the local supply of hides and skins exceeds that of other European countries, the quantity of the manufactured article is less. Most of the tanneries are small, numbering in 1872 nearly 13,000, and producing over 10,000,000 dressed skins. A greater impulse has been given to

* In the government of Saratov 128 villages redistribute annually, 22 every 2, 13 every 3, 21 every 5, 20 every 6, 2 every 8, 32 every 10 years; in 32 there has been no division since 1858; in 23 none since 1862; and in one division has been theoretically abolished.

the tallow industry, and the preparation of stearine and chemicals derived from animal fats now ranks fourth in importance of all the Russian manufactures. Notwithstanding its great economic value, the flour industry has made little progress, steam being slightly employed, though there are some 80,000 small mills

Fig. 250.—Commercial Movement between St. Petersburg and Astrakhan.

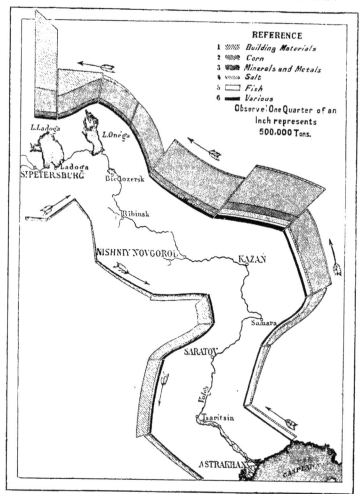

REFERENCE
1. Building Materials
2. Corn
3. Minerals and Metals
4. Salt
5. Fish
6. Various

Observe: One Quarter of an Inch represents 500,000 Tons.

at the rapids, dams, and races along the streams and canals. The distilleries, which supplied the most reliable source of revenue, were also very numerous; but the fiscal regulations, very burdensome to small establishments, have encouraged large speculators, though in 1876 there were still 3,900 distilleries, yielding

and upwards of 154,000,000 gallons of brandy. In
et-root sugar-mills, of which half were in Kiev, the rest
ds," and the produce rose from 122,700 tons in 1868
The production of machinery has also made fair
y as 212 of the 1,150 Russian locomotives had been
same year the total value of the large industries was
)0, or about one-fifth of the agricultural produce.
progress dates from the epoch of the emancipation,
productiveness has increased, on an average, fivefold.
he value of the smaller industries, which may exceed
6,500,000 hands, of whom 6,000,000 reside in the
ter months only, when field operations are suspended.
e thought that this so-called *kustarnaya*, or "planta-
ne nation from the proletariate, but the hope is vain,
to centralization is manifest. Thus in the Shuya
at home were fivefold those working in the factories
proportion was reduced to double, and in 1872 the
equal.

TRADE AND SHIPPING.

varied from £110,000,000 to £142,000,000 between
verland movement Germany naturally takes the lead,
traffic, France ranking third, and Austria-Hungary
their exchanges with the empire. The exports consist
ural produce and raw materials, the imports mainly
nd cotton : between 1872 and 1876 the latter averaged
reely £61,000,000, showing a difference of about
i imports. The total exchanges are about 40s. per
than those of France.
s naturally absorbed by the Baltic, on which are
westernmost outports of the most densely peopled
trade of the Euxine and Sea of Azov is far inferior,
I White Sea combined scarcely equals the movement

fly in the hands of foreign shippers, and many of the
Russian flag really belong to Greeks. Excluding
often wrongly included in that of Russia, this state
European countries for its tonnage, being in this
Spain, Holland, and Sweden. The proportion of
the Russian commercial marine is still only as 1 to
of the gross tonnage.
n since the expansion of the railway system, still
er 53,600 river craft of all kinds were built between
Some on the Volga have a carrying capacity of over

2,000 tons, though the mean burden is only from 150 to 160 tons. Most of
these boats are built for a single voyage, after which they are broken up and
sold as timber or fuel. Two-thirds of the river steamers ply on the Volga,
representing three-fourths of the collective tonnage of all the inland navigation.

CANALS.—RAILWAYS.

Of the 23,240 miles of navigable highways no more than one-sixtieth was
represented by canals in 1870; but some of these canals have a commercial

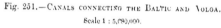

Fig. 251.—CANALS CONNECTING THE BALTIC AND VOLGA.

Scale 1 : 5,080,000.

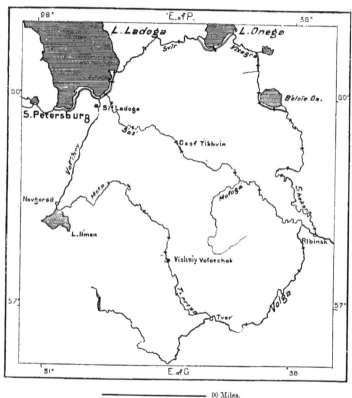

90 Miles.

importance equal, or even superior, to that of main rivers. Their construction has
been greatly facilitated by the marshy or lacustrine nature of the regions about
the upper courses of the main streams. The cuttings required to connect the
Volga and Dwina by their head-streams have a total length of scarcely 9 miles;
while the Tikhvinskiy Canal, the shortest of those connecting St. Petersburg with
the Volga, crosses the crest of the water-parting by a cutting less than 10 miles

all these facilities, the Volga still remains unconnected
iemen, Dnieper, and Don, while the last named is also
. All these canals are, moreover, constructed with
depth of scarcely 6 feet.

the railway system, notwithstanding the prejudices of
er, so far yielded to the pressure of public opinion as
the rigidly straight line between St. Petersburg and
n 1855 the total length of rails has risen from 600 to

g. 252.—RUSSIAN RAILWAY SYSTEM.
Scale 1 : 247,000 000.

500 Miles.

ive of the Caucasus and Finland, and representing a
)0. The system has thus proved far more costly than
nature of the land, where the only great engineering
· the rivers, and the causeways across the marshy
inies, being protected by the Government security for
ents, have been able to dispose of a lavish expenditure,
ital belongs to the State. The Russian network at
, being exceeded only by those of the United States,
France; but taking area and population into account,

it is surpassed even by those of such countries as Portugal and Rumania. The system also labours under the commercial inconvenience of using different gauges, though this presents the anticipated strategic advantage of preventing the German trains from penetrating into the country. On the main lines the five-foot gauge prevails. Railway accidents are, on an average, more frequent in Russia than elsewhere in Europe.

The centre of the system is at Moscow, where the five main lines converge, and where the passenger and goods traffic is about double that of St. Petersburg.* From this centre the lines run westwards to the Central European network,

Fig. 253.—Growth of Attendance in the Russian Gymnasia and Universities from 1808 to 1877.

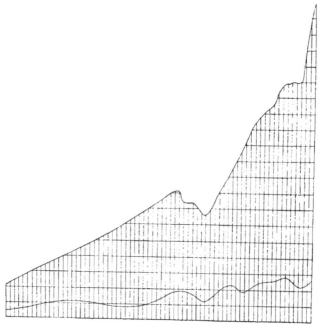

Upper Curve: Students of the Gymnasia. Lower Curve: University Students.

southwards to Odessa, Taganrog, and the Caucasus, eastwards to the Volga at various points, and beyond it to Orenburg. But they do not yet reach the Caspian or Central Asia, though an isolated line connects Perm with Yekaterinenburg, beyond the Urals. The northern lines stop at Vologda, so that the Finnish and Scandinavian systems advance much nearer to the Arctic Ocean than does the Russian. Some of the great trunk lines also pass from 3 to 8 miles from such important places as Tver, Orol, Kursk, to the great inconvenience of passengers.

* Moscow (1873), 1,903,954 passengers; 3,034,000 tons; St. Petersburg, 1,050,213 passengers; 1,287,000 tons. Total passengers (1879), 38,000,000; merchandise, 4,510,000 tons.

EDUCATION.

IN education the same contrast presents itself as in her industrial life. The high schools were already organized, and scientific teaching fully developed in all the Universities, while primary instruction was still almost entirely neglected. As long as 20,000,000 of the people were in a state of thraldom mental culture was necessarily regarded as dangerous. But few primary schools existed, and those of Little Russia and of the Ukrainian communities in Great Russia were closed in the present century. Ever since 1830 the Government was taking steps to prevent even the free children of the " lower orders" from entering the middle schools reserved for the nobles. But the emancipation of the serfs was accompanied by a general movement in favour of popular education. Under private action Sunday schools were opened at first in Kiev, and they had already 20,000 pupils in the empire when they were closed by a Government decree in 1862. Since then private citizens have rarely been able to promote the cause of national instruction. Normal training schools have long been wanting, notwithstanding the petitions of the Zemstvos, on whom the State wished to impose teachers trained in the ecclesiastical seminaries. But since the Franco-German war a few normal schools have been founded under pressure of public opinion.* In her general educational system Russia is thus still behind even such countries as Japan and Egypt.

In 1876 there were only 24,456 primary schools in European Russia, and 1,019,488 children, including 177,900 girls, in attendance—a proportion of little over 1 to 80 of the population. Yet rudimentary instruction is obligatory for all soldiers in the regimental schools, so that the War Department may be said to do more for education than the Minister of Public Instruction.

The secondary establishments—gymnasia, " real " schools, military gymnasia, ecclesiastical seminaries, and boarding schools—were attended in 1877 by 88,400 students, besides 41,630 following the courses of the special ministerial institutes. Since the end of the last century the academies for girls of noble birth have given courses analogous to those of the gymnasia, and in 1876 there were 320 middle schools for girls of all classes, with 55,620 pupils. Since 1861 women began also to attend the University and medical courses, but in 1863 they were excluded from the former, and for some years past they have frequented foreign Universities, especially those of Zurich, Berne, and Geneva.

The eight Russian Universities, modelled on those of Germany, have a comparatively much smaller attendance.† Entrance is rendered very difficult by excessively hard examinations, while many of the students are periodically expelled in consequence of their tendency to become imbued with the new ideas. But such is the love of learning that, in spite of all coercive measures, candidates continue to present themselves in ever-increasing numbers. This real devotion to

* Normal schools in 1877 :—For towns, 7 ; for villages, 61 ; pupils, 4,596, of whom 727 were girls.

† Russian Universities (1878) :—Students, 6,250 ; history, philology, letters, 721 ; physics and mathematics, 1,365 ; law, 1,641 ; medicine, 2,865 ; free auditors, 449 ; professors, 635 ; yearly budget, 2,529,470 roubles

science is also shown by the fact that serious historical, ethnological, and other scientific works are comparatively more read than light literature. The number of newspapers is still inferior to that of the Western states, and none of them enjoy the enormous circulation of certain popular papers in England, France, and America; but the solid reviews containing original articles of real value are eagerly read by thousands of subscribers. And while making comparatively little use of her postal and telegraphic systems, Russia actually publishes more works than either Great Britain or Austria-Hungary.* Except in Roman Catholic Warsaw few of these works are devoted to theology.

GOVERNMENT AND ADMINISTRATION.

RUSSIA is the only European state in which the sovereign is absolute master. In the legal formula the " Emperor of all the Russias" is an autocratic ruler, to

Fig. 254.—HOUSE OF THE ROMANOVS NEAR KOSTROMA.

whom, according to the divine ordinance itself, all owe fealty, " not through fear alone, but also for conscience' sake." This theory of absolute autocratic power has been gradually developed. As in France the royal authority ended by prevailing over that of the vassal states and the great nobles, so in Muscovy autocracy replaced the old institutions of the communes and " free orders," and assumed a

	Telegrams.		Per 1,000 Inhabitants.	Letters.		Per Head.
• United Kingdom (1876) .	21,820,023	or	638	1,116,688,000	or	33·4
Germany	,,	13,456,728	,,	286	594,994,000	,, 16·6
France	,,	11,412,161	,,	275	366,506,000	, 10·2
Russia	.,	4,275,000	,,	53	86,612,000	, 1·

Publications:—France (1875), 14,195; Germany (1878), 13,912; Russia (1877), 7,500; England (1877), 5,095; Austria (1876), 1,902.

.eice of the Mongolians and of the ecclesiastical and
yzantium. The Czarism of the Romanovs received
eaucratic institutions introduced from Germany, and
legitimacy elevated by Joseph de Maistre into State
c power of the Emperor is, in theory at least, to some
from the French Revolution, consolidated by the
.ionalities in the empire, especially since the annexa-
ice the time of Peter the Great the autocracy of the
. absolute principle.   Peter drew up reports for the
had been created by himself, while in 1730 the
arter limiting the autocracy by a council of the chief
1 this charter was afterwards torn up, yet Alexander I.
ls, to found a " Committee of Public Safety " for the
is of " bridling the despotism of his Government."
ve, judicial power flow all from the Czar, who is
laws guaranteeing the predominance of the National
cession to the throne. But based as they are on the
ise laws themselves be modified by the same right ?
only at the pleasure of the Czar.
t to avail himself of the council of the boyards (duma
Geieral, created in 1711 an assembly which has
ose functions have been frequently changed. This is
nate " (pravitelst-royushchiy senat), which he is said
:o the Dutch States General.   To it " everything was
to obey it as the Czar himself; " and its decrees were
ated from him.   But the appointment of its members
four years afterwards, named a " Reviser-General "
ed every month an officer of the guard to watch over
them to prison if they failed in their duty. He
breats of forfeiture, disgrace, and death, in case of
institution is merely a court for registration and
lecrees, and in judicial matters a Supreme Court of
eper of the Laws," " Controller of all the Adminis-
nder of every Russian citizen's legal rights," are
nly in trivial matters.
led, by virtue of their office, to share in the legisla-
sters.   By their reports, issued with the imperial
y their explanatory circular notes, they enjoy very
leven ministerial departments—Imperial Household,
Home Office, Public Instruction, Finance, Justice,
General Control.   Each minister depends directly on
erial functions are, moreover, confided to the head of
erial Chancery—that is, the secret police ; to the
rho, in the name of the reigning Empress, administers

the charitable institutions and various educational establishments, especially girls'
schools; to the head of the Government stud; lastly, to the Procurator General
of the Holy Synod, or Minister of Public Worship.

The absolute system also confers undoubted authority on its direct representatives, ministers, provincial governors, *isprarnik*, or heads of districts, cantonal
provosts (*stanoroï pristar*), agents of local safety (*uriadnik*). All these functionaries

Fig. 255.—Proportion of Exiles in Siberia according to Provinces before the Judicial Reforms.

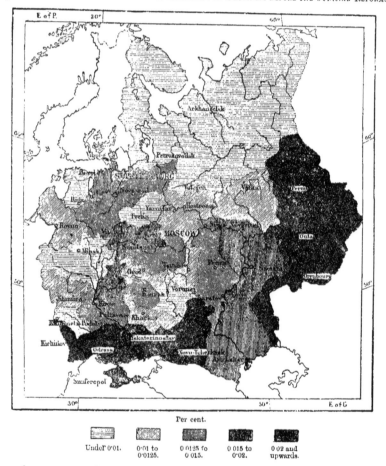

Per cent.

Under 0·01. 0·01 to   0 0125 to   0 015 to   0 02 and
           0·0125.   0 015.   0·02.   upwards.

have far more extensive powers than the corresponding officials in other European
states. Before the emancipation of the serfs the governor was a real "master of
his province," possessing the right of intervening in all matters, and personally
controlling the tribunals and finances of his territory. The Zemstvo, judicial
reforms, and the creation of Chambers of Control for a time curtailed his authority,
but the right of veto enjoyed by him in all appointments to the Zemstvo and

municipalities has restored his power. Even since the judicial reforms of 1864 he has not been deprived of an exorbitant privilege analogous to that of the *lettres de cachet* under the old French régime.

Except in Poland and in the Baltic Provinces, already possessing their feudal Diets, an instalment of representative government was everywhere introduced in 1864 by the creation of the *Zemstvo*, or provincial Parliaments, analogous to the German Landtag. In these Zemstvo (from *Zemlia land*) the deputies belong to all classes—nobles, citizens, traders, peasants—and hold their sittings in three distinct chambers, or curiæ. The first, or assembly of landed proprietors, consists of owners of estates averaging from 550 to 1,780 acres according to the provinces, of delegates of proprietors with at least 55 acres, and of representatives of the clergy as holders of ecclesiastical lands. The second, or assembly of burgesses, comprises merchants, traders, and manufacturers doing a trade of at least 6,000 roubles, house owners, and representatives of industrial associations disposing of a fixed capital. Lastly, the third curia, that of the rural communes, includes the delegates of the peasantry, elected at second hand—that is, by the members of the bailiwicks, who are themselves named by the peasantry at the rate of one for every group of ten families. The presidency of the first curia belongs to the marshal of the nobles, of the second to the mayors of the towns, of the third to the officials employed in the administration of rural affairs, the president in chief of the Zemstvo being always the marshal of the nobles, except where specially appointed by the Czar. The first two bodies are chosen from their respective classes, but the peasantry may elect nobles or priests to represent them, and the number of deputies is everywhere calculated in such a way as to leave the rural classes in a minority.

The sessions are very short, the district Zemstvo sitting only for ten days in the year, while the delegates chosen by it for the provincial Zemstvo meet annually in the chief town for twenty days only. But every three years the assemblies name an *uprava*, or administrative committee, whose president must be approved by the governor or minister, who has also the power of suspending all decisions of the Zemstvo which he may consider contrary to the laws or the good of the state. The range of subjects of which these bodies can take cognisance is, moreover, very limited, and becomes yearly more restricted by ministerial injunctions.

The municipal institutions have passed through an evolution analogous to that of the Zemstvo. Under the old régime the interests of the boroughs were looked after by a *duma*, or council, chosen by the traders and burgesses (*mieshchane*), and in certain grave contingencies the general electoral body decided. Now the Government has attempted to fuse all classes in the urban, as it has in the provincial administration. St. Petersburg, Moscow, Odessa, first received a new municipal organization resembling that of the Zemstvo, and in 1870 the municipalities were reconstituted in nearly all the towns of the empire. The urban electors are divided into three curiæ, according to the amount of their taxes, each group appointing for four years an equal number of deputies (*glasniye*, from *glas*, "vote"). These, constituting the duma, name in their turn, also for four years, the *uprava*, or administrative committee, and the mayor (*golova*, or "head"). The latter, being intrusted with the executive, has powers superior to those of the body naming him.

The new judicial organization is considered as the most liberal of all the reforms introduced since the emancipation, and in certain respects this is the case. Hitherto the courts, and especially the juries, have given proof of that clemency which is so characteristic of the nation. During the trials the accused are not bound to incriminate themselves, and the procurator, or president, generally sums up with great impartiality. Nevertheless the direct interference of the State is already making itself felt, and the procedure followed in political cases has been frequently modified since the first of these charges was heard in 1871.

No definite law regulates the press, which is still mostly subject to the provisional enactment of 1865. Preventive censure has been abolished in the two capitals in favour of certain original works, translations, journals, and compilations, authorised to appear after depositing from 3,000 to 5,000 roubles as a security for their good behaviour.

THE CHURCH.—ARMY AND NAVY.—FINANCE.

THE Czar is not, as is generally supposed, the spiritual head, but only the protector, of the Græco-Russian Church. Of all the Czars Paul I. alone attempted to assume the priestly functions, and even to say mass. But he was prevented from doing so by the remark that he had been twice married. According to the Russian catechism the only head of the Church is Jesus Christ. Nor can the legislative and executive departments be united in one person, the first being expressly reserved to the councils, the second to the national synods and the bishops. Since the time of Peter the Great the government of the Church, formerly intrusted to a patriarch, has been confided to a "Most Holy Synod," appointed, however, by the Czar. This assembly, presided over by the St. Petersburg and Novgorod Metropolitan, is composed of a few prelates, succeeding each other by rotation ; but a lay procurator, sometimes a general named by the Czar, is the medium of the sovereign pleasure, proposes and promotes all questions, sees to the execution of all decisions. No synodal act is valid without his confirmation, and he further possesses the right of veto against all decrees of the assembly opposed to the will of the prince. Under such an administration the Russian Church has been thoroughly centralized. The bishops have become simple ecclesiastical prefects, proposed by the Synod—that is, the imperial procurator—and named by the Emperor from a list of three candidates, the first of whom is nearly always chosen. Each bishop is, moreover, assisted by a diocesan consistory, whose members are appointed by the Synod. To facilitate the action of the central power the Church has been divided into eparchies, or bishoprics, whose limits nearly always coincide with those of the civil governments. Of the sixty eparchies those of St. Petersburg-Novgorod, Kiev, and Moscow alone bear the title of metropolis, nineteen that of archbishoprics. The "popes," or priests, are allowed to marry but once only, and widowers mostly retire from the "white," or secular, to join the "black," or regular clergy, and from amongst this class the Czar now usually chooses the higher Church dignitaries. The ecclesiastical body numbers at present about 254,000, of whom 70,000 officiate in the 625 cathedrals, the 39,400 churches, and the 13,600 chapels of the empire. Of monasteries there are 480, and of nunneries 70 only—a proportion characteristic of the Greek as compared with the Latin Church.

The Government has imposed on other religions the administrative and bureaucratic forms of the dominant creed. Thus the Latin Catholics are directed by a college independent of the Vatican, seated in St. Petersburg, under the presidency of the Primate and Archbishop of Mogilov-Gubernskiy, and under the control of a lay representative of the imperial authority. So also the Lutherans, Calvinists, and

Fig. 256.—THE MEJIBOJ LINES.

Scale 1 : 250,000.

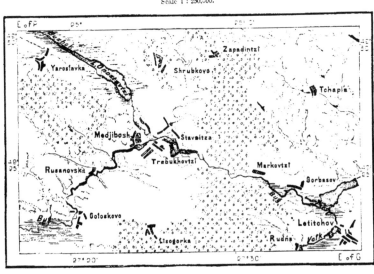

6 Miles.

Armenians are controlled by an outward organization more or less analogous to that of the State Church, and superior to the individual hierarchy of each. The Mussulman communities of the Tatars and Kirghiz as well as the Buddhist Kalmuks have also been obliged to conform their worship to the hierarchical forms imposed on them by St. Petersburg. Each of these confessions has its central authority; each of them has been subjected to the pleasure of procurators or secretaries representing the imperial power; each of them has its consistories intrusted with functions analogous to the Orthodox consistories. All these religions are tolerated, but are not allowed to proselytize, while the Raskolniks, regarded as apostates from the National Church, have not even the right of public worship.

The Russian army has been completely reorganized since the Franco-German

war.   Before 1874 it was recruited by conscription amongst the peasantry and
artisans of the towns, besides volunteers and the sons of soldiers.   But according
to the new law all able-bodied men twenty-one years old are bound to military
service, without the option of substitution.   The use of the rod is abolished

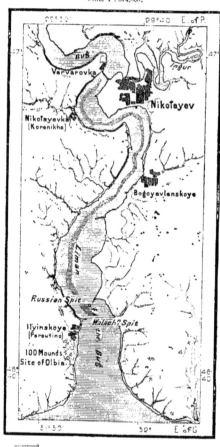

Fig. 25.—NIKOLAYEV AND THE BUG LIMAN.
Scale 1 : 574,000.

0 to 32 Sheep-folds. Mounds. Farmsteads.   32 Feet and
  Feet.   upwards.

except in the disciplinary corps.
The normal period of service is
fifteen years, of which six are in
the active army and nine in the
reserve.   But the population being
too great to need all available
recruits, about two-thirds of them
are at once dismissed to the militia,
while the rest, chosen by lot, are
retained in the active service only
till the cadres are completed, when
they receive their congé for an
indefinite period.   In the reserve
they are bound to serve only during
war-time, but must present themselves yearly at the nearest camp
for a few days' exercise.   Few
regiments except those of the
Guards are complete, although of
the 650,000 yearly recruits about
200,000 are retained for active
service.   The Don, Ural, Terek,
Kuban, and Siberian Cossacks
have preserved their special military organization, each of their
"armies" furnishing a certain
number of fully equipped regiments always ready to take the
field at ten days' notice.   In 1877
the European Cossacks in the
active service numbered 40,000,
which on the first call might be
raised to 100,000.

In the same year the whole
army numbered over 1,000,000, but it varies from 710,000 to 1,200,000 according
to circumstances.   Till recently the mortality was excessive, but since 1872 has
been much reduced by improved sanitary arrangements, and a more vigilant
control over the army contractors.   In time of peace it now amounts only to
1 per cent., and even in war the proportion has been considerably lessened by
more efficient ambulance and hospital service.

Some of the Russian fortresses are amongst the strongest in Europe. Kronstadt, Sveâborg, and Viborg, in the Gulf of Finland, seem to have been rendered impregnable from the sea, while Modlin, or Novo-Georgievsk, and Warsaw in the Polish quadrilateral, may serve as entrenched camps for large armies. In the Black Sea, Sebastopol, rising stronger than ever from its ruins, and Kertch no less formidable, serve as the advanced bulwarks of Nikolayev, the "Russian Portsmouth."

Russia lacks the advantage enjoyed by England of being able to unite her fleets. The two squadrons maintained in the Baltic and Euxine have no access to the open sea except through narrow straits held by foreign powers; still both combined would form a first-class navy, both for the number of ships and the strength of their armaments. About two-thirds of the most formidable are concentrated in the Baltic, where they cover St. Petersburg and Kronstadt, while menacing the coasts of Sweden, Denmark, and Germany. The European fleet consists altogether of 28 ironclads and 304 steamers, including two so-called *popocka* in the Euxine, which sail with circular motion, but which proved of little service during the last war. The naval forces, numbering about 29,000, are recruited, like the army, by conscription, but the length of service, formerly twenty-two years, has been reduced to nine, of which seven are in the active and two in the reserve.

These formidable armaments, together with the land forces and the interest on the national debt, absorb nearly two-thirds of the imperial revenue, although this has been considerably increased since the middle of the century. The first regular Budgets date only from 1866, but notwithstanding fictitious balances, the expenditure seems to have exceeded the income every year since 1832, except in 1871 and 1875. The chief sources of income are the indirect taxes, and of these the most willingly paid is that on alcoholic drinks, the national vice guaranteeing over one-third of the State expenditure. The taverns are daily crowded with copious consumers of coarse brandies mixed with extract of belladonna and other baneful ingredients, and since the financial reforms of 1865 yielding an annual addition to the revenue of about £1,280,000. In 1876 the Budget received from this source about £30,000,000, while the salt and tobacco duties figured for £1,432,000 and £1,683,000 respectively.

The spirit excise is about sufficient to meet the army expenditure in time of peace. In 1876 the Minister of War demanded over £30,000,000, and in the same year the cost of the navy was over £4,300,000. In ordinary times the army and navy together involve an outlay of about £32,000,000, but this sum has been doubled and even trebled during the great wars. The last campaign in the Balkans cost, down to November 13th, 1878, £142,000,000. In 1876 the interest of the national debt exceeded £17,000,000, a burden which has since been considerably increased. The expenses of the imperial family are relatively greater than in any other sovereign European state, except in Turkey and Montenegro. But it is difficult to ascertain the actual sum, as the Czar has no civil list in the ordinary sense. However, the total income drawn from the imperial domain—forests, mines, and arable lands—is known to exceed £2,000,000, and the Minister of the Imperial Household expends from £1,850,000 to £2,000,000. As in all absolute monarchies, the Budget is also burdened with heavy charges for pensions and grants to the bureaucracy.

The Russian Budget slightly exceeds that of Great Britain, but, including the special budgets, it is inferior to those of France and Germany. The debt, however, is still smaller than those of the three great Western powers, and the vast natural resources of the country have been little utilised. On the other hand, the future is too overcast, and the stability of the present order too doubtful, to restore the credit of Russia in the European money market. Hence its foreign loans are still contracted under great disadvantages, and the national debt amounted in 1879 to £606,453,000, including the paper money in forced circulation, and now greatly below par.* Though still bearing the name of "silver rouble," this paper, which in 1879 amounted to £120,000,000, is really worth little more than one-half of its nominal value.

NOTE.—The social state of the empire as above depicted receives a startling illustration from the tragic fate of Czar Alexander II., the news of whose assassination on Sunday, March 12th, 1881, was received after these sheets had passed through the press. The significance of this event seems to lie in the fact that the conspirators found it possible to make such arrangements for carrying out their fell purpose in the very heart of the capital as must have, on this occasion, precluded the possibility of failure. It is stated that even had the second bomb proved ineffective, the road to the Winter Palace was lined with so many other determined "Nihilists," all armed with the same terrific engine of destruction, that escape must have in any case been impossible. What has been done once may, of course, be done again ; and should the attempt be repeated, as is already threatened, the whole fabric of society as at present constituted may be shaken to its very foundations.

The late Emperor was born on April 19th, 1818, married the Princess Marie of Hesse-Darmstadt in 1841, and succeeded his father, Nicholas I., in March, 1855, towards the close of the Crimean war. His reign must be ever memorable for the emancipation of the serfs, one of the most momentous events in the history of humanity. The credit of this wise and bold act is due entirely to the Czar, who, in the face of much opposition, not only gave freedom to 40,000,000 serfs, but made them communal proprietors, and released them from the penal jurisdiction of the great landowners. Nevertheless the later years of his reign were passed amidst wars, threatened revolutions, frequent attempts on his life, and domestic troubles, the latter due especially to his unhappy relations with the Princess Dolgoruki. He is succeeded by his son Alexander III., who was born on March 10th, 1845, and married November, 1866, to Maria Dagmar, daughter of the King of Denmark, and sister of the Princess of Wales and of the Queen of the Hellenes. His eldest son and heir apparent is Nicholas, born May 18th, 1868.

* This is the value in paper roubles: in silver the debt only amounts to £306,000,000. Included in the latter amount are £116,000,000 advanced to railway and manufacturing companies.

APPENDIX.

｜TISTICAL TABLES.

DENMARK

—ARFA AND POPULATION.

Area. Miles.	Population. 1889.	Towns.
8	， 235.254	Copenhagen, 235,254.
468	121,487	Roskilde, 5.895 ; Kjöge, 3.122 : Frederiksberg (Suburb of Copenhagen), 26.510.
522	83,356	Helsingör (Elsinore), 8,978 ; Hilleröd, 3,059.
627	93.340	Helbek, 3.265.
569	87,477	Ringsted, 2,127 ; Sorö, 1,464 ; Slagelse, 6,076 ; Corsör, 3,959.
644	101.169	Præstöe, 1.460 ; Nestved, 4,792.
225	35.365	Rönне, 6,472.
640	97,008	Nakshov, 5,278 ; Maribo, 2,403.
681	128,947	Odense, 20,804 : Assens, 3,196.
633	117,559	Svendborg, 7,185 : Nyborg, 5,402 ; Rudkjöbing (Langeland), 3,197.
1,070	100,548	Hjörring, 4,308 ; Frederikshavn. 2,891.
650	63,991	Thisted, 4,182 ; Nykjöbing, 4,560.
1.132	96,205	Aalborg, 24,152.
1,170	93,371	Viborg, 7,653,
938	104,313	Randers, 13.457.
956	140,888	Aarhus, 24.831 ; Horsens, 12,652.
901	108,513	Vejle, 7.145 ; Fredericia, 8,275 ; Kolding, 7,141.
1,747	87,408	Ringkjöbing, 2,035.
1,174	73,255	Ribe, 3.932.
514	11,221	Thorshavn, 900.
4,769	1,980,675	
3,400	72,000	Reykjavik, 2,000.
4,000	9,800	Sukkertoppen, 359 ; Julianehaab, 223.
139	37,600	St. Thomas.
9,308	2,000,075	

II.—COMMERCE.

MERCANTILE MARINE (1879).—Sailing vessels, 3,096 of 210,768 tons; steamers, 190 of 46,651.

SHIPPING (1878).—Entered, 41,652 vessels of 1,356,943 tons; cleared, 40,718 vessels of 667,335 tons. Total tonnage, including coasting trade, 2,024,178.

Railways open (1878), 849 miles. Telegraph lines, 2,097 miles, with 5,602 miles wire. Telegraphic dispatches (1878), 939,322. Letters forwarded through the lines (1878), 25,483,599. Journals, packages, &c. (1878), 24,879,891.

Total imports from—

				1874 (£).	1877 (£).	1878 (£).
Germany	.	.	.	4,621,611	4,716,000	3,516,020
Great Britain	.	.	.	3,226,222	2,975,500	2,329,700
Sweden	.	.	.	1,458,888	1,355,277	1,138,600
Norway	.	.	.	541,722	302,166	291,310
Russia	.	.	.	549,000	637,166	410,400
All other countries	.	.	.	2,557,557	2,534,891	2,377,430
Total	.	.	.	12,955,000	12,521,000	10,663,460

Total exports to—

Germany	.	.	.	3,303,200	2,960,000	2,598,060
Great Britain	.	.	.	3,949,222	3,538,333	3,535,340
Sweden	.	.	.	1,349,000	1,452,777	1,275,140
Norway	.	.	.	821,944	664,500	592,700
Russia	.	.	.	48,440	51,000	44,050
All other countries	.	.	.	511,174	460,390	935,140
Total	.	.	.	9,983,000	9,127,000	8,980,430

SWEDEN.

I.—AREA AND POPULATION.

Provinces (Län).	Area. Sq. Miles.	Population. 1878.	Principal Towns, with Population.
Stockholm, Town	13	169,429	Stockholm, 169,429.
„ Län	2,995	143,763	Södertelge, 3,156.
Upsala	2,063	108,844	Upsala, 14,069; Enköping, 2,342.
Södermannland	2,631	144,821	Nyköping, 4,896; Eskilstuna, 7,467; Malmköping.
Öster Gotland	4,232	270,328	Linköping, 4,520; Norrköping, 27,410; Wadstena, 2,440; Motala.
Jönköping	4,464	195,323	Jönköping, 15,037; Esksjö, 2,860.
Kroneberg	3,841	169,890	Vexjö (Wexiö), 4,615.
Kalmar	4,437	243,600	Kalmar, 10,469; Vestervik, 6,007; Oskarshamn, 6,348.
Gotland	1,203	55,011	Wisby, 6,994.
Blekinge	1,164	135,639	Karlskrona, 18,276; Carlshamn, 6,330.
Kristianstad	2,507	232,116	Kristianstad, 9,125; Engelholm, 2,056.
Malmöhus	1,847	345,927	Malmö, 35,626; Lund, 13,611; Landskrona, 9,560; Helsingborg, 10,986; Ystad, 6,903.
Halland	1,899	135,411	Halmstad, 8,000; Warberg, 2,863.
Göteborg and Bohus	1,952	257,466	Göteborg, 74,418; Uddevalla, 6,718; Strömstad, 2,205.
Elfsborg	4,947	290,763	Wenersborg, 5,571; Alingsås, 2,343; Borås, 4,389; Amål, 2,125; Trollhättan.
Skaraborg	3,306	258,901	Mariestad, 2,646; Lidköping, 4,520; Skara, 2,991; Sköfde, 2,954; Falköping, 2,494; Carlsborg.
Wermland	7,347	269,583	Karlstad, 7,218; Kristinehamn, 4,826; Filipstad, 2,928.
Örebro	3,520	181,473	Örebro, 11,008.
Westmanland	2,623	127,586	Westerås, 5,906; Köping, 2,860; Arboga, 3,631.
Kopparberg	11,420	190,299	Falun, 6,887.
Gefleborg	7,419	172,577	Gefle, 18,526; Söderhamn, 7,333; Hudiksvall, 4,331.
Wester Norrland	9,530	162,514	Hernösand, 5,010; Sundsvall, 8,476.
Jemtland	19,594	79,764	Östersund, 2,437.
Westerbotten	21,943	103,151	Umeå, 2,720.
Norrbotten	40,563	87,681	Luleå, 2,960; Haparanda, 1,011; Piteå, 2,215.
Large Lakes	3,516	—	
Total	170,976	4,531,863	

The total area includes 13,940 square miles of lakes.

Nationalities (1870).—6,711 Lapps, 14,932 Finns, 70 Gipsies, 12,015 natives of foreign countries.

Creeds (1870).—All Lutherans, with the exception of 3,999 Protestant Dissenters, 1,836 Jews, 573 Roman Catholics, and 30 Greek Catholics.

II.—COMMERCE.

ng vessels (458,789 tons) and 761 steamers (86,103 tons).
's in operation, 370 miles building.

	1875 (£).	1877 (£).	1878 (£).
. .	5,104,000	4,848,944	3,563,670
. .	3,067,000	3,742,000	3,176,910
. .	2,715,000	2,746,000	2,336,600
. .	1,012,000	2,143,000	1,477,840
. .	835,000	828,000	646,000
. .	2,160,000	2,748,056	2,211,680
. .	14,893,000	17,056,000	13,412,700
. .	6,069,000	6,478,000	5,212,800
. .	719,000	758,000	647,470
. .	1,370,000	1,236,000	1,174,710
. .	313,000	217,000	269,700
. .	394,000	371,000	397,320
. .	2,610,000	2,961,000	2,619,920
. .	11,475,000	12,049,000	10,321,920

NORWAY.

—AREA AND POPULATION.

s.	Population. 1876.	Principal Towns, with Population.
1	107,804	Frederikshald, 9,792; Frederiksstad, 9,616; Moss, 4,509.
1	116,335	Fidsvold, 300; Drøbak, 2,047.
	76,054	Cristiania, 76,054 (with suburbs, 106,781).
4	120,618	Congsvinger, 941; Hamar, 2,188.
9	115,814	Lillehammer, 1,551; Gjövik, 1,051.
5	102,186	Drammen, 18,643; Congsberg, 4,357.
2	87,506	Laurvik, 7,737; Holmestrand, 2,147; Tönsberg, 4,913; Sandefiord, 2,307.
7	83,171	Skien, 5,362; Porsgrund, 3,457; Brevik, 2,252; Kragerö, 4,669.
1	73,415	Arendal, 4,132; Osterrisör, 2,390; Grimstad, 1,657.
1	75,121	Christiansund, 11,766; Flekkefiord, 1,651; Mandal, 3,885.
7	110,965	Stavanger, 19,004.
0	119,303	No town.
	33,830	Bergen, 33,830.
2	86,208	Florö, 503.
0	117,220	Christianssund, 8,251; Aalesund, 5,603; Molde, 1,672.
1	116,804	Troidhjem, 22,152.
2	82,271	Levanger, 955; Namsos, 1,529; Steikjaer, 1,427.
7	104,151	Bodb, 1,519.
4	54,019	Tromsö, 5,409.
)	24,075	Hammerfest, 2,102; Vadsö, 1,764; Vardö, 1,322.
3	1,806,900	

II.—COMMERCE.

1875 (£).	1876 (£).	1877 (£).	1879 (£).
0 9,836,000	9,294,000	10,544,000	7,404,660
0 5,596,000	6,468,000	6,100,000	4,996,430

34 vessels of 1,493,041 tons, inclusive of 273 steamers of 46,869 tons.
48   „   1,526,689   „   308   „   51,674 „

THE RUSSIAN EMPIRE.

1.—AREA AND POPULATION.

Governments.	Area. Sq. Miles.	Population.	Principal Towns, with Population.
BALTIC PROVINCES			
St. Petersburg . . .	20,760	1,326,875	St. Petersburg. 667,963 ; Kronstadt, 47,166 ; Tzarkoïe-Selo, 14,465.
Esthonia (Esthland) .	7,818	323,961	Revel, 31,269.
Livonia (Livland) . .	18,158	1,000,876	Riga, 103.000 ; Dorpat. 20,540 : Pernau, 9,568.
Kurland	10,535	619,154	Mitau, 22,185 ; Libau, 10,767.
GREAT RUSSIA.			
Moscow	12.858	1,913,699	Moscow, 601.969 ; Sergiyevskiy. 27.741 ; Kolomna, 18.808 ; Serpukhov. 16,720 ; Podolsk. 10,973.
Tver	25.224	1,528,881	Tver. 38.248 ; Rjov, 18.732 ; Vishni-Volochok, 17,408 ; Torjok, 12,910 ; Ostashkov, 10.806.
Yaroslav	13,750	1,001,718	Yaroslav, 26,429 ; Ribinsk, 15,047 ; Uglich. 13.069.
Kostroma	32,702	1,176,097	Kostroma, 27,178.
Vladimir	18,864	1,259,923	Vladimir, 16,422 ; Ivanovo, 17,000 ; Murom, 10,700 ; Shuya, 16,440.
Nijni-Novgorod . . .	19,797	1,271,564	Nijni-Novgorod (Nishegorod), 44,190 ; Arzamas, 10,406.
Razan	16,254	1,477,433	Razan, 19,990 ; Kasinov. 14,100.
Tula	11,956	1,253,037	Tula, 57,374.
Kaluga	11,940	936,252	Kahlga, 38,608 ; Jizdra, 11,703 ; Borovsk. 9,491.
Tambov	25,684	2,150,971	Tambov, 26,403 ; Kozlov, 25,522 ; Morshansk, 19,500 ; Lipetzk, 14,213 ; Borisoglebsk, 12,610.
Voronej	25,439	2,153,696	Voronej, 42,112 ; Ostrogohsk, 9,900.
Kursk	17,937	1,951,807	Kursk. 31,754 ; Belgorod, 16,097 ; Miropolye, 10,754 ; Rilsk, 9,445.
Orol	18,041	1,596,881	Orol. 44,280 ; Yeletz, 30 540 ; Bolkhov, 19 224 ; Bransk, 14,657 ; Mtzensk, 14,159 ; Livni, 12,178 ; Karachev, 11,267.
Smolensk	21.638	1,140,015	Smolensk, 24,352 ; Vazma, 11,827 ; Dorogobuj. 9,100.
Pskov	17.069	755,701	Pskov, 18.331.
Novgorod	47.235	1,011,445	Novgorod, 17,093 ; Staraya Rusa, 14,756.
Olonetz	57.638	295,392	Petrozavodsk, 10,901 ; Olonetz. 1.341.
Archangel	331,500	281,112	Archangel (Arkhangelsk), 18,268.
Vologda	156,496	1,003,039	Vologda, 17,223.
EAST RUSSIA.			
Perm	128,250	2,198,666	Perm. 22,288 ; Yekaterinburg, 25,138 ; Kungur, 10,894.
Viatka	9,116	2,406.024	Viatka, 21.240 ; Ishovsk, 21,500 ; Sarapul, 7,688.
Kazan	54,601	1,789,980	Kazan, 86,262 ; Chistopol, 13,030.
Penza	24,996	2,198,666	Penza, 34,334 ; Saransk, 9,369.
Simbirsk	19,110	1,205,881	Simbirsk. 26,822 ; Sizran, 19.443.
Samara	10,200	1,837,081	Samara, 51,217 ; Pokrovskaya, 12,950 ; Buzuluk, 14,878.
Ufa	47,033	1,364,925	Ufa, 20,917 ; Zlatoust, 16,629.
Orenburg	73,887	900,547	Orenburg, 35 623.
Saratov	32,623	1,751,268	Saratov, 85,220 ; Volsk. 31,269 ; Kamishin, 15,700 ; Khvalinsk, 15.630 ; Atkarsk, 15,200 ; Kuznetzk, 14,185 ; Dubovka, 12,737 ; Serdobsk, 12,200 ; Tzaritzin. 11.800 ; Petrovsk, 10,770.
Astrakhan	86,670	601,514	Astrakhan, 48,220.
LITTLE RUSSIA.			
Kharkov	21,040	1,698,015	Kharkov, 101,175 ; Akhtirka, 17,820 ; Sumi, 14.126 ; Izum, 12,962 ; Starobelsk, 12.960 ; Bielopolye, 12,256 ; Lebeda 1. 11,897 ; Slavansk. 11,690.
Poltava	19,265	2,102,614	Poltava, 33.979 ; Kremenchug. 30,472 ; Kobelaki, 12,989 ; Priluki, 12,878 ; Zenkov, 10 589.
Chernigor	20,233	1,659,600	Chernigov, 16,174 ; Nejin, 21,590 ; Glukhov, 13,398 ; Starodub, 12,333.
Kiev (Kiyev)	19,688	2,175,132	Kiev, 127,251 ; Berdichev, 52,563 ; Belaya-Zerkov, 18,700 ; Vasilkov, 16,597 ; Uman, 15,393 ; Cherkasi, 13.914 ; Tarashtcha, 11,420 ; Zvenigorodka,11,375 ; Skvira, 10,061.

Population of the Russian Empire (continued).

les.	Population.	Principal Towns, with Population.
13	1,086,264	Novo-Cherkask, 33,397.
18	1,352,300	Yekaterinoslav, 24,267; Taganrog, 48,186; Rostov, 44,453; Bakhmut, 17,399; Azov, 16,791; Nakhichevan, 16,258; Pavlograd, 11,891; Novo Moskovsk, 10,575; Lugan, 10,050.
39	704,997	Kertch, 22,459; Simferopol, 17,130; Sebastopol, 13,260; Berdansk, 12,223; Karasubazar, 11,669; Bakhchisarai, 10,528.
23	1,596,809	Odessa, 184,819; Nikolayev, 82,805; Kherson, 46,320; Yelisavetgrad, 35,179; Tiraspol, 16,692; Ananyev, 15,383; Alexandriya, 10,5,1; Novo-Georgievsk (Krilof), 10,225.
26	1,205,932	Kichinov, 102,427; Akkerman, 39,201; Bender, 24,625; Ismail with Tuchkov, 21,000; Khotin, 18,148; Bolgrad, 15,000.
23	1,933,188	Kamenetz-Podolskiy, 22,611; Balta, 18,842; Vinnitza, 18,780; Moghilov-Podolskiy, 18,129; Proskurov, 11,751.
38	1,719,890	Jitomir, 43,047; Staro-Constantinov, 15,805; Lutzk, 11,838; Kremenetz, 11,819; Novgorod-Volinsk, 9,340.
74	1,182,230	Milsk, 35,563; Bobruisk, 26,872; Pinsk, 17,718; Slutsk, 9,922.
65	1,008,521	Grodno, 31,060; Brest-Litovskiy, 22,132; Belostok, 17,658; Stonim, 11,596.
92	1,156,041	Kovno, 33,050; Shavli, 13,343; Vilkomir, 11,118; Rossieni, 10,700.
12	1,001,909	Vilna, 64,217; Disna, 6,111.
51	947,625	Moghilov, 40,431; Gomel, 13,030.
39	888,727	Vitebsk, 31,182; Dunaburg, 29,613; Polotzk, 11,928.
90	65,864,910	
26	1,968,626	Helsingfors, 33,602; Åbo, 21,794; Wiborg, 12,009; Tammerfors, 12,000; Uleaborg, 8,679; Björnaborg, 7,350; Kuopio, 6,050; Nikolaistad (Vasa), 5,060.
57	6,528,017	Warsaw, 308,548; Lodz, 39,078; Lublin, 26,708; Plock, 19,189; Kalisz, 16,957; Piotrkow, 16,949; Suwalki, 15,585; Czestochowa, 14,830; Lomza, 13,335; Wloclawek, 12,445; Zgierz, 11,468; Radom, 11,339; Augustowo, 10,660; Kalwarya, 10,200; Siedlce, 10,013.
25	5,628,300	Tiflis, 104,024; Yeisk, 26,276; Nukha, 25,000; Shemakha, 24,500; Stavropol, 24,000.
10	3,440,000	Irkutsk, 32,300; Tomsk, 25,000; Tobolsk, 17,500.
8	4,400,000	Tashkend, 86,233; Khokand, 50,000; Samarkand, 30,000; Omsk, 30,000.
0*	87,829,853	

ive of the Caspian Sea (169,663 square miles).

GIONS IN THE RUSSIAN EMPIRE.

According to Schwanebach, 1876.

RUSSIA IN EUROPE.		POLAND.		FINLAND.	
Number.	Per Cent.	Number.	Per Cent.	Number.	Per Cent.
55,846,630	85·0	285,546	4·7	347,37	1·8
2,897,560	4·4	4,596,956	76·3	830	0·2
2,355,488	3·7	327,815	5·4	1,821,468	98·0
38,720	0·05	—		—	
1,944,378	2·9	815,433	13·5	—	—
2,363,658	3·6	426	0·0	—	—
258,125	0·4	245	0·0	—	—
65,704,559	100·0	6,026,421	100·00	1,857,035	100·0

RELIGIONS IN THE RUSSIAN EMPIRE (*continued*).

	CAUCASUS.		SIBERIA.		CENTRAL ASIA.	
	Number.	Per Cent.	Number.	Per Cent.	Number.	Per Cent.
Greek Catholics	2,114.991	44·4	3,016.174	88·6	320,506	9·5
Roman Catholics	25,915	0·54	24,316	0·7	1,396	0·0
Protestants	7,825	0·16	5,563	0·1	418	0·0
Armenians	595,310	12·5	—	—	—	—
Jews	22,732	0·40	11.941	0·3	3,396	0·1
Mohammedans	1,987.213	41·70	61,059	1·8	3,016,302	89·8
Pagans	4,683	0·10	286,016	8·4	14,740	0·4
Total . . .	4,763,809	100·0	3,105,069	100·0	3,356,758	100·0

III.—POPULATION OF RUSSIA IN EUROPE, POLAND, AND FINLAND

According to Nationalities (estimated for 1879).

Slavs	Great Russians . . .	40,000,000
	Little „ . .	16,370,000
	White „ &c. . .	3,600,000
	Bulgarians, Servians, &c. .	150,000
	Poles	5,000,000
Lithuanians		1,900,000
Letts		1,100,000
Rumanians		750,000
Greeks		75,000
Gipsies		15,000
Germans		1,000,000
Swedes		236,000
Mongols	Kalmuks	120,000
	Samoyeds . . .	4,000
Jews		3,000,000
Armenians		36,000

FINNS, LAPPS, AND UGRIANS.

Finlanders	1,840,000
Karelians in Russia . . .	300,000
Ehstes and Lives . . .	800,000
Lapps	4,000
Permians and Ziryanians . .	150,000
Mordvinians	1,000,000
Chuvashes	700,000
Cheremissians	260,000
Votyaks	240,000
Meshtcheryaks and Tepyaks .	270,000
Voguls	2,000
Vepses and others . . .	28,000

TURKS.

Kazan Tatars	1,050,000
Bashkirs	750,000
Crimean Tatars . . .	80,000
Kirghiz	180,000

IV.—AGRICULTURE.

DISTRIBUTION OF THE LAND ACCORDING TO ITS CULTIVATION:—
40·3 per cent. of the total area covered with forests. 20·6 per cent. is under the plough. 12·3 per cent. consists of meadows and pastures, 26·8 per cent. is covered with lakes or lies waste.

LIVE STOCK (1872):—
19,266,267 horses, 24,000,000 cattle, 49,000,000 sheep, 9,804,000 pigs, 1,180,000 goats, 340,200 reindeer, 29,300 camels.

DISTRIBUTION OF THE LAND ACCORDING TO PROPRIETORSHIP:—

	Before the Emancipation of the Serfs.	After the Emancipation of the Serfs.
Crown Lands	64·6 per cent.	45·6 per cent.
Noblemen's Estates . . .	30·6 „	22·6 „
Imperial Domains . . .	3·3 „	1·8 „
Small Freeholds . . .	1·5 „	30·0 „
Total	100·0 „	100·0

V.—PRODUCE OF MINES (INCLUDING ASIA).

	1860.	1876 or 1877.
Gold	77,580 lbs.	87,480 lbs.
Silver	31,248 „	24,588 „
Platinum	4,284 „	3,456 „
Copper	110,299 centals.	93,525 centals.
Lead	— „	15,660 „
Nickel Ore	— „	3.808
Zinc	— „	101,590
Cobalt Ore	— „	153
Cast Iron	6,501,102 „	8,986,620 „
Coal	7,387,700 „	36,933,000 „
Salt	9,671,150 „	14,166,600 „
Petroleum	568,100 „	3,000,000 „
Sulphur	— „	6,126 „
Plumbago	— „	2,360 „

ᵗ NOT LIABLE TO THE PAYMENT OF LICENSE DUTIES (1873).

ztories.	Number.	Operatives.	Value of Products (£).
. . .	833	164.000	16.000,000
. . .	634	88,500	8.960,000
. . .	402	35,600	2.080,000
. . .	121	9.600	1.248,000
. . .	175	11,800	1.610,400
. . .	388	13,500	1,920,000
. . .	333	5,000	1.040,000
. . .	3,297	17.620	4.680,000
. . .	893	5,330	3.200,000
. . .	132	33 910	4,672,000
. . .	193	13.370	861,130

ities there were (1876) 3,913 distilleries and 261 sugar refineries.

VII.—MILITARY FORCES.

ᴿSSIAN ARMY ON A WAR FOOTING.

	Combatants.
CAUCASUS):—	
luding a total of 292 regiments, or 768 battalions . .	758,000
otal of 28 battalions	28,000
ra total of 56 regiments of 4 squadrons (to be raised to 6)	33,600
a total of 15 battalions	13,000
Railway battalions, 2 Torpedo companies, 9 Telegraph	
. . . .	10,000
y, with 246 batteries and 1,968 guns)	
ry (204 guns) }	60,000
. . . .	912,600
intended for service in the field) :—	
2 battalions of Infantry (485 battalions)	485,000
. . . .	16.800
field batteries, with 768 guns)	23,000
. . . .	5,000
. . . .	529,800
ND CAUCASUS):—	
attalions	16,000
tillery	70,000
. . . .	86,000
us) :—	
id 7 battalions of Rifles	3 0,000
. . . .	11,200
l and 3 batteries of Horse Artillery (18 guns) . . .	12,800
. . . .	5,000
. . . .	329,000
ᴵSUS) :—	
s of Cavalry, 15 batteries of Horse Artillery (90 guns) .	60,000
ents of Cavalry, 5 battalions, 5 batteries of Horse Artillery	
. . . .	33,000
ats of Cavalry. 2 batteries of Horse Artillery (8 guns) .	7,000
ments of Cavalry	1,200
giments of Cavalry, 9 battalions of Infantry, 3 batteries	
. . . .	19,000
of Cavalry	5 400
. . . .	125,000
s, 12 Local battalions, 27 Line battalions, 10 Cossack	
. . . .	50,000
ossacks	10,000
es (88 guns), 3 Horse Batteries (18 guns), 4 Garrison	
. . . .	3,200
. . . .	63,200

with 3,610 field guns.
numerous escort detachments, &c., scattered over the empire, and
the Militia troops in the Caucasus and elsewhere, and in Finland
ᵈ which it is proposed to summon in case of a national war.
ardly exceeds 730,000 men, but on January 1st, 1877, there were
,005,825 men (784,161 Infantry, 70,925 Cavalry, 125,927 Artillery,

THE RUSSIAN NAVY.

	Baltic.	Black Sea.	Caspian.	Lake Aral.	Pacific.
28 Ironclads, with 201 guns	24	4	—	—	—
199 Steamers, with 545 guns	115	25	13	6	10
162 Transports	85	58	4	—	15

The navy is manned by 1,200 officers and 27,000 men.

VIII.—COMMERCE.

Commercial Marine (1880):—

	Sailing and Steam.		Steamers only.	
	Vessels.	Tons.	Vessels.	Tons.
White Sea	586	31,056	11	1,832
Baltic	638	106,620	63	19,078
Black Sea	2,135	211,310	171	59,128
Caspian	1,040	110,294	36	10,982
Pacific Ocean	15	20,000	15	20,000
Total	4,414	479,280	296	111,020

Imports from —

	1873 (£).	1877 (£).	1878 (£).
Germany	27,122,000	23,355,600	30,760,080
England	19,496,000	14,849,800	23,752,480
Austria	2,972,000	3,193,000	7,971,200
France	4,021,240	1,624,600	3,935,680
Holland	849,680	1,018,700	6,199,360
Sweden and Norway	478,200	323,300	3,201,600
Finland	1,526,000	1,440,800	1,562,000
Turkey	716,900	712,000	571,680
Asia (overland)	3,273,000	3,287,400	4,496,800
All other countries	10,419,980	5,730,800	12,842,400
Total	70,875,000	55,536,000	95,293,280

Exports to—

Germany	17,599,000	31,477,000	27,734,000
England	20,668,300	23,762,000	30,596,000
Austria	4,074,300	7,931,200	6,446,000
France	4,244,500	3,845,400	13,786,100
Holland	2,554,300	6,199,300	5,174,000
Sweden and Norway	1,033,200	3,201,600	2,253,300
Finland	1,412,000	2,040,300	1,973,000
Turkey	2,418,500	661,800	2,675,800
Asia (overland)	1,561,100	1,104,300	1,476,500
All other countries	2,945,200	4,226,900	6,791,700
Total	58,310,400	84,469,800	98,906,400

In the above tables the rouble is supposed to be worth about 3·8 shillings, but allowing for the discount on paper roubles at various periods, the average annual imports and exports appear to have been as follows :—

Periods.	Imports (£).	Exports (£).	Total (£).
1820—1839	10,500,000	12,000,000	22,500,000
1840—1849	15,000,000	18,000,000	33,000,000
1850—1859	21,000,000	23,000,000	44,000,000
1860—1869	33,000,000	32,000,000	65,000,000
1870—1878	69,000,000	59,000,000	128,000,000

INDEX.

END OF VOL. V.